MASS SPECTROMETRY OF BIOLOGICAL MATERIALS

Second Edition, Revised and Expanded

MASS SPECTROMETRY OF BIOLOGICAL MATERIALS

Second Edition, Revised and Expanded

edited by

BARBARA S. LARSEN
CHARLES N. McEWEN
E.I. du Pont de Nemours and Co., Inc.
Wilmington, Delaware

CRC Press
Taylor & Francis Group
Boca Raton London New York

CRC Press is an imprint of the
Taylor & Francis Group, an **informa** business

CRC Press
Taylor & Francis Group
6000 Broken Sound Parkway NW, Suite 300
Boca Raton, FL 33487-2742

First issued in paperback 2019

© 1998 by Taylor & Francis Group, LLC
CRC Press is an imprint of Taylor & Francis Group, an Informa business

No claim to original U.S. Government works

ISBN-13: 978-0-8247-0157-4 (hbk)
ISBN-13: 978-0-367-40058-3 (pbk)

The publisher offers discounts on this book when ordered in bulk quantities. For more information, write to Special Sales/Professional Marketing at the address below.

Library of Congress Cataloging-in-Publication Data

Mass spectrometry of biological materials / edited by Barbara S. Larsen. Charles N. McEwen. --2nd ed., rev. and expanded.
    p.   cm.
    Includes bibliographical references and index.
    ISBN 0-8247-0157-7 (alk. paper)
    1. Mass spectrometry. 2. Biomolecules--Analysis. 3. Proteins-Analysis. I. Larsen, Barbara Seliger. II. McEwen, Charles N.
QP519.9.M3M38   1998
572 '.36--dc21

                                                  97-52814
                                                     CIP

Visit the Taylor & Francis Web site at
http://www.taylorandfrancis.com

and the CRC Press Web site at
http://www.crcpress.com

Preface

Rapid advances in the biological sciences have led to an increased demand for chemical and structural information from biological systems. Mass spectrometry plays a pivotal role in the structural characterization of biological molecules, especially those present in minute quantities in complex matrices. Innovations in ionization methods, improvements in instrumentation, the use of tandem mass spectrometers, and improved interfaces with separation techniques are providing the capabilities needed to analyze a wide variety of biological materials. This volume provides fundamental information about the technology used by the leading research groups in applying mass spectrometry to the study of biological systems.

Background chapters are provided on the ionization methods and mass spectrometers used in biological analyses and on how databases are used in the identification of proteins from mass spectral data. This book also provides in a clear and concise manner the essential parameters necessary for the analysis of extremely low levels of complex biological materials. Included in these chapters are methods for isolating and identifying biologically important target molecules, methods for identifying proteins and modified proteins and determining their tertiary structures and noncovalent interactions, and methods for characterizing oligonucleotides and DNA. These chapters provide experiences from the authors' own laboratories, giving answers to why and how the experiments are performed and with what equipment, as well as the "oh by the ways" without which duplicating the work of others is often difficult. This book is intended to serve as a guide for applying mass spectrometry to the analysis of biological materials and as a cross-fertilization source of ideas and methods to help advance the field.

Chapter 1 is a review of ionization methods and instrumentation used in the analysis of biological materials. Chapters 2 and 3 discuss new mass spectrometers and the capabilities of these instruments for solving fundamental problems in the biological sciences. Chapters 4–6 discuss methods used in isolating and identifying target molecules often present in trace quantities in complex mixtures. Chapters 7–12 discuss methods for identifying proteins and related molecules, frequently at very low levels of material. Chapters 13 and 14 discuss the study of noncovalent interactions and the use of hydrogen/deuterium exchange for tertiary structure determination. Chapters 15 and 16 discuss the characterization of oligonucleotides and DNA, and Chapter 17 contains an in-depth discussion on screening protein databases using mass spectrometry-generated data.

This volume is a tribute to the contributing authors, who, writing about cutting-edge techniques and developments in their own laboratories, produced chapters of exceptionally high quality. Our intent was to create an instructional volume rather than a series of review articles, and we applaud the skill and effort put forth by the authors to achieve this objective.

ACKNOWLEDGMENT

These are wonderful times to be involved in mass spectrometry and the biological sciences. Many people have contributed to the developments in mass spectrometry that have made this such an exciting field in which to be working, but two research groups stand out for their exceptional contributions to the advancement of science. We wish to express our appreciation to Professor John Fenn for his outstanding work in combining electrospray ionization and mass spectrometry to produce a powerful new analytical method, and to Professor Franz Hillenkamp for the work he and Dr. Michael Karas have accomplished in introducing the very powerful analytical technique of matrix-assisted laser desorption/ionization mass spectrometry. These outstanding scientists have been excellent advocates for the continued development of the technology they introduced. It has been our great pleasure to have been associated with John Fenn since 1984 and, more recently, to have benefited from interactions with Franz Hillenkamp. Many of the advances since the first edition of *Mass Spectrometry of Biological Materials* are the direct result of developments in electrospray ionization and matrix-assisted laser desorption/ionization mass spectrometry.

Barbara S. Larsen
Charles N. McEwen

Contents

Contributors

Ruedi Aebersold, Ph.D. Associate Professor, Department of Molecular Biotechnology, University of Washington, Seattle, Washington

Philip C. Andrews, Ph.D. Professor, Department of Biological Chemistry, University of Michigan, Ann Arbor, Michigan

Bruno Antonsson, Ph.D. Research Scientist II, Department of Biochemistry, Geneva Biomedical Research Institute, Glaxo Wellcome Research and Development S.A., Geneva, Switzerland

Jian Bai, Ph.D.* Visiting Scientist, Department of Chemistry, University of Michigan, Ann Arbor, Michigan

Carthene R. Bazemore Walker, Ph.D. Research Associate, Department of Chemistry, University of Virginia, Charlottesville, Virginia

Ronald C. Beavis, Ph.D.† The Skirball Institute of Biomedical Research, New York University Medical School, New York, New York

---

Current affiliations:
*Hardware Design Engineer, California Analytical Division, Hewlett-Packard Company, Palo Alto, California.
†Senior Research Scientist, Department of Biopharmaceutical Development, Eli Lilly & Company, Indianapolis, Indiana.

R. Kevin Blackburn, M.S.   Scientist, Department of Analytical Chemistry, Glaxo Wellcome, Inc., Research Triangle Park, North Carolina

Rose Boyle, M.S.   Graduate Student, Department of Molecular Biotechnology, University of Washington, Seattle, Washington

Pamela F. Crain, Ph.D.   Research Associate Professor, Department of Medicinal Chemistry, University of Utah, Salt Lake City, Utah

Chhabil Dass, Ph.D.   Associate Professor, Department of Chemistry, The University of Memphis, Memphis, Tennessee

Anne Dell, Ph.D.   Professor, Department of Biochemistry, Imperial College of Science, Technology and Medicine, London, England

Axel Ducret, Ph.D.*   Senior Research Fellow, Department of Molecular Biotechnology, University of Washington, Seattle, Washington

Ricky D. Edmondson   Laboratory for Biological Mass Spectrometry, Department of Chemistry, Texas A&M University, College Station, Texas

David Fenyö, Ph.D.   National Resource for the Mass Spectrometric Analysis of Biological Macromolecules, Rockefeller University, New York, New York

Daniel Figeys, Ph.D.   Scientist, Department of Molecular Biotechnology, University of Washington, Seattle, Washington

Ming Gu, Ph.D.   Senior Research Fellow, Department of Molecular Biotechnology, University of Washington, Seattle, Washington

Paul A. Haynes, Ph.D.   Senior Fellow, Department of Molecular Biotechnology, University of Washington, Seattle, Washington

Jack D. Henion, Ph.D.   Professor, Department of Analytical Toxicology, New York State College of Veterinary Medicine, Cornell University, Ithaca, New York

---

* *Current affiliation*: Senior Research Biologist, Department of Biochemistry and Molecular Biology, Merck Frosst Canada, Inc., Montréal, Quebec, Canada.

Donald F. Hunt, Ph.D.   Professor, Departments of Chemistry and Pathology, University of Virginia, Charlottesville, Virginia

Robert L. Johnson   Research Investigator I, Department of Analytical Chemistry, Glaxo Wellcome, Inc., Research Triangle Park, North Carolina

Daniel B. Kassel, Ph.D.   Director, Analytical Chemistry, CombiChem, Inc., San Diego, California

Barbara S. Larsen, Ph.D.   Research Associate, Central Research and Development, E.I. du Pont de Nemours and Co., Inc., Wilmington, Delaware

Yan-Hui Liu, Ph.D.*   Research Assistant, Department of Chemistry, University of Michigan, Ann Arbor, Michigan

Joseph A. Loo, Ph.D.   Senior Research Associate, Department of Chemistry, Parke-Davis Pharmaceutical Research, Division of Warner-Lambert Company, Ann Arbor, Michigan

David M. Lubman, Ph.D.   Professor, Department of Chemistry, University of Michigan, Ann Arbor, Michigan

Roy McDowell   Research Officer, Department of Biochemistry, Imperial College of Science, Technology and Medicine, London, England

Charles N. McEwen, Ph.D.   Research Fellow, Central Research and Development, E.I. du Pont de Nemours and Co., Inc., Wilmington, Delaware

Barbara M. Merrill, Ph.D.   Research Investigator, Department of Analytical Chemistry, Glaxo Wellcome, Inc., Research Triangle Park, North Carolina

Howard R. Morris, Ph.D.   Professor, Department of Biochemistry, Imperial College of Science, Technology and Medicine, London, England

M. Arthur Moseley III, Ph.D.   Research Investigator II, Department of Analytical Chemistry, Glaxo Wellcome, Inc., Research Triangle Park, North Carolina

---

* Current affiliation: Senior Scientist, Schering-Plough Research Institute, Kenilworth, New Jersey.

Rachel R. Ogorzalek Loo, Ph.D.   Research Associate, Department of Biological Chemistry, University of Michigan, Ann Arbor, Michigan

Ron Orlando, Ph.D.   Assistant Professor, Complex Carbohydrate Research Center and Departments of Biochemistry and Molecular Biology, and Chemistry, University of Georgia, Athens, Georgia

Maria Panico, Ph.D.   Research Fellow, Department of Biochemistry, Imperial College of Science, Technology and Medicine, London, England

Thanai Paxton   Research Student, Department of Biochemistry, Imperial College of Science, Technology and Medicine, London, England

Carol V. Robinson, Ph.D.   Research Fellow, Oxford Center for Molecular Sciences, New Chemistry Laboratory, Oxford, England

David H. Russell, Ph.D.   Professor, Department of Chemistry, and Director, Laboratory for Biological Mass Spectrometry, Texas A&M University, College Station, Texas

Kristin A. Sannes-Lowery, Ph.D.*   Postdoctoral Fellow, Department of Chemistry, Parke-Davis Pharmaceutical Research, Division of Warner-Lambert Company, Ann Arbor, Michigan

Jeffrey Shabanowitz, Ph.D.   Principal Scientist, Department of Chemistry, University of Virginia, Charlottesville, Virginia

Douglas M. Sheeley, Sc.D.   Research Investigator, Department of Analytical Chemistry, Glaxo Wellcome, Inc., Research Triangle Park, North Carolina

Nicholas E. Sherman, Ph.D.   Research Associate, Department of Microbiology, University of Virginia, Charlottesville, Virginia

David Siemieniak   Howard Hughes Medical Institute, University of Michigan, Ann Arbor, Michigan

Jannavi R. Srinivasan, Ph.D.   Research Fellow, Department of Chemistry, University of Michigan, Ann Arbor, Michigan

---

* Current affiliation: Research Chemist, Wacker Silicones Corporation, Adrian, Michigan.

Richard B. van Breemen, Ph.D. Associate Professor, Department of Medicinal Chemistry and Pharmacognosy, University of Illinois at Chicago, Chicago, Illinois

Patrick J. Venta, Ph.D. Assistant Professor, Department of Microbiology, Michigan State University, East Lansing, Michigan

Duane L. Venton, Ph.D. Professor, Department of Medicinal Chemistry and Pharmacognosy, University of Illinois at Chicago, Chicago, Illinois

Julian D. Watts, Ph.D. Research Scientist, Department of Molecular Biotechnology, University of Washington, Seattle, Washington

Ray Wieboldt, Ph.D. Postdoctoral Associate, Department of Analytical Toxicology, New York State College of Veterinary Medicine, Cornell University, Ithaca, New York

Charles P. Woodbury, Ph.D. Associate Professor, Department of Medicinal Chemistry and Pharmacognosy, University of Illinois at Chicago, Chicago, Illinois

Yi Yang, Ph.D.* Complex Carbohydrate Research Center and Department of Biochemistry and Molecular Biology, University of Georgia, Athens, Georgia

Yanni Zhang, Ph.D. Postdoctoral Fellow, Department of Molecular Biotechnology, University of Washington, Seattle, Washington

Yongdong Zhu Research Assistant, Department of Chemistry, University of Michigan, Ann Arbor, Michigan

Jerry Zweigenbaum, M.S. Department of Analytical Toxicology, New York State College of Veterinary Medicine, Cornell University, Ithaca, New York

---

Current affiliation: Drug Metabolism & Pharmacokinetics, SmithKline Beecham Pharmaceuticals, King of Prussia, Pennsylvania.

1

Instrumentation and Ionization Methods for the Analysis of Biological Materials by Mass Spectrometry

Charles N. McEwen and Barbara S. Larsen
E.I. du Pont de Nemours and Co., Inc., Wilmington, Delaware

Mass spectrometry has become an integral part of research in the biological sciences as a result of recent developments in ionization methods and instrumentation. In this chapter we provide a brief description of the variety of instruments and ionization methods that are used in the analysis of biological materials. Chapters 2 and 3 discuss in more detail new mass spectrometers that are being developed to have an important impact on the analysis of biopolymers and related materials. While these chapters will be of value to experienced mass spectroscopists, Chapter 1 is written for those less familiar with mass spectrometry so that they can better appreciate the ionization methods and mass analyzers discussed in subsequent chapters of this book.

I. VOLATILE AND THERMALLY STABLE MATERIALS

Mass spectrometers separate gas-phase ions, either positive or negative, according to their mass-to-charge (m/z) ratio. Because gas-phase ionization generally produces ions with a single positive or negative charge, the calibrated m/z scale provides the mass of the ion in atomic mass units (daltons). The two most common ionization techniques that are used to ionize gas-phase molecules are electron impact ionization (EI) [1] and chemical ionization (CI) [2]. In either case, ionization is initiated by electrons that are "boiled" from a hot filament and passed through a potential gradient before interacting with gas-phase molecules.

EI [1] in its most common form ionizes gas-phase molecules through the interaction of neutral molecules with high-energy electrons (approximately 70 eV of kinetic energy). The interaction causes a secondary electron to be ejected to form a molecular ion and fragment ions:

$$e^- + M \rightarrow M^{+\cdot} + 2e^- \tag{1}$$

$$M^{+\cdot} \rightarrow f_1^+ + f_2^+ + f_n^+ \tag{2}$$

where M represents neutral molecules and f_n^+ represents the fragment ions. Negative ions can be formed under electron impact conditions but with poor sensitivity and only for certain compound types. Electron attachment and dissociative electron capture occur with higher efficiency for analyte molecules capable of stabilizing an electron. Electron attachment conditions are obtained by increasing the pressure in the ion source using a gas such as nitrogen or methane (see chemical ionization section that follows), which produces a large flux of near thermal electrons by Eq. 1 but few negative ions from the buffer gas.

The fragmentation that occurs under EI conditions (Fig. 1) provides information that can be used to confirm or determine an unknown structure. Search routines can be used to identify compounds in an electronic library of mass spectra, or the fragmentation pattern can be used to aid in the interpretation of the mass spectrum. Unfortunately, many biologically important molecules, even if volatile, produce a small or nonexistent molecular ion under EI conditions (see Fig. 1). Also, mixtures of compounds can produce very complex mass spectra under EI conditions because of the abundance of fragment ions.

Sample introduction can be achieved in a number of ways. Direct vaporization can be used to introduce a gaseous sample into the ionization volume for relatively pure materials. For solids and low-vapor-pressure liquids, this is done by placing the sample in a heated sample probe that can be inserted near the ion volume. The sample is vaporized by heating and the vapors migrate into the ion source volume. For mixtures, a separation method may be needed prior to introduction of the sample components into the mass spectrometer.

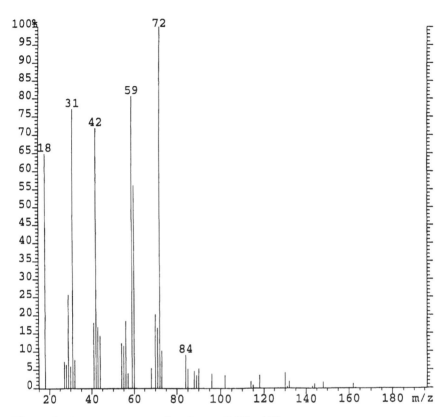

Figure 1   EI mass spectrum of D-glucose (MW = 180).

Separation of compounds ionized by EI is normally done by gas chromatography (GC). The drawback of gas chromatography/mass spectroscopy (GC/MS)
is the requirement that compounds be sufficiently volatile and thermally stable
to be vaporized and passed through a GC column. A great majority of biological materials do not fit this profile. However, a number of important biological
compounds can be made more volatile and thermally stable by derivatization.
In this way amino acids, nucleosides, steroids, bile acids, amines, many drugs
and their metabolites, and other classes of compounds can be analyzed by GC/
MS with high sensitivity [3].

   An approach for obtaining EI mass spectra for compounds separated by
liquid chromatography (LC) is through a particle beam interface [4]. In this
interface, the liquid stream from the mass spectrometer is nebulized and the
resulting spray passes through a series of skimmers with pumping regions be-

tween each skimmer pair to remove the solvent from the nebulized droplets. Particles of analyte enter the hot ion source and are thermally vaporized. The gas-phase molecules are then ionized by EI or CI. Compounds that can be successfully run by this method must have some volatility but cannot be so volatile that they are pumped away with the solvent.

CI [2] is a secondary process of ionization initiated by high-energy electrons. In chemical ionization, a so-called reagent gas is added to the source at a typical pressure of 0.3–1.0 torr. To achieve this higher pressure in the ionization volume while maintaining a good vacuum in the mass spectrometer, the CI source is built to be much tighter toward gas leaks than an EI source. Nevertheless, the principle of operation is the same in the two sources. In the CI source, an electron beam ionizes the reagent gas, which is in much higher concentration than the analyte, producing primary ions. Because of the high gas pressure in the source, the primary ions undergo ion molecule reactions to produce the reagent ions that are stable to further reaction with the reagent gas. Reagent ions in turn ionize the gas-phase analyte molecules. Equations 3–7 show the reactions that occur when methane is used as the reagent gas under CI conditions (A = analyte).

$$CH_4 + e^- \rightarrow CH_4^{+\cdot} + 2e^- \tag{3}$$

$$CH_4^{+\cdot} \rightarrow CH_3^+ + H^\cdot \tag{4}$$

$$CH_4^{+\cdot} + CH_4 \rightarrow CH_5^+ + CH_3^\cdot \tag{5}$$

$$CH_3^+ + CH_4 \rightarrow C_2H_5^+ + H_2 \tag{6}$$

$$CH_5^+ \text{ or } C_2H_5^+ + A \rightarrow AH^+ + CH_4 \text{ or } C_2H_4 \tag{7}$$

Chemical ionization is frequently used with EI to help identify molecular ions or to simplify complex mass spectra. The degree of fragmentation can be controlled in CI by judicious selection of the reagent gas (Fig. 2). Use of argon as the reagent gas results in mass spectra that are similar to EI mass spectra, while use of ammonia as the reagent gas generally provides abundant molecular ions and/or adduct ions. Argon ionizes by a charge-transfer mechanism to produce odd-electron molecular ions [$M^{+\cdot}$] and copious fragmentation. Ammonia generally produces even-electron protonated molecular ions [$M + H$]$^+$ or addition complexes [$M + NH_4$]$^+$ without significant fragmentation (Fig. 2b) [2].

Sample introduction in CI is similar to EI. However, it is possible to obtain molecular ion information on less volatile materials by using a technique called direct-introduction chemical ionization (DCI) [2]. In this method, sample is placed on a wire that can be inserted into the CI ion region and heated. Molecules vaporized from the filament undergo collisional stabilization and soft ionization through collisions with the reagent gas and reagent ions. A similar approach that further reduces fragmentation is potassium ionization of desorbed

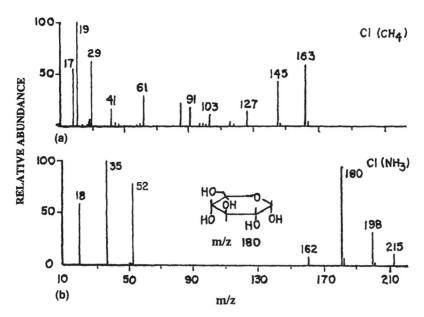

Figure 2 CI mass spectrum of D-glucose using (a) methane and (b) ammonia as the reagent gases. The reagent ions are (a) m/z 19, 29, 41, and (b) m/z 18, 35, 52.

species (KIDS) [5]. This method is similar to DCI except that a special glass is coated onto the wire filament and emits K^+ ions when heated. The sample is placed directly on the specially prepared filament, which is then placed in the ionization volume and heated. Molecules vaporize from the surface with potassium ions and are ionized by capture of K^+.

Separation combined with chemical ionization is most frequently accomplished using GC/MS, or particle beam interface LC/MS. The same conditions of volatility as discussed earlier for EI and particle beam ionization apply.

Field ionization (FI) [6] is a method for producing essentially only molecular ions from stable volatile materials. In this method, a high field is generated at the tips of fine dendrites grown from the surface of a thin wire. When a gaseous molecule comes near a dendrite tip, an electron is removed from the molecule, forming a "cool" molecular ion. FI is infrequently used in mass spectrometry.

Instrumentation used with EI and CI ionization is not limited to a single analyzer type. However, because these are continuous ionization processes as opposed to pulsed ionization, the gas-phase ions are commonly analyzed using quadrupole or magnetic sector mass analyzers. For some applications, other analyzer types such as ion traps or time-of-flight will have advantages. These analyzers are discussed in later sections of this chapter.

Magnetic sector [7] mass spectrometers separate charged particles in a magnetic field according to their mass-to-charge ratio. For this to occur, all gas-phase ions must enter the magnetic field having approximately the same kinetic energy (eV). The kinetic energy of any ion depends on a number of factors such as the direction of thermal motion at the moment of ionization, but the primary source of kinetic energy is derived from the voltage (V) used to accelerate the ions from the source into the magnetic field (B). The magnetic field exerts a force perpendicular to the movement of the charged particle according to the equation $m/z = B^2 r^2/2V$, where r is the radius of curvature of the ions traveling through the magnetic field. Ions of low mass will travel in a trajectory having a smaller radius than the higher mass ions for a given magnetic field and acceleration voltage. In practice, the magnetic field is scanned, frequently from high to low field, so that ions of successively lower mass-to-charge ratios pass through a fixed slit and strike an electron multiplier detector. The pattern of detected ion signals when displayed on a calibrated scale is called a *mass spectrum.*

Electrostatic analyzers (ESA) [7,8] are composed of two curved plates onto which opposite electric fields are placed to cause the ion beam to follow the curvature of the plates. Positive ions, for example, are repelled from the positively charged plate toward the negatively charged plate. The exact path any ion takes is determined by its momentum. Ions with lower momentum travel a shorter path than ions of higher momentum, thus producing a focusing effect. An ESA can be placed before or after the magnetic sector to provide energy resolution and directional focusing of the ion beam. An ESA combined with a magnetic sector allows higher mass resolution and more accurate mass measurement (Fig. 3). The mass accuracy that can be obtained with a double-focusing magnetic sector mass spectrometer is sufficient in many cases, for ions below about 700 Da, to specify a unique elemental composition [9]. A knowledge of the elemental composition of each ion in the mass spectrum greatly simplifies interpretation of mass spectra of unknowns.

Quadrupole mass analyzers [7,8] use four precisely machined metal rods that the ions must pass between to reach the detector. Ions are accelerated to a few volts prior to entering the quadrupole analyzer. The quadrupole rods act as mass filters when rf and dc fields are applied to diagonally opposed rods. By sweeping the rf and dc voltages in a fixed ratio (e.g., low to high voltages), ions of successively higher masses follow a stable path to the detector. Although quadrupole mass spectrometers are not capable of high resolving power typical of double-focusing magnetic sector mass spectrometers, it is possible to

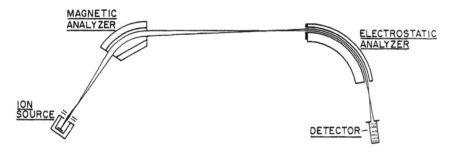

Figure 3 A double-focusing magnetic sector mass spectrometer of B/E geometry.

design quadrupole instruments to obtain good mass accuracy for well-resolved peaks [10].

MS/MS instruments [11,12] are mass analyzers connected in tandem to provide additional structural information through collision-induced dissociation (CID) of analyte ions with an inert gas such as helium or argon, or a surface (SID). A variety of analyzer combinations can be employed to obtain MS/MS results including tandem sector instruments, tandem quadrupole instruments, hybrid sector/quadrupole, hybrid quadrupole/time-of-flight, etc. Multiple steps of fragmentation (MS/MS/MS) or (MS)^n can be performed on a single ion trap mass analyzer or on a Fourier transform mass spectrometer (discussed later). Currently, the most common MS/MS instrument is the triple quadrupole mass spectrometer, which consists of two scanning quadrupole mass analyzers separated by an rf-only quadrupole (Fig. 4). The rf quadrupole transmits ions between the first and second mass analyzers [11].

MS/MS instruments can be used to perform a variety of experiments. For example, the sensitivity and specificity of an analysis can be increased by using the first mass analyzer as a mass filter to pass only ions having a particular mass (e.g., the molecular ion of the compound of interest) and then using the second mass analyzer to determine if characteristic product (daughter) ions are produced from the mass selected precursor (parent) ion (Fig. 5). A number of factors affect the collisionally induced fragmentation of the precursor ion, including the mass of the collision gas, the gas pressure, and the collision energy. Because the first mass analyzer only passes the mass of interest, chemical noise that normally limits the level of detection is virtually eliminated in the fragment ion spectrum. Tandem mass analyzers can also be operated to determine the precursor ion of a mass-selected fragment ion. In either of these methods, one analyzer is set to pass ions of a selected mass while the other either scans or jumps between mass-selected ions. Alternatively, both mass analyzers can be

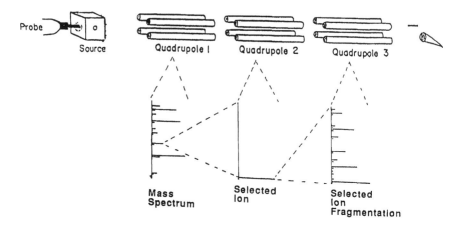

Figure 4 Diagram of a triple-quadrupole mass spectrometer showing MS/MS capabilities of selecting a precursor ion using the first quadrupole analyzer, collisionally fragmenting the mass-selected ion in the rf-only center quadrupole, and mass analyzing the fragment ions in the third quadrupole analyzer.

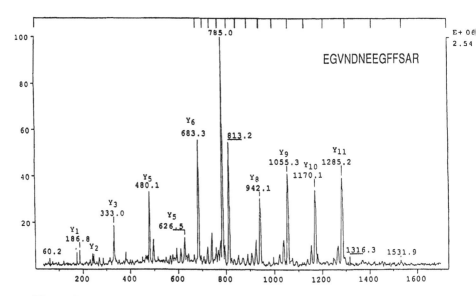

Figure 5 MS/MS daughter ion spectrum of the doubly charged molecular ion of Glu-fibrinogen peptide. See Chapter 11 for details on nomenclature of fragmentation.

scanned with an offset in mass to pass ions only when a mass-selected neutral fragment is lost. This experiment is used to identify compound classes. For example, acetates could be selected by using an offset of 60 amu [3]. Ion trap instruments are only able to perform full-scan parent-to-daughter experiments, but they have the advantages of high sensitivity as well as the ability to perform several stages of parent/daughter fragmentation $(MS)^n$. Combining MS/MS with a separation method gives the selectivity of the separation method plus mass selection and the specificity of determining if characteristic product ions are produced in the correct ratio from the selected precursor ion.

II. NONVOLATILE AND HIGH-MASS MATERIALS

Most biological materials cannot be vaporized without decomposition. For this reason EI, CI, and other ionization methods requiring vaporization before ionization fail to produce molecular ions of the compound of interest. Many biological molecules have high molecular weights, and thus the mass spectrometer must also be capable of mass analyzing and detecting these high-mass ions. Ionization of nonvolatile materials can occur in the solid or liquid phases with simultaneous or subsequent desorption into the gas phase for mass analysis. Alternatively, rapid desorption of neutrals can occur followed by gas-phase ionization. The result of ionization must be gas-phase ions that are stable on the time scale that allows mass measurement to occur.

Field desorption (FD) [6] was the first ionization method to extend mass spectrometry to nonvolatile and higher mass compounds. The method is similar to FI (discussed earlier) except that analyte is applied to the FD emitter from solution. The emitter, which is a thin wire onto which fine carbon dendrites are grown, is placed into the ion volume and a high voltage is applied between the emitter and the ion exit plate of the ion source. Heating the emitter causes a migration of the analyte to the tips of the dendrites, where the electric field is in excess of 10^8 V/cm. Under these conditions electrons are stripped from the analyte molecules with subsequent desorption of the ions for mass analysis. This method is seldom used today for biological materials.

Plasma desorption mass spectrometry (PDMS) [13] was the first particle desorption method of producing gas-phase ions from a nonvolatile, thermally labile, and high-mass solid analyte. In this method, massive, highly charged, MeV fission fragments from radioactive decay of ^{252}Cf strike the surface of a thin gold foil onto which analyte has been applied. Because this is a pulsed ionization technique, the ions that are desorbed during each ionization event are mass measured using time-of-flight mass spectrometry. This technique has mostly been replaced by other methods of ionization.

Thermospray ionization (TSI) [14] is a liquid introduction method that can readily be used with liquid chromatography and is capable of ionizing nonvolatile and thermally unstable compounds over a moderate mass range. In TSI, an aqueous solution containing the analyte and a volatile electrolyte such as ammonium acetate is forced through a restricted-diameter tube with a flow rate as high as 2 ml/min. The tube is heated to a sufficient temperature to cause nebulization and partial vaporization of the liquid. Under certain conditions of temperature and flow rate, ions are produced by ion evaporation from very small droplets in the nebulized vapor. Ion evaporation is believed to occur due to excess charge, either positive or negative, that builds up from an imbalance in the electrolyte ions as the vapor droplets evaporate. The vapor and ions from the spray pass through an open tube that goes through the ion source region to a rotary vacuum pump. A small orifice in the tube allows some of the ions that are produced to be sampled in the high-vacuum region of the mass analyzer. Primarily molecular ions are produced in TSI. Alternatively, using the same ion source design, ions can be produced by a corona discharge or from electrons "boiled" from a hot filament. This method produces a greater abundance of ions than TSI, but is not capable of ionizing nonvolatile materials and is more prone to production of fragment ions. TSI has mostly been replaced by electrospray/atmospheric pressure chemical ionization (discussed later).

Fast atom bombardment (FAB) or liquid secondary ion mass spectrometry (LI-SIMS) [15,16] was a popular technique for ionizing biological materials having molecular weights less than about 6000. In this method analyte dissolved in an appropriate solvent is added to a matrix, which is typically a viscous, low-vapor-pressure liquid such as glycerol or 3-nitrobenzyl alcohol. The matrix vaporizes slowly and does not overwhelm the vacuum pumps of the mass spectrometer. Bombardment of the liquid surface with KeV ions or neutral molecules/atoms results in desorption of analyte molecular ions, usually with minimal fragmentation. The liquid surface refreshes itself after each ionization event so that a continuous beam of analyte ions is produced. However, because FAB is a surface technique, in mixtures the more surface-active molecules will produce the most abundant ions. Because the most surface-active molecules can be changed by appropriate selection of the liquid matrix, it is often possible to change the relative abundances of ions from mixtures (Fig. 6).

Sample introduction with FAB/LISIMS is accomplished by introducing the analyte dissolved in a matrix such as glycerol or nitrobenzyl alcohol into the high vacuum of the ion source and in the path of the KeV ions or neutral particles. FAB is therefore a batch process, but can be adapted to liquid flow methods. Continuous-flow FAB (CfFAB) [17,18] is the interface that connects liquid chromatography with FAB ionization mass spectrometry. The technique uses flow rates of a few microliters per minute, which is compatible with capillary-column liquid chromatography, but because CfFAB is more technically

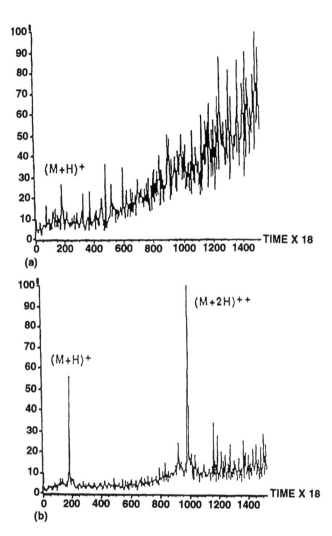

Figure 6 FAB mass spectrum of two peptides run using (a) glycerol and (b) 3-nitro-benzyl alcohol showing the effect of surface activity on ionization. Figure taken in part from Ref. 16.

challenging, frequently less sensitive, and more limited in the scope of compounds that can be ionized than electrospray ionization, it is infrequently used.

Atmospheric-pressure chemical ionization (APCI) [19] combined with liquid chromatography is an excellent method for the analysis of biological materials that have some degree of volatility. Because the same basic source design is used for APCI and electrospray ionization (discussed later), the methods are complementary and allow analysis of a wide range of compounds. APCI works well for most materials with molecular weights below 1000 but, unlike electrospray ionization, frequently produces fragment ions that can aid interpretation. The technique is compatible with flow rates from about 0.2 to 2 ml/min.

APCI is a special case of chemical ionization and is easily adapted to liquid separation methods. As the name applies, ionization takes place at atmospheric pressure with a corona discharge being used to produce ionization. The primary ions formed at atmospheric pressure undergo sufficient ion molecule collisions to give abundant H_3O^+ ions from traces of water, or NH_4^+ ions from traces of ammonia. These reagent ions protonate any compound more basic than water (using H_3O^+) or ammonia (using NH_4^+) or form adduct ions. Analyte can enter the ionization region from a liquid stream that exits an LC column. A coaxial stream of nebulizing gas is used to produce a fine spray. APCI requires the application of sufficient heat to the ionization region by way of a drying gas (usually nitrogen) to vaporize the solvent droplets and analyte. The analyte ions that are formed at atmospheric pressure are then sampled through a small orifice, and usually differentially pumped skimmers, before entering the analyzer region of the mass spectrometer. Negative ions can also be formed under APCI conditions. Negative ionization is by reagent ions such as solvated O^{2-} rather than electron attachment reactions.

Electrospray ionization (ESI) [20] is currently a very popular method for analysis of biological materials. ESI is capable of producing molecular ions from vanishingly small amounts of material over an exceedingly high mass range (this volume). Combined with APCI, most biological materials below about 150,000 Da can be ionized. ESI has been shown to ionize materials with molecular weight equivalents to 10^9 Da, but special charge detection mass spectrometry is required and the mass accuracy is no better than about $\pm 10\%$ [21].

Electrospray dates back to Zeleny's work in 1917 [22]. The ESI process produces charged liquid droplets from a liquid stream in the presence of an electric field. The liquid droplets formed during electrospray are highly charged, having either excess positive or negative charges. Evaporation of these droplets leads to a state of instability due to the excess charge on the droplet surface. The surface charge on the droplet is reduced by increasing the surface area, either by formation of a so-called Taylor cone, which expels smaller droplets, or by coulombic explosion, when the surface charge reaches the Raleigh limit [23]. In either case smaller droplets are formed, which in turn undergo the same process, until either the field strength at the droplet surface allows field evaporation of analyte ions, or evaporation of the solvent results in only the solvated analyte remaining as a multiply charged ion.

The first attempt to interface ESI with mass spectrometry was by Dole and coworkers in the late 1960s and early 1970s [24,25]. Fenn and coworkers [26] reported the first successful interface of ESI to a quadrupole mass spectrometer in 1984 using a glass capillary as an orifice for the ions formed at atmospheric pressure to enter the vacuum region of the mass spectrometer while limiting the gas load on the vacuum system (Fig. 7). In the Fenn source design, heated nitrogen gas is used to aid drying of the liquid droplets formed in the electrospray process. Other source designs include a small aperture or a heated metal capillary for transporting ions from atmospheric pressure into the vacuum of the mass spectrometer.

In ESI, a series of molecular ions differing only in the number of charges is observed for most compounds with molecular weights above about 1000. Figure 8 is an example of an ESI mass spectrum of a protein. As a result of multiple charging, the molecular ion is much higher in mass than the mass-to-charge limit of the mass spectrometer used to obtain the mass spectrum

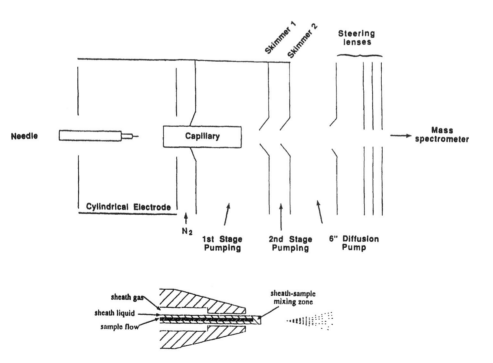

Figure 7 Electrospray source design for a magnetic sector mass spectrometry showing ESI nebulizer.

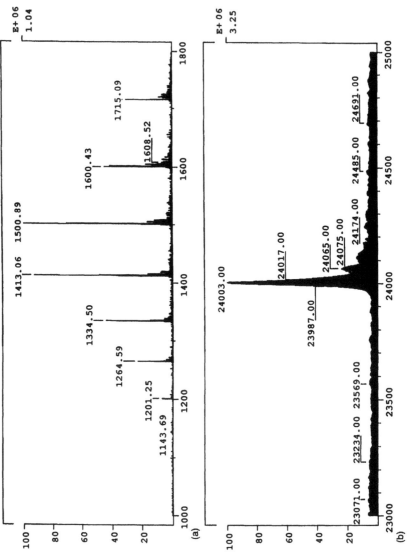

Figure 8 ESI mass spectrum of dethiobiotin synthetase. (a) Multiply charged spectrum; (b) computer-calculated molecular ion.

[27]. Thus, the molecular ion of a protein of MW 24,000 having 10 protons, and thus 10 positive charges, would be observed at m/z 2401 ($[24,000 + 10]/10 = 2401$). In the mass spectrum, only the m/z ratio is observed. Both m and z are unknowns but can be determined by simultaneously solving two equations with two unknowns. The number of charges on an ion can be determined from the m/z of two peaks that differ by one charge (two adjacent multiply charged molecular ion peaks) using the following equations:

$$n_j = (m_k - X)/(m_k - m_j) \quad \text{or} \quad n_k = (m_j - X)/(m_k - m_j)$$

where $X = 1$ for ionization by H^+, 23 for ionization by Na^+, etc. Ions having n_j charges have $m/z = m_j$, and the next lower charge state, $n_k = n_j - 1$, has $m/z = m_k$. While it is relatively simple to calculate the number of charges and determine the molecular weight for each m/z measured, most commercial mass spectrometers with ESI have computer algorithms that determine MWs from the multiply charged mass spectra even for complex mixtures.

ESI is inherently very sensitive for polar and highly functionalized compounds. Full-scan mass spectra are readily obtained for most compounds that can be ionized by electrospray at a concentration $<10^{-5}$ M and frequently $<10^{-7}$ M. Because ESI is a concentration-sensitive ionization method, mass spectra can be obtained consuming only attomole amounts of sample by using micro- or nanoflow techniques, which use fine needles to create stable sprays over several minutes while consuming only nanoliters of solution (see Chapters 6, 7, 8, and 11) [28]. One problem in handling very small quantities of material lies in making a solution sufficiently concentrated to produce a mass spectrum. With micro- and nanoflow it is possible to sum numerous scans while consuming only nanoliters of solution. Thus, peaks that may not be observed in a single scan are readily observed after summing 50 or more scans. Summing the ion current over 10 min consuming 10 nl/min theoretically uses 1/1000 the sample consumed spraying the same solution for 10 min at 10 μl/min.

Sample introduction is by infusion in a solvent that is compatible with ESI. Because ESI is generated from a flowing liquid at atmospheric pressure, it is readily interfaced to mass spectrometers through an atmospheric pressure inlet similar to APCI. Miscible solvents can be mixed to enhance the ability to electrospray a solution. Alternatively, nestled tubes can be employed (Fig. 7) to provide a sheath flow that achieves the same goal without mixing solvents until they reach the point where the electrospray process occurs. Further enhancement of the electrospray process can be achieved, especially for high flow rates, by forcing a gas such as nitrogen through the outermost tube as a means of enhancing nebulization of the droplets. A loop injection valve can be used to introduce the sample into the liquid flow. This can be done manually, or automatically using an autoinjector. With an autoinjector, large numbers of samples can be run without operator intervention. Combined with computer programs APCI and ESI are well suited for analysis of combinatorial libraries.

Separation methods can also be used to handle small quantities of sample as well as complex mixtures. Commonly used gradients such as acetonitrile/ water and methanol/water with acetic acid, trifluoroacetic acid, or formic acid are also excellent solutions for ESI, although 100% water or addition of trifluoroacetic acid increases the difficulty of forming a stable spray. Frequently, a sheath liquid flowing in a concentric tube is used to minimize the effect of a gradient on ESI operating conditions.

As noted earlier, ESI is concentration rather than flow dependent. Thus, a packed capillary column operating at 2 μl/min flow will require 0.2% as much sample as a conventional 4.6-mm column operating at 1 ml/min, assuming equivalent resolution. In practice, capillary LC/MS requires much more attention to detail for proper operation, but the payoff in addition to enhanced sensitivity is increased resolution and much lower solvent consumption. One danger of ESI is the aerosols that are generated during the electrospray process.

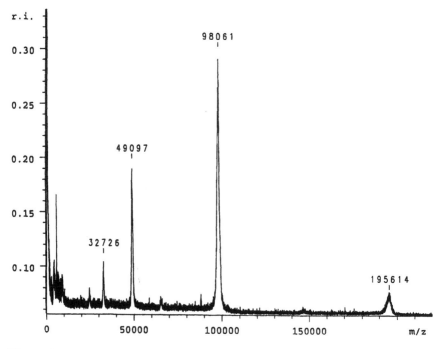

Figure 9 MALDI mass spectrum of phosphorylase B (MW 98,328) showing multiply charged and dimer molecular ions.

Systems where the aerosols are not efficiently trapped potentially expose the laboratory occupants to solvents and dangerous compounds. Clearly, operation at low flow rates decreases the hazards.

Matrix-assisted laser desorption/ionization (MALDI) [29] mass spectrometry is also a successful technique for analyzing biological materials because it is applicable over a wide mass range (see Fig. 9), requiring femtomoles or less of material while consuming only a small fraction of the analyte applied to the target. Like FAB, MALDI is a batch process where samples are prepared and introduced into the vacuum system of the mass spectrometer. Most MALDI mass spectrometers allow a number of samples to be prepared on a target thus making the batch process reasonably efficient. The sample is cocrystallized with a matrix compound in about a 1/1000 molar ratio. Ionization occurs when the dried matrix is irradiated with a focused laser beam in the mass spectrometer ion source region (Fig. 10). Most of the energy from the laser is adsorbed by the matrix, resulting in a rapid expansion from the solid to the gas phase in the micrometer-sized area receiving the radiation. Ionization of the analyte is believed to primarily occur in the high-pressure region just above the irradiated surface and may involve ion–molecule reactions or reaction of excited state

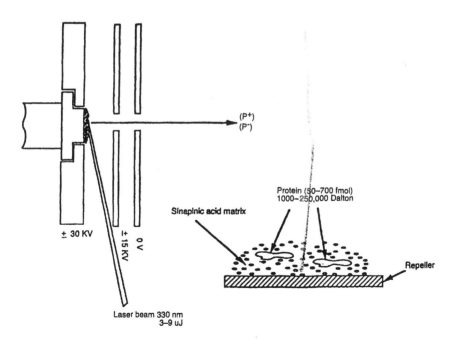

Figure 10 Pictorial representation of MALDI process.

species with analyte molecules. Nitrogen lasers operating at 337 nm are the most common because they are inexpensive and work well, but other ultraviolet (UV) and even infrared (IR) lasers can be employed with the properly selected matrix materials.

MALDI is especially forgiving of impurities and mixtures. Because the method is robust, applicable to a large variety of compounds, and relatively easy to operate, it is often the method of choice for determining molecular weights.

Sample preparation involves dissolving the sample in a solvent at low micromolar or even nanomolar concentration and mixing the sample solution with the matrix solution. The matrix is typically 10^3 to 10^4 times more concentrated than the analyte. The mixing can be done on the target or prior to spotting the sample solution on the target. Typically less than a microliter of solution is applied to the target. Water-soluble samples such as peptides and proteins are prepared by dissolving the analyte in water or water–acetonitrile

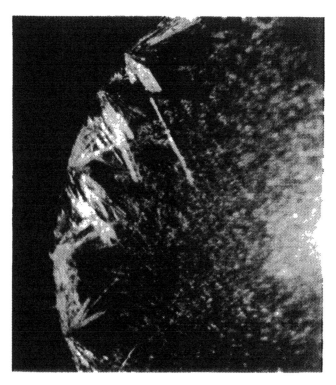

Figure 11   MALDI matrix preparation showing crystalline and amorphous regions.

(or water–methanol) with 1% HOAc or 0.1% trifluoroacetic acid (TFA). The matrix is typically dissolved in an approximately 1:1 water:acetonitrile solution at 5–10 mg/ml. Commonly used matrix materials are 2,5-dihydroxybenzoic acid, nicotinic acid, sinapinic acid, and α-cyanocarboxylic acid. When the solutions of matrix and analyte are mixed and dried on the target, a crystalline outer ridge and an amorphous inner area are typically produced (Fig. 11). The outer crystalline region usually provides the best results for hydrophilic materials. The use of volatile organic solvents will produce a more uniform target spot with smaller and more reproducible crystals. The latter type of sample preparations usually gives more uniform ionization (Chapter 2).

Separation methods are not readily interfaced online with MALDI. The best approach appears to involve spraying the LC eluent onto a target precoated with a matrix material. The spray can be confined to a small area and eliminates the problem of liquid spreading when eluent is applied directly to the target. Moving the target over the course of the LC separation provides modest resolution. Because MALDI is robust to impurities and mixtures, it is usually not necessary to achieve high chromatographic resolution.

Instrumentation used for nonvolatile samples is somewhat dependent on the ionization method employed. FD, TSI, and FAB are mostly employed on either magnetic sector or quadrupole mass spectrometers, which were discussed in the first section. PDMS has been exclusively used with time-of-flight mass spectrometers because of the pulsed nature of the ionization. ESI and MALDI have been interfaced to a variety of mass spectrometers. Fundamentally, ESI is a continuous ionization method and can be advantageously interfaced to magnetic sector and quadrupole mass spectrometers. However, it has also been interfaced successfully to time-of-flight and ion trap instruments.

Time-of-flight (TOF) [30] mass spectrometers have seen a recent resurgence in popularity driven by ionization methods that require high mass capabilities, high speed, or high duty factors (percent of ions formed that are detected). Recent developments in TOF-related technologies have led to a mass analyzer capable of high resolution, fast scans, high sensitivity, accurate mass measurement, and essentially unlimited mass range (Chapter 2) [31]. In TOF-MS, ions are pulsed from the ionization region using an electric field and require a finite time before the ions reach the detector and a subsequent draw-out pulse can occur (Fig. 12). Unlike quadrupole and magnetic sector instruments, which work by eliminating all ions except those of the mass being detected, the TOF instrument detects all of the ions in a draw-out pulse, thus producing a full spectrum with each pulse. The pulsed nature of TOF-MS provides exceptional sensitivity for pulsed ionization techniques such as MALDI. However, for a continuous-ionization technique such as ESI, there is a delay time during the mass measurement of the ions from a draw-out pulse when ions cannot be sampled. To minimize loss of ions formed in the continuous ESI process, the

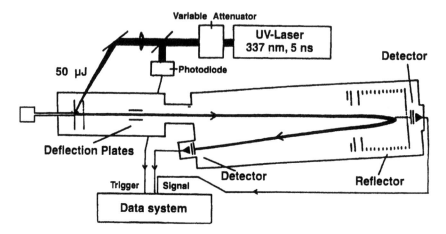

Figure 12 Schematic of a reflectron time-of-flight mass spectrometer with laser ionization, ion mirror, and linear and reflectron detectors.

direction of the ions entering the TOF ion region is orthogonal to the instrument flight direction. Because TOF instruments have an open geometry (no beam-limiting slits or apertures for the beam to pass through), a time slice of the ESI beam can be effectively sampled by the TOF instrument. Thus, while a set of ions is being mass analyzed in the TOF instrument, ions from the ESI source are filling the sampling region. This is possible because of low acceleration of the ESI beam, which travels a small distance; simultaneously, those ions being mass measured are accelerated by a high voltage and travel a much longer distance. Using this approach, the duty cycle for ESI can be made to approach 100%.

The TOF analyzer approach is not readily adapted for MS/MS capabilities, although metastable ions formed in the field free region before the ion mirror in a reflectron TOF instrument can be mass analyzed using a suitable calibration after passing through the ion mirror (Fig. 13). MS/MS spectra can be obtained, albeit at greater expense, by adding a mass analyzer such as a quadrupole or magnetic sector prior to an orthogonal TOF analyzer [31]. Ions can be mass selected (precursor/parent ions) in the first analyzer, caused to undergo collisional fragmentation, and the fragment (daughter) ions mass analyzed using the TOF analyzer. The advantage of this approach over a triple quadrupole is that good mass resolution and mass accuracy can be achieved for the daughter ions with excellent sensitivity. The disadvantage, especially for MALDI, is that the initial mass analyzer can limit the mass range that is achievable (Chapter 3).

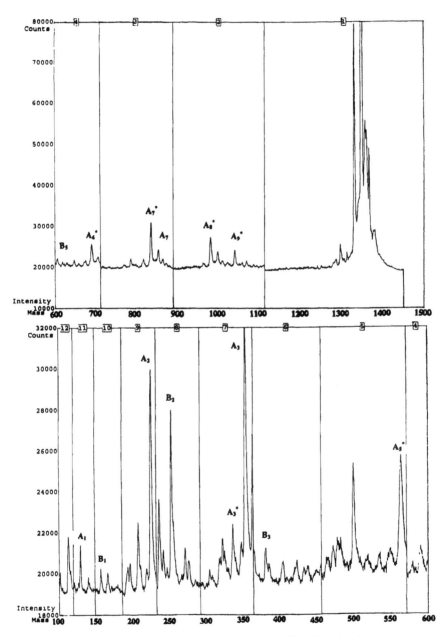

Figure 13 Metastable mass spectrum of substance P. The lines in the spectrum are where different mass ranges were acquired using different ion mirror voltages. The spectrum is "stitched" together automatically under computer control.

Ion trap devices such as the quadrupole ion trap (IT) and the ion cyclotron resonance (ICR) cell also have a duty function because ions are not being sampled during mass analysis. These devices are therefore well suited for a pulsed technique such as MALDI, except that the instruments have a limited practical mass range. One means of minimizing the loss of ions from a continuous ionization source is to use an external ion source with a linear ion guide, which can also be used to store ions for injection into the trap. In this way the duty cycle can be improved. Hexapoles have been employed as ion guides.

Fourier-transform ion cyclotron resonance mass spectrometers (FTICR-MS) [32,33] have the advantage of very high mass resolution and good mass accuracy along with $(MS)^n$ capabilities. In the FTICR-MS, ions are trapped by a combination of electric and magnetic fields inside a small cell. In the magnetic field, ion packets precess with a cyclotron motion that is mass dependent. The frequency of the ion motion is detected by an electric circuit that connects opposite walls of the ion cell. As ions near the detector walls, electrons in the circuit are either attracted or repelled by positive or negative ions, respectively. The mass-dependent frequency measurements are transformed to a mass-to-charge versus ion intensity scale using computer algorithms.

The mass range of the instrument and the mass resolution that can be achieved are dependent on the field strength of the magnet. The higher the magnetic field, the better the performance that can be achieved but at a higher cost. FTICR instruments are capable of resolving the isotope envelope of multiple charged ions with molecular weights beyond 100,000 [34]. Resolving the isotopes in an electrospray experiment allows direct determination of the number of charges on the ion being observed because the number of isotope peaks observed per mass-to-charge unit is directly related to the number of charges on the ion. A singly charged ion will have only one isotope per m/z unit, a doubly charged ion, two, a triply charged ion three, etc. (Fig. 14). Thus, an ion observed at mass 1500 having 20 isotope peaks per m/z unit will have a mass of $20 \times 1500 = 30,000$. While FTICR can determine the mass of any single isotope peak to within a few parts per million, it becomes increasingly difficult with increasing mass to determine how any one isotope peak relates to the monoisotopic mass or the average chemical mass of a compound.

The $(MS)^n$ capabilities of FTICR make it possible to obtain structural information from the same set of ions that was trapped for molecular weight determination. High resolution and mass accuracy also apply to the fragment ions that are trapped in the MS/MS experiment, making FTICR a powerful tool for structure elucidation. While FTICR is especially well suited for ESI because the multiply charged ions fall within the mass range of the FTICR instrument, it is more limited when interfaced to MALDI where singly charged ions are dominant. The ability to obtain high resolving power with FTICR is inversely proportional to the mass-to-charge ratio of the ion being measured and directly

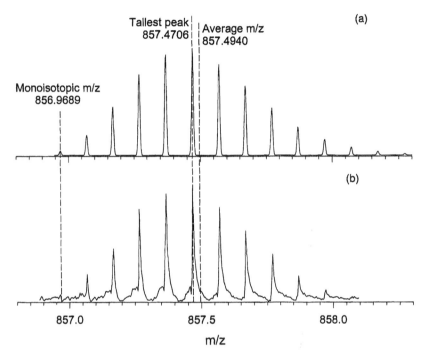

Figure 14   ESI mass spectrum of ubiquitin +10 charge state. (a) Theoretical and (b) experimental isotope distribution obtained using an FTICR mass spectrometer.

proportional to the magnetic field strength. A singly charged ion above about *m/z* 10,000 is difficult to observe using current commercial FTICR instruments.

Quadrupole ion trap mass spectrometers (ITMS) [35,36] do not use a magnetic field to trap ions and are therefore much less expensive than FTICR instruments. The advantages of the ITMS are the high mass capabilities, high sensitivity for full-scan data, and the ability to obtain $(MS)^n$ spectra (daughter ion scans only). The ion trap can produce high resolution over a limited mass window, but this has not yet translated to accurate mass measurement. These instruments are readily interfaced to ESI.

The ion trap analyzer consist of a baseball-size chamber made up of three hyperbolic electrodes, a ring and two end caps. Ions can either be injected into or formed in the trap. For biological applications, ions are generated externally, by, for example, ESI, and transmitted into the ion trap volume. The ions are focused into the center of the trap by an oscillating electric field (rf field) applied to the ring electrode. A time slice of ions can be trapped and subsequently m/z-selected ions can be made unstable in the axial direction (between the end caps) by applying an oscillating field to the end caps that is in resonance with the secular frequency of the selected ions. In this way, ions can be ejected through a hole in the end cap and detected by an electron multiplier.

ITMS instruments are commercially available interfaced to gas and liquid chromatographs. In addition, hybrid mass spectrometers are being developed which incorporate an ion trap analyzer. An example of a hybrid ion trap mass spectrometer is one in which a high-resolution magnetic sector mass spectrometer precedes the ion trap. A principle advantage of this instrument is high-resolution mass selection of the parent ion using the double-focusing magnetic sector and $(MS)^n$ capabilities and high resolution of daughter ions provided by the ion trap.

III. SUMMARY OF IONIZATION METHODS USED FOR BIOLOGICAL ANALYSES

Electron impact (EI): Usually provides molecular weight and structurally useful fragmentation for volatile materials. Sometimes molecular ions are weak or not observed. Large libraries of EI mass spectra (e.g., NIST and Wiley) are available for computer searching. Readily combined with gas chromatography for mixture analysis.

Chemical ionization (CI): Various chemical ionization gases are available to tailor fragmentation. Usually provides strong molecular ions for volatile materials. Frequently combined with gas chromatography.

Negative ion chemical ionization (NCI): For compounds that can stabilize a negative charge. Electron capture can be as much as $1000\times$ more sensitive than positive ionization CI for electronegative compounds that can stabilize an added electron.

Atmospheric-pressure chemical ionization (APCI): Chemical ionization at atmospheric pressure. Sensitive method of ionization that requires minimal volatility. Liquid introduction at atmosphere enhances ease of coupling to liquid chromatography.

Fast atom bombardment (FAB): Ions are sputtered in vacuum from a liquid surface having low volatility. Nonvolatile and thermally labile molecules produce primarily molecular ions. Can be coupled to a liquid chromatograph.

Electrospray ionization (ESI): An atmospheric-pressure liquid flow ionization method for producing molecular ions from a wide variety of compounds over a wide mass range. Peptides, proteins, organic salts, etc. are readily analyzed. Coupling to a liquid chromatograph is relatively straightforward.

Matrix-assisted laser desorption/ionization (MALDI): The analyte is mixed with a suitable matrix material and a laser is used for ionization. The method produces molecular ions over a wide mass range. Metabolites, proteins, and synthetic polymers are among the compound types that can be analyzed using MALDI-MS.

REFERENCES

1. JR Chapman. Mass spectrometry ionization methods and instrumentation. In: JR Chapman, ed. Protein and peptide analysis by mass spectrometry. Totowa, NJ: Humana Press, 1996.
2. AG Harrison. Chemical ionization mass spectrometry. 2nd ed. Boca Raton, FL: CRC Press, 1992.
3. FG Kitson, BS Larsen, CN McEwen. Gas chromatography and mass spectrometry, a practical guide. San Diego, CA: Academic Press, 1996.
4. GG Jones, RE Pauls, RC Willoughby. Analysis of styrene oligomers by particle beam liquid chromatography/mass spectrometry. Anal Chem 63:461, 1991.
5. D Bombick, JD Pinkston, J Allison. Potassium ion chemical ionization and other uses of an alkali thermionic emitter in mass spectrometry. Anal Chem 56:396, 1984.
6. L Prokai, ed. Field desorption mass spectrometry. New York: Marcel Dekker, 1990.
7. C Dass. Mass Spectrometry: instrumentation and techniques. In: DM Desiderio, ed. Mass spectrometry: Clinical and biomedical applications. Vol. 2. New York: Plenum Press, 1994.
8. RJ Anderegg. Mass spectrometry: an introduction. In: CH Suelter, JT Watson, eds. Biomedical applications of mass spectrometry. Vol. 34 of Methods of biochemical analysis. New York: Wiley, 1990.
9. RA Zubarev, P Hakansson, B Sundqvist. Accuracy requirements for peptide characterization by monoisotopie molecular mass measurements. Anal Chem 68:4060–4063, 1996.
10. AN Tyler, E Clayton, BN Green. Accurate mass measurement of polar molecules at low resolution using electrospray ionization and a quadrupole mass spectrometer. Proceeding of the 44th ASMS Conference on Mass Spectrometry and Allied Topics, Portland, OR, May 12–16, 1996, p 74.
11. RA Yost, CG Enke. Tandem quadrupole mass spectrometry. In: FW McLafferty, ed. Tandem mass spectrometry. New York: Wiley, 1983, pp 175–195.
12. KL Busch, GL Glish, SA McLuckey. Mass spectrometry/mass spectrometry: Techniques and applications of tandem mass spectrometry. New York: VCH, 1988.
13. RD MacFarlane. ^{252}Cf-Plasma desorption mass spectrometry using polymer surfaces. Trends Anal Chem 7:179, 1988.

14. CR Blakley, ML Vestal. Thermospray interface for liquid chromatography/mass spectrometry. Anal Chem 55:750, 1983.

15. M Barber, RS Bordoli, RD Sedgwick, AN Tyler. Fast atom bombardment of solids: a new ion source for mass spectrometry. J Chem Soc Commun 325–327, 1981.

16. BS Larsen, Tips for the analysis of peptides using FAB. In: CN McEwen, BS Larsen, eds. Mass spectrometry of biological materials. New York: Marcel Dekker, 1990, pp 197–214.

17. RM Caprioli, T Fan. High sensitivity mass spectrometric determination of peptides: Direct analysis of aqueous solutions. Biochem Biophys Res Commun 141:1058, 1986.

18. RM Caprioli, T Fan, JS Cotrell. Continuous-flow sample probe for fast atom bombardment mass spectrometry. Anal Chem 58:2949, 1986.

19. AP Bruins. Mass spectrometry with ion sources operating at atmospheric pressure. Mass Spectrom Rev 10:53, 1991.

20. RB Cole, ed. Electrospray ionization mass spectrometry: Fundamentals, instrumentation, & applications. New York: Wiley, 1997.

21. SD Fuerstenau, WH Benner. Molecular weight determination of megadalton DNA electrospray ions using charge detection time-of-flight mass spectrometry. Rapid Commun Mass Spectrom 9:1528, 1995.

22. Z Zeleny. Instability of electrified liquid surfaces. Phys Rev 10:1,1917.

23. M Mann, JB Fenn. Electrospray mass spectrometry principles and methods. In: DM Desiderio, ed. Mass spectrometry: Clinical and biomedical applications, Vol. 1. New York: Plenum Press, 1992.

24. M Dole, LL Mack, RL Hines, RC Mobley, LD Ferguson, MB Alice. Molecular beams of macroions. J Chem Phys 49:2240, 1968.

25. M Dole, HL Cox Jr., J Geiniec. Electrospray mass spectrometry. In: RF Gould, ed. Adv Chem Ser No 125, Polymer Molecular Weight & Methods, 1973, pp 73–84.

26. M Yamashita, JB Fenn. Electrospray ion source. Another variation on the free-jet theme. J Phys Chem 88:4451, 1984.

27. C-K Meng, M Mann, JB Fenn. Electrospray ionization of some polypeptides and small proteins. Proceedings of the 36th ASMS Conference on Mass Spectrometry and Allide Topics, San Francisco, CA, June 5–10, 1988, p 771.

28. MS Wilm, M Mann. Electrospray and Taylor-Cone theory, Dole's beam of macromolecules at last? Int J Mass Spectrom Ion Processes 136:167, 1994.

29. M Karas, F Hillenkamp. Laser desorption ionization of proteins with molecular masses exceeding 10000 daltons. Anal Chem 60:2299, 1988.

30. RJ Cotter, ed. Time-of-flight mass spectrometry: Instrumentation and applications in biological research. Washington, DC: American Chemical Society, 1997.

31. M Guilhaus, V Mlynski, D Selby. Perfect timing: Time-of-flight mass spectrometry. Rapid Commun Mass Spectrom 11:951, 1997.

32. GA Valaskovic, NL Kelleher, FW McLafferty. Attomole protein characterization by capillary electrophoresis-mass spectrometry. Science 273:1199, 1996.

33. B Asamoto, ed. FT-ICR MS: Analytical applications of Fourier transform ion cyclotron resonance mass spectrometry. Weinheim, Germany: VCH, 1991.

34. NL Kelleher, MW Senko, MM Siegel, FW McLafferty. Unit resolution mass spec-

tra of 112 kDa molecules with 3 Da accuracy. J Am Soc Mass Spectrom 8:380, 1997.

35. RE March, RJ Hughes. Quadrupole storage mass spectrometry. New York: Wiley, 1989.

36. KR Jonscher, JR Yates III. The quadrupole ion trap mass spectrometer—A small solution to a big challenge. Anal Biochem 244:1, 1997.

2

High-Resolution Mass Spectrometry and Accurate Mass Measurements of Biopolymers Using MALDI-TOF

Ricky D. Edmondson and David H. Russell
Texas A&M University, College Station, Texas

Biological mass spectrometry has undergone many changes since the introduction of matrix-assisted laser desorption ionization (MALDI) by Karas and Hillenkamp [1] and electrospray ionization (ESI) by Fenn [2]. Although ^{252}Cf-plasma desorption mass spectrometry [3], laser desorption ionization [4], and fast atom bombardment mass spectrometry (FAB-MS) [5] provided many opportunities for identification and characterization of polar biologically important molecules, such experiments were difficult and the sensitivity was marginal. Conversely, the sensitivity of MALDI and ESI is excellent, low picomoles to attomoles [6], the sample preparation requirements are minimal, both molecular weight and structural information can be obtained, and the instrumentation requirements are modest [7,8]. In this chapter we discuss requirements for obtaining high-resolution MALDI–time-of-flight (TOF) mass spectra and show how high mass resolution can facilitate accurate mass measurements. In turn we illustrate the utility of and the need for accurate mass measurements in the analysis of peptide maps generated by enzymatic digestion of proteins.

I. MASS RESOLUTION AND TIME-OF-FLIGHT MASS ANALYSIS

TOF mass spectrometers are widely used for both MALDI and ESI analysis of peptides and proteins as well as many other classes of biomolecules. The analytical advantages of TOF include relatively low initial cost, low maintenance cost, increased sensitivity over scanning instruments, the ability to record a complete spectrum from each pulsed ionization event (so-called Felgett advantage), a large dynamic range, and a large mass range (in excess of 100 kD).

The major disadvantage of TOF has traditionally been low mass resolution. The mass resolution of MALDI-TOF is especially poor because the ions are formed with a broad kinetic energy distribution [9] and a mass-independent initial velocity [10,11]. The kinetic energy distribution of MALDI-formed ions gives rise to a broad ion arrival time distribution at the detector. Because the kinetic energy distribution increases as the molecular weight increases, the peak widths for high m/z ratio ions are greater than those for low m/z ratio ions [9]. As a result of the low mass resolution, the carbon isotope peaks of the ions are rarely resolved in the MALDI mass spectra. Consequently, mass measurements of proteins are made using average mass (M_r) values. For proteins with $M_r >$ 5000, the distribution of the ^{13}C isotopes yields a peak shape for the $[M + H]^+$ ion that is nearly Gaussian [12], and Gaussian peak profiles can be accurately centroided and calibrated in terms of m/z ratios [13]. In theory, any peak shape based on isotopic abundances can be accurately centroided provided the entire peak profile is used in the centroiding algorithm [12]. However, peaks in the MALDI-TOF mass spectra can be asymmetrically broadened due to metastable ion decay [14] or delayed ion formation [15–17], and this makes it difficult to accurately centroid the entire peak. To avoid errors due to peak tailing, the centroid is often taken using only the upper portion of the peak (e.g., centroiding only the top 50% or 30% of the peak profile). Accurate centroids can be calculated for proteins with $M_r > 5000$ because the top 50% of the peak is characteristic of the theoretical peak profile. However, the ^{13}C isotope distribution for peptides with $M_r < 4000$ are non-Gaussian, and centroiding peaks in this mass range at a level other than using the entire peak profile results in mass measurement errors [13].

Factors other than the kinetic energy distribution can lead to line broadening in the MALDI-TOF mass spectrum. For example, (1) unresolved adduct ions $[M+Na]^+$, $[M+K]^+$, etc.; (2) unresolved matrix–analyte adduct ions $[M+(matrix)_n]^+$; (3) unresolved fragment ions $[M+H-NH_3]^+$, $[M+H-H_2O]^+$; and (4) heterogeneity of the sample (oxidation of Met or Cys, glycosylation, phosphorylation) contribute to the peak width for the $[M+H]^+$ ion measured at low mass resolution (e.g., mass resolution of 200–500). In each of these cases, the error associated in centroiding the unresolved peak results in significant errors for the measured mass.

The performance of MALDI-TOF in terms of ion yields and mass resolution can be improved by the appropriate choice of MALDI matrix, careful sample preparation, and optimizing the laser fluence and spot size [18]. A trade-off exists between the advantages and disadvantages of "hot" MALDI matrices such as α-cyano-4-hydroxycinnamic acid that typically produce high ion yields but result in more analyte ion fragmentation [19]. Sample preparation techniques that produce a thin film of matrix and analyte crystals have shown improvements in mass resolution [20–23] as well as durability when washing the crystals to remove excess salts.

Further improvements in mass resolution can be obtained by instrument modification. For example, an ion mirror, or reflectron, compensates for the kinetic energy distribution of the ions and greatly increases the mass resolution [24]. The enhanced mass resolution of the reflectron is greatest for peptides with $M_r < 4000$, and there are few examples where the reflectron markedly improves performance of high-m/z protein ions. Metastable decay of ions in the flight tube leads to a broadening of the peak profile due to unresolved peaks generated by the loss of small neutrals (NH_3, H_2O, CO_2) [15]. For proteins with $M_r > 10,000$ the broadening due to metastable ions is usually greater than the broadening due to the kinetic energy distribution, and thus the peaks observed in the reflected mass spectra are broader than peaks in the linear-TOF mass spectra. For this reason, the most effective mass range of a reflectron TOF mass spectrometer is usually limited to the analysis of peptides or small proteins.

Alternatively, peak broadening due to the ions' kinetic energy distribution can be reduced by applying the principles of Wiley and McLaren's time-lag focusing [25]. For MALDI-TOF, this is commonly referred to as delayed extraction (DE) [23,26–29]. DE involves forming the ions in a field-free region and applying a high-voltage pulse to accelerate the ions. The extraction pulse is delayed with respect to firing the laser by a few hundred nanoseconds. DE results in a significant increase in the mass resolution for linear-TOF instruments. For example, using DE it is possible to resolve the carbon isotope peaks of peptides up to $M_4 \sim 4000$. A difficulty in using DE with a linear-TOF instrument is that the energy focusing is mass dependent. Thus, it is difficult to analyze peptide mixtures at high mass resolution if the desired mass range is greater than approximately 2000 Da. For mixtures of analytes having a broad range of molecular weights, it is advantageous to use DE in conjunction with a reflectron-TOF instrument. The analysis of peptides using DE-MALDI-reflectron TOF results in high mass resolution ($m/\Delta m = 10,000–15,000$) and mass measurement accuracy in the low ppm range (1–10 ppm) [23]. The mass dependence in the energy focusing of DE observed in the linear mode is not as evident when DE is coupled with a reflectron, because the reflectron is performing the majority of the energy focusing.

The major advantage of DE-linear TOF comes in the analysis of proteins with M_r 10–30 kD. Because the delay between the laser firing and the application of the extraction voltage pulse is typically short (100 ns–1 μs), peak broadening problems of proteins associated with metastable ion decay in DE linear mass spectra are not as severe as the peak broadening observed in reflectron mass spectra. As a result, DE-linear-TOF provides better-quality mass spectra for proteins than does reflectron-TOF mass spectra. For proteins with M_r 10–30 kD the mass resolution for DE-linear-TOF is not sufficient to resolve the individual carbon isotopes, but the width of the peak profile is approximately that of the unresolved isotopic envelope. An example of a high-resolution MALDI DE-linear-TOF mass spectrum of horse heart myoglobin (M_r 16951.5) is given in Fig. 1. The $[M+H]^+$, $[M+2H]^{2+}$, $[M+3H]^{3+}$, and $[M + matrix]^+$ ions are observed in the mass spectrum. The inset shows the expanded $[M+H]^+$ ion region. The full width at half maximum (FWHM) of the $[M+H]^+$ ion signal is ~8.4 Da, corresponding to a mass resolution ($m/\Delta m$) of 2000. In the continuous linear-TOF mass spectrum of this protein a mass resolution of 200 is obtained. The peak profile contained in Fig. 1 agrees well with the theoretical peak width of the unresolved isotopic envelope of myoglobin (8 Da) (Fig. 2). Because the measured peak width shows no signs of artificial broadening and the peak shape is Gaussian, centroiding this peak allows for an accurate mass assignment. In fact, we observe average deviations of ~0.2–0.3 Da (10–20 ppm) on multiple measurements of proteins (<20 kD) when the measured peak width approaches the resolution limit set by the calculated unresolved isotopes [30]. Accurate mass measurements of unresolved peak profiles are ultimately limited to ~10 ppm due to the varying $^{12}C/^{13}C$ abundance. The naturally varying abundance of ^{13}C causes the average mass of carbon to range from 12.0107 to 12.0111 Da [31].

Even when accurate mass measurements of an unknown protein can be made, the identity of an unknown protein cannot be unambiguously determined solely on the basis of accurate mass measurement. The method we utilize for protein identification involves digesting the unknown protein using specific endoproteases, and analyzing the resulting peptide map of the protein. The enzymatically generated peptides typically fall in the mass range of a few hundred

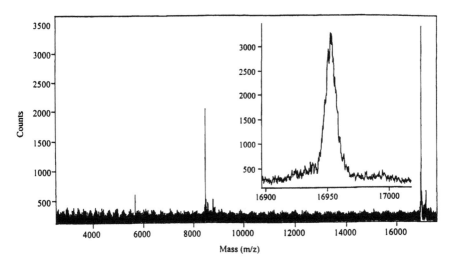

Figure 1   MALDI/DE-LTOF mass spectrum of horse heart myoglobin.

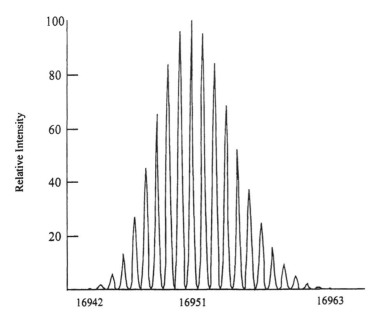

Figure 2   Calculated isotope distribution of horse heart myoglobin.

daltons to 3–4 kD. The mass resolution required to resolve isotope peaks over the mass range of the digest fragments (m/z up to ~5000) can be achieved by using DE-reflectron-TOF. Mass measurement accuracy of <10 ppm can be obtained on the basis of mass assignment using the all-^{12}C isotope. Using the all-^{12}C isotope peak removes the effect of varying ^{12}C/^{13}C abundance ratios as well as problems associated with centroiding a non-Gaussian isotope distribution. The abundance of the all-^{12}C isotope is sufficiently high to give good signal-to-noise ratios (S/N) up to M_r ~5000, but it is difficult to assign accurate masses using the all-^{12}C isotope for mass values larger than 5000 Da due to the low abundance of the all-^{12}C in this mass range.

II. APPLICATIONS OF HIGH-RESOLUTION TIME-OF-FLIGHT MASS SPECTROMETRY

High-resolution mass spectrometry has traditionally been used to determine the elemental composition of low-M_r (<500 Da) ions [32]. Up to approximately m/z 500 unique elemental compositions are separated by ~1 ppm, and when combined with complimentary chemical information (e.g., infrared spectroscopy, nuclear magnetic resonance, or some chemical knowledge of the compound or its origin) the mass measurement accuracy required for determination of elemental composition is on the order of 5 ppm [33]. On the other hand, accurate mass measurements of 1 ppb are required to determine the elemental composition of molecules with M_r 1–5 kD. Analysis of low-M_r species (<500 Da) is difficult using MALDI-TOF due to the high abundance of low-mass matrix ions and the narrow peak profiles for the low-mass ions. Ultimately, the solution to accurate mass measurements at low m/z can be overcome by using faster digitizers (GHz) and fast detector response (<1 ns).

Recently, the first use of accurate mass MALDI-TOF in the determination of an elemental composition was reported [34,35]. The compound studied was a naturally occurring substituted flavin adenine dinucleotide (FAD), a cofactor for nitroalkane oxidase isolated from *Fusarium oxysporum*. The mass of the substituted flavin was 887.2143 ± 0.0068 Da, and as mentioned earlier, the mass measurement accuracy required to unambiguously identify the elemental composition required MMA of ~1 ppb. In this study the identity of the substituent group on the flavin was of interest. Because the flavin and substituted flavin differed only by the mass of the substituent, mass difference (substituted flavin minus unsubstituted flavin) was used to determine the mass of the substituent (102.0572 Da). An elemental composition search using 102.0572 ± 0.0068 mass resulted in only one possibility ($C_4H_8NO_2$, $M_r$calc − $M_r$meas = 0.0017 Da), after eliminating nitrogen rule violators. $C_4H_8NO_2$ corresponds to nitrobutane bound to the N(5) position of the isoalloxazine ring of FAD [34,35].

Accurate mass measurements of peptides having M_r 1–5 kD can be performed at the <10 ppm level because the naturally occuring amino acids limit the number of possible elemental combinations. For example, the nominal masses for lysine and glutamine are both 128 Da, but their exact mass values differ by 0.03639 Da [36]. To determine a lysine for glutamine substitution in a 1000-M_r peptide requires mass measurement accuracy of 36 ppm, whereas the same substitution in a 5000-M_r peptide requires mass measurement accuracy of 7 ppm.

Highly accurate mass measurements have practical utility in the analysis of peptides in the rapidly expanding area of protein database searching [37]. Proteins can be identified using the mass of peptides that result from enzymatic digestion because the peptide fragments serve as "fingerprints" for a particular protein [38,39]. The number of peptides required to correctly identify a protein is directly related to the accuracy with which the peptide masses are measured [40]. If all of the predicted digest fragments are observed in the mass spectrum, accurate mass measurements are not required for correct protein identification; however, it is rare that all of the peptide fragments are observed in the mass spectrum. When many digest fragments are used in the database search and the required number of matched peptides is high, the mass window for matching peptides can be large (~1 Da). Several proteins in the database have peptides with similar (within 1 Da) mass values to the observed digest fragments, resulting in a large number of possible proteins, but the correct protein is identified by having the highest number of matched peptides.

Large error limits ($\pm$ 1 Da) often used in database searching cannot be tolerated when the number of observed peptide fragment ions is small. Several scenarios can lead to the observation of only a few peptide fragments in the mass spectrum after a protein has undergone enzymatic digestion. Certain proteins resist enzymatic digestion; this is especially true for proteins that have a large number of cysteine residues that form intramolecular disulfide bonds. To facilitate digestion, the disulfide bonds can be broken by adding a reducing agent, such as dithiothrietol, followed by alkylation of the cysteines to prevent recombination. Note that complete reduction/alkylation is not practical when the aim of the study is to locate the cross-linked residues (discussed later) [41]. Even when a complete map of peptide fragments is generated from the enzymatic digestion, the number of observed peptides in the mass spectrum can be small due to discrimination in the ionization process. The peptide fragments that contain acidic residues may not be observed in the positive ion MALDI mass spectra. The presence of buffers and salts used in protein isolation/purification can also lead to suppression of analyte ion signals. The choice of the MALDI matrix and the pH of the matrix/analyte solution also strongly influence peptide ion yields [42]. Suppression of analyte ion signal is especially daunting when the amount of peptides being produced is near the detection limit (a few femtomoles). Under the conditions just outlined, accurate mass measurements of the peptide fragments are crucial to correct protein identification. In cases where only a few peptide fragments are observed, the mass error limits for matching peptides must be narrow in order to reduce the number of possible proteins. The number of proteins that yield peptide fragments within 1 Da of the experimentally determined mass value is quite large, whereas the number of proteins that yield peptide fragments within 20 ppm of the measured m/z is remarkably small. In a recent report by Mann and coworkers [37], a database search using mass limits of 30 ppm resulted in a unique protein match, whereas searching with a 1-Da mass limit resulted in 163 proteins each having at least 10 matching peptides.

The utility of enzymatic digestion in conjunction with high-resolution/ accurate mass measurement MALDI-TOF in the analysis of proteins can be illustrated by the protein horse heart cytochrome C ($M_r = 12360.1$). Figure 3a contains a MALDI/DE-RTOF mass spectrum of a tryptic digest of horse cytochrome C. The inset (Fig. 3b) shows an expansion of the ion at m/z 1633.620. A predicted tryptic fragment ion (Ile^9-Lys^{22}) has a calculated [M + H]$^+$ ion mass of 1633.820 Da, similar to the measured value. But assigning the ion at m/z 1633.620 to the digest fragment (Ile^9-Lys^{22}) results in a mass error of 122 ppm. Conversely, assignment of the observed m/z 1633 peak to the digest fragment (Cys^{14}-Lys^{22}) with the heme group covalently bound to Cys^{14} and Cys^{17} results in a mass error of only 3 ppm. The assignment of the observed fragment to the heme containing peptide is further supported by comparing the observed isotopic distribution to the calculated distribution. The insets of Fig. 3 contain the isotope distribution of the observed fragment ion (Fig. 3b) and the calculated isotopic distribution of the heme containing digest fragment (Fig. 3c). The uncharacteristic isotope distribution is due to the iron isotopes in the heme group. Thus, accurate mass measurements in conjunction with analyzing the isotope distribution at high mass resolution can aid in the correct assignment of digest fragment ions.

Protein identification via database searching with accurate mass measurements of a peptide map is illustrated by the following example. A small amount of horse heart myoglobin was subjected to a 24-hr enzymatic digestion using trypsin. A representative mass spectrum of the tryptic digest of horse heart myoglobin is given in Fig. 4. The observed [M + H]$^+$ ions were entered in an online protein database search platform (http://rafael.ucsf.edu/MS-Fit.html). Eleven observed peptides matched with the database entry for myoglobin,* and

---

*Most database entries for myoglobin contain the incorrect sequence. The sequence commonly found incorrectly lists residue 122 as asparagine [41] instead of aspartic acid [42]. The correct M_r for myoglobin is 16951.5, but the incorrect sequence gives rise to the often reported $M_r = 16950.5$ [36,43], in spite of this sequence error being addressed by several researchers [42,44]. The problem is especially pertinent to mass spectrometrists who often use horse heart myoglobin as a mass calibration standard.

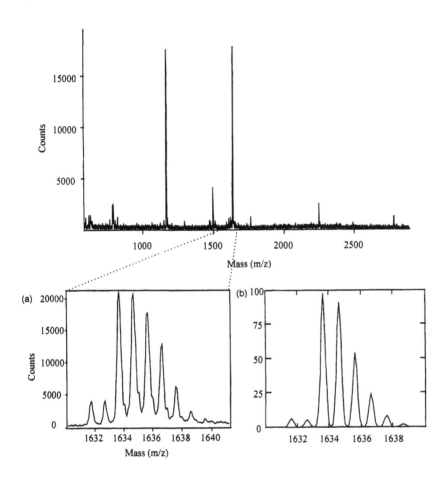

Fragment Sequence	Measured m/z	Calculated m/z	Δ m/z	m/z error (ppm)
heme CAQCHTVEK 14 22	1633.620	1633.615	0.005	3
IFVQKCAQCHTVEK 9 22	1633.620	1633.820	-0.200	-122

Figure 3 MALDI/DE-RTOF mass spectrum of a tryptic digest of horse heart cytochrome C. (a) Expansion of the ion at m/z 1633. (b) Calculated isotope distribution of the tryptic fragment Cys^{14}-Lys^{22} + the heme group.

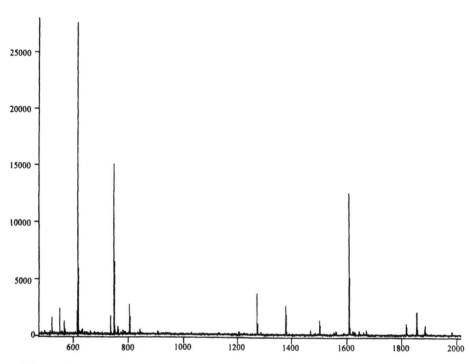

Figure 4 MALDI/DE-RTOF mass spectrum of a tryptic digest of horse heart myo-globin.

the observed and calculated mass values for the peptide $[M+H]^+$ ions are contained in Table 1. A peak corresponding to the heme group was also identified (Table 1), although the mass of the heme peak was not recognized by the database search. The average error for the observed peptides is ~6 ppm with the largest error less than 15 ppm. Figure 5 contains expanded regions of the $[M+H]^+$ ions at m/z 748 and 1885. These ions are shown to illustrate two points: (1) high mass resolution observed over the entire mass range of the digest fragments, and (2) the effect of S/N on mass measurement accuracy. The mass assignment of the ion at m/z 1885 results in a mass error of ~10 ppm, whereas the mass assignment of the ion at m/z 748 results in a mass error of 1 ppm. The S/N of the latter ion is ~15 times greater than the former. The observed fragment ions with the lowest observed abundances, m/z 397, 735, and 1885, have the highest observed mass errors, whereas the most abundant ion, corresponding to the heme group, had the lowest observed mass error (0.1 ppm). The 11 observed tryptic fragments contain 120 of the 153 amino acids

Table 1 Identification of Tryptic Digest Fragments of Horse Heart Myoglobin

Fragment sequence	Measured m/z	Calculated m/z	$\Delta m/z$	m/z Error (ppm)
Gly^1-Lys^{16}	1815.9143	1815.9030	0.0113	6.2
Val^{17}-Arg^{31}	1606.8618	1606.8553	0.0065	4.0
Leu^{32}-Lys^{42}	1271.6619	1271.6636	−0.0017	−1.3
His^{48}-Lys^{50}	397.2513	397.2563	−0.0050	−12.6
His^{64}-Lys^{77}	1378.8493	1378.8422	0.0071	5.1
Gly^{79}-Lys^{96}	1982.0649	1982.0572	0.0077	3.9
Gly^{80}-Lys^{96}	1853.9666	1853.9622	0.0044	2.4
His^{97}-Lys^{102}	735.4773	735.4881	−0.0108	−14.7
Tyr^{103}-Lys^{118}	1885.0419	1885.0224	0.0195	10.3
His^{119}-Lys^{133}	1502.6627	1502.6698	−0.0071	−4.7
Ala^{134}-Arg^{139}	748.4349	748.4357	−0.0008	−1.0
Heme	616.1774	616.1773	0.0001	0.1

of myoglobin. Covering 78% of the amino acids in a single enzymatic digest definitely allows for the identification of the protein; however, our objective would not have been achieved if we had intended to observe full coverage of the protein, that is, each residue observed in a peptide fragment.

A quick glance at the digest spectra in Figs. 3 and 4 reveals a vast difference in observed ion abundances. Assuming the trypsin digestion went to completion (cleaving at each lysine and arginine), there should be equal concentrations of each enzymatically generated peptide fragment, but this is not apparent from the mass spectrum. The peptides are not observed at the same intensity (or not observed at all), due to discrimination effects in the ionization process. The spectrum in Fig. 4 was acquired from a spot on the sample plate that contained approximately 1–2 pmol of digested protein. Because of the low intensity of several signals, it is evident that if only a few femtomoles of sample had been applied to the probe, the low-intensity signals would not have been seen at all. If only the most intense signals had been identified and the database had been searched using only four mass values and requiring a minimum of three peptides for a match, the following results would be obtained: Searching with peptide mass error limits of 1 Da results in 378 possible proteins, searching with a 100-ppm window results in 30 possible proteins, and searching with a 20-ppm window results in 5 possible proteins. One of these 5 proteins from the last search corresponds to the correct myoglobin sequence and matched 4/4 peptides, the second matched protein is the database entry for the incorrect sequence of myoglobin and matched 4/4 peptides (the incorrect sequence at residue 122 was not covered by the 4 peptides used in the search), and the

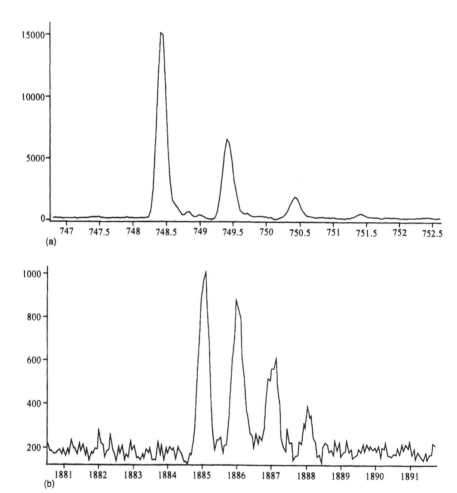

Figure 5 MALDI/DE-RTOF mass spectra of myoglobin tryptic digest expanding the [M + H]$^+$ ions at (a) m/z 748 and (b) m/z 1885.

remaining 3 proteins matched 3/4 peptides and are myoglobin mutants with single amino acid substitutions.

   The preceding example was performed on a relatively pure, commonly available protein that is well suited for MALDI-MS. The question remains: Can comparable results, in terms of mass resolution and mass measurement accuracy, be obtained for a small amount of protein isolated from biological origins? To investigate this issue we examined recombinant and isolated native acyl coenzyme (Acyl-CoA) binding proteins (ACBP) from the mouse and rat liver. Native and recombinant rat ACBP were analyzed to confirm the sequence of the recombinant protein. The mouse liver ACBP, isolated from tumor cells that overexpress the protein, was analyzed to determine if there were any amino acid substitutions, deletions, or posttranslational modifications (and if these modifications existed, which amino acid residues are modified) in the tumor-cell-isolated ACBP as compared to ACBP isolated from normal mouse liver cells.

As mentioned earlier, the observed peptide fragment ions from a single enzymatic digest rarely cover the entire sequence of a protein; therefore, multiple enzymes are used in order to observe full coverage. Figure 6 contains linear MALDI-DE/TOF mass spectra of native rat ACBP, recombinant rat ACBP, and mouse liver ACBP. The observed molecular mass for recombinant rat ACBP (9896.2 Da) agreed well with the calculated M_r (9896.3 Da), but the measured molecular mass for the native ACBP (9938.2 Da) is ~42 Da higher than the calculated M_r. A 42-Da mass difference corresponds to several possible single amino acid substitutions, such as Ala>Leu/Ile, Asp>Arg, Gly>Val, or Ser> Glu. Even more possibilities exist if more than one amino acid is substituted. The measured M_r of mouse liver ACBP (9911.4 Da) is also observed to be ~42 Da higher than the calculated M_r (9869.3 Da). Each protein was divided into aliquots and digested using trypsin and Asp-N. Native and recombinant rat ACBP were also subjected to a digest by chymotrypsin. The resulting peptide fragments were mass analyzed using MALDI-DE/RTOF. A mass spectrum of the chymotrypsin digest of native rat ACBP is given in Fig. 7. The insets illustrate the high mass resolution that is obtained using MALDI-DE/RTOF over the entire mass range of peptide fragment ions. Along with this high mass resolution we also observe high mass measurement accuracy. The average mass error for the observed enzymatically generated peptides from the native rat ACBP is ~7 ppm. Table 2 contains the observed and calculated $[M+H]^+$ ions that resulted from the trypsin and Asp-N digests of native rat ACBP. Neither digest on its own yielded sufficient peptide ions to obtain full coverage of the protein, but a combination of the observed peptides from the two enzymes resulted in full coverage. The 42-Da mass difference between the native and recombinant rat ACBP was found to occur in the N-terminal peptide fragment. The mass difference was determined to arise from a posttranslational acetylation of the N-terminal serine. Acetylation of the N-terminus was also observed for the isolated mouse liver ACBP, explaining the 42-Da difference between its measured and calculated M_r. The observance of the posttranslational N-terminal acetylation is a common occurrence in both rat and mouse ACBP.

The results of the trypsin and Asp-N digests of mouse ACBP are contained in Table 3. The mass measurement accuracy of the observed trypsin and Asp-N fragment ions of mouse ACBP is ~6 ppm. As in the case of the rat

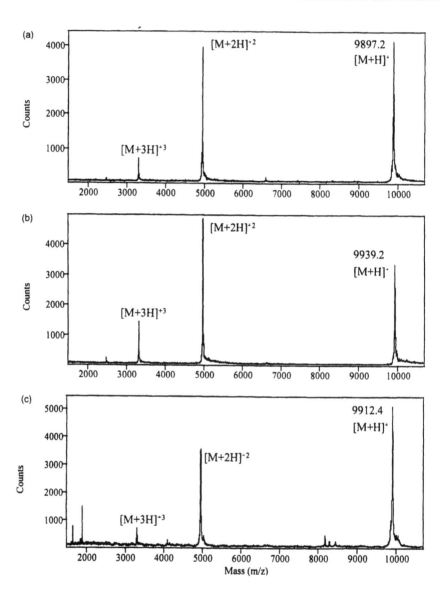

Figure 6   MALDI/DE-LTOF mass spectra of Acyl-CoA binding protein (ACBP) from (a) recombinant rat, (b) native rat, and (c) native mouse.

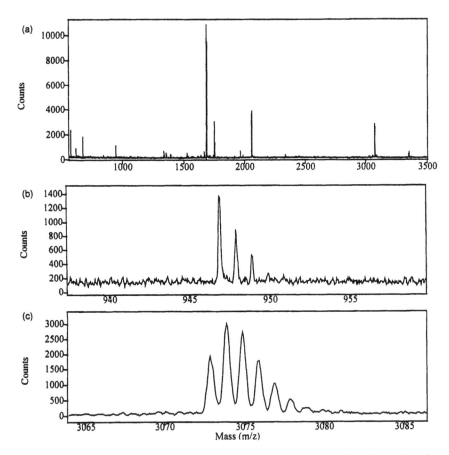

Figure 7 (a) MALDI/DE-RTOF mass spectrum of a chymotryptic digest of native Rat ACBP, (b) expansion of *m/z* 947, and (c) expansion of *m/z* 3073.

ACBP, the combined digests provide full coverage of each amino acid in the mouse ACBP. This confirms that the sequence of the ACBP isolated from tumor cells is the same as normal ACBP. The coverage for both the native rat ACBP and the isolated mouse ACBP is illustrated in Fig. 8, with the solid lines representing the observed tryptic fragments and the dashed lines representing the observed Asp-N fragments.

In addition to molecular weight measurements, mass spectrometry can provide structural information. To show the utility of high-resolution MALDI-TOF in studying secondary protein structure, we examined the disulfide bonds of bovine insulin. Bovine insulin is a heterodimeric protein composed of an *A*

Table 2 Identification of Enzymatic Digest Fragments of Native Rat Liver ACBP

Fragment sequence	Measured m/z	Calculated m/z	$\Delta m/z$	m/z Error (ppm)
Trypsin fragments				
Ac-Ser^1-Arg^{14}	1635.812	1635.798	0.014	8.6
Asp^8-Arg^{14}	802.458	802.442	0.016	19.9
Leu^{15}-Lys^{32}	2227.101	2227.111	−0.010	−4.5
Tyr^{17}-Lys^{32}	1985.934	1985.932	0.002	1.0
Gln^{33}-Lys^{50}	1912.023	1912.014	0.009	4.7
Gln^{33}-Lys^{52}	2097.108	2097.130	−0.022	−10.5
Ala^{53}-Lys^{62}	1275.690	1275.685	0.005	3.9
Trp^{55}-Lys$^{60.}$	835.374	835.374	0.000	0.0
Trp^{55}-Lys^{62}	1076.542	1076.553	−0.011	−10.2
Tyr^{72}-Lys^{81}	1237.661	1237.668	−0.007	−5.7
Tyr^{72}-Lys^{82}	1365.759	1365.763	−0.004	−2.9
Tyr^{72}-Lys^{83}	1493.834	1493.858	−0.024	−16.1
Asp-N fragments				
Asp^6-Thr^{20}	1713.959	1713.950	0.009	5.3
Asp^{21}-Gly^{37}	2014.978	2014.958	0.020	9.9
Asp^{42}-Leu^{47}	670.387	670.389	−0.002	−3.0
Asp^{48}-Trp^{55}	945.550	945.552	−0.002	−2.1
Asp^{56}-Ile^{86}	3644.950	3644.936	0.014	3.8

chain ($M_r = 2339$) and a B chain ($M_r = 3395$). Two disulfide bonds link the A and B chains, and an additional disulfide bond is located between Cys^6 and Cys^{11} in the A chain as shown in Fig. 9. Intact bovine insulin was subjected to a partial reduction using dithiothrietol, and then the reduced cysteines were alkylated using iodoacetamide. Note that each alkylation results in the addition of CH_2CONH_2 (58 Da). The partially alkylated insulin was separated using high-performance liquid chromatography (HPLC), and the reduced, alkylated bovine insulin with four cysteines alkylated (and one disulfide bond remaining intact) was analyzed by MALDI-DE/RTOF. The MALDI-TOF mass spectrum confirmed the presence of a single species with M_r of 5966 [M_r of insulin (5734) + 4 alkylations]. The alkylated protein was then digested using α-chymotrypsin and the peptide map was analyzed using MALDI-DE/RTOF. A mass spectrum of the peptide digest mixture is contained in Fig. 10. The insets illustrate the mass resolution that is obtained for the peptide fragments. Five mass spectra of the digest fragments were acquired and the mass values were averaged; the results are given in Table 4. A high degree of mass measurement

Table 3 Identification of Enzymatic Digest Fragments of Tumor Cell Mouse Liver ACBP

Fragment sequence	Measured m/z	Calculated m/z	Δm/z	m/z Error (ppm)
Trypsin fragments				
Ac-Ser[1]-Lys[13]	1493.709	1493.712	−0.003	−2.0
Ac-Ser[1]-Arg[14]	1649.827	1649.814	−0.013	−7.9
Ac-Ser[1]-Lys[16]	1891.001	1890.993	0.008	4.2
Asp[8]-Arg[14]	802.457	802.442	0.015	18.7
Leu[15]-Lys[32]	2227.122	2227.111	0.011	4.9
Tyr[17]-Lys[32]	1985.927	1985.932	−0.005	−2.5
Gln[33]-Lys[50]	1912.014	1912.014	0.000	0.0
Gln[33]-Lys[52]	2097.147	2097.13	0.017	8.1
Ala[53]-Lys[60]	1034.502	1034.506	−0.004	−3.9
Ala[53]-Lys[62]	1275.694	1275.685	0.009	7.1
Trp[55]-Lys[60.]	835.377	835.374	0.003	3.6
Trp[55]-Lys[62]	1076.554	1076.553	0.001	0.9
Tyr[72]-Lys[81]	1223.646	1223.652	−0.006	−4.9
Tyr[72]-Lys[82]	1351.748	1351.747	0.001	0.7
Tyr[72]-Lys[83]	1479.84	1479.842	−0.002	−1.4
Lys[82]-Ile[86]	608.38	608.377	0.003	4.9
Asp-N fragments				
Asp[6]-Thr[20]	1713.940	1713.950	−0.010	−5.8
Asp[21]-Gly[37]	2014.939	2014.958	−0.019	−9.4
Asp[42]-Leu[47]	670.390	670.389	0.001	1.5
Asp[48]-Trp[55]	945.550	945.552	−0.002	−2.1
Asp[56]-Val[77]	2529.275	2529.302	−0.027	−10.7
Asp[78]-Ile[86]	1093.622	1093.626	−0.004	−3.7

accuracy was obtained, with the average observed mass error for the peptides only ~10 ppm. The peak shown in the middle inset of Fig. 10 has an uncharacteristic isotope profile due to overlapping fragment ions. The ion at m/z 1886 is identified as a fragment of the B chain with the Cys[7] alkylated, and m/z 1890 is identified as an A-chain/B-chain fragment attached via the disulfide between A20 and B19. High mass resolution allows both of the overlapping peaks to be correctly identified, but if the mass spectrum of the digest fragments were acquired at low mass resolution it would be very difficult, if not impossible, to correctly identify these two fragment ions.

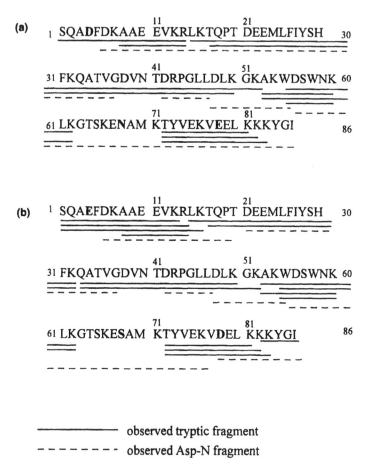

Figure 8   Amino acid sequence of ACBP from (a) rat and (b) mouse.

III. MASS SPECTROMETRY AND DATABASE SEARCHING

A serious caveat regarding database searching for protein identification is directly related to the issue of mass measurement accuracy. Although mass spectrometrists strive to achieve the highest level of mass measurement accuracy for the intact protein, this objective is not always of practical utility. For instance, accurate mass measurements can be obtained for proteins (see Fig. 1), but the practical uses of such data may be limited. For example, the protein being analyzed often has undergone posttranslational modifications (discussed

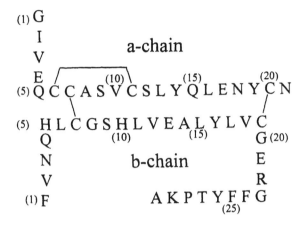

Figure 9 Amino acid sequence of bovine insulin showing location of the disulfide bonds.

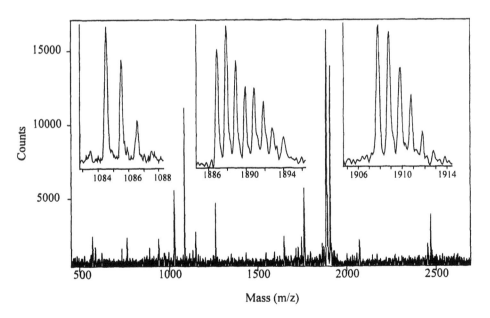

Figure 10 MALDI/DE-RTOF mass spectrum of a chymotryptic digest of the reduced/ alkyated bovine insulin with insets demonstrating resolution.

Table 4 Identification of Chymotrypsin Digest Fragments of Reduced/Alkylated Bovine Insulin

	LVCGERGFF, B17-B25	LVCGERGFF \Alkylation. B17-B25 + CH$_2$CONH$_2$	CN / LVCGERGFF, (B17-B25) / (A20-A21)	QLENYCN / LVCGERGF, B17-B24 / (A15-A21)	FVNQHLCG-SHLVEALY \Alkylation, (B1-B16) + CH$_2$CONH$_2$	QLENYCN / LVCGERGFF-17. (B17-B25) / (A15-A21)-17	QLENYCN / LVCGERGFF, (B17-B25) / (A15-A21)
1	1027.509	1084.536	1260.565	1760.812	1886.923	1890.867	1907.843
2	1027.481	1084.534	1260.579	1760.763	1886.926	1890.885	1907.826
3	1027.523	1084.535	1260.573	1760.802	1886.961	1890.858	1907.850
4	1027.525	1084.537	1260.558	1760.788	1886.929	1890.881	1907.854
5	1027.515	1084.554	1260.565	1760.788	1886.952	1890.860	1907.853
Average deviation	0.013	0.006	0.007	0.013	0.015	0.010	0.008
Average observed m/z	1027.511	1084.539	1260.568	1760.791	1886.938	1890.870	1907.845
Calculated m/z	1027.504	1084.525	1260.551	1760.774	1886.922	1890.839	1907.842
Error (Da)	0.007	0.014	0.018	0.017	0.016	0.031	0.003
Error (ppm)	7	13	14	10	8	16	2

earlier), and thus the resulting measured M_r may differ from the M_r contained in the database (calculated from cDNA). If the database is searched with a narrow M_r window, an incorrect identification may result. If, on the other hand, the database is searched with a larger M_r window, the protein may then be correctly identified. It is important to remember that the search is performed using not only the M_r of the protein, but more importantly, the accurate masses of the peptide digest fragments.

In addition to database searching for protein identification, accurate mass measurements of enzymatic digest fragments can be used in the analysis of known proteins. By using multiple enzymes, overlapping fragments can be produced and used to confirm the sequence of recombinant proteins. This method can also be used to identify and locate the sites of posttranslational modifications such as oxidized methionine or cysteine; glycosylation; phosphorylation; or N-terminal acetylation. Accurate mass measurements can also be used to search for amino acid substitutions between protein isoforms. Often these modifications are substitutions of Asp for Asn, or Glu for Gln, which result in a 1-Da mass shift.

IV. CONCLUSIONS

In this chapter we have dealt with issues relating to mass measurement accuracy such as mass resolution, peak shape, and peak centroiding, but it must be stressed at this point that the ultimate limit of mass measurement accuracy often lies in the mass calibration that is used [45]. In order to obtain the level of mass measurement accuracy illustrated herein (low ppm), careful attention must be given to several factors in the MALDI-TOF experiment. One of these is the high-voltage power supplies. It is our experience that the high-voltage power supplies are stable only after warming up for an hour prior to data acquisition. Power-supply stability is especially important if an external calibration is used, as the power-supply voltage will drift between the acquisition of the sample and the calibration standard. Another crucial factor in obtaining an accurate mass calibration is the homogeneity of the matrix/analyte crystals or, more importantly, the sample support. A deviation in the sample support or matrix/analyte crystal of only 10 μm corresponds to a mass deviation of 5 ppm [23], illustrating the need for highly polished surfaces and microcrystalline sample preparation techniques. For highly accurate external calibration, it is advantageous to place the sample spot containing the calibration standard as close as possible to the sample spot containing the analyte, thus minimizing the effects of deviations in the sample support.

A limitation to using accurate mass data in assigning enzymatic digest fragments occurs when different fragment ions with different amino acid compositions have the same elemental composition [37]. For example, the elemental composition of the peptides Gly-Gly-Lys and that of Asn-Lys are identical $(C_{10}H_{20}N_4O_4)$, and therefore their mass values are identical and cannot be distinguished from one another using only accurate mass data. Such discrepancies can be resolved by acquiring postsource decay (PSD) [46] mass spectra to confirm the sequence (i.e., the identity) of the peptide.

V. EXPERIMENTAL

The mass spectra shown herein were acquired on a Voyager Elite XL TOF mass spectrometer (PerSeptive Biosystems, Framingham, MA). The instrument has been previously described in detail [23,27]. Briefly, the instrument has a two-stage gridded ion source equipped with delayed extraction. A variable-voltage wire ion guide (typically biased at -10 to -1 V) extends from the ion source to the entrance of the single-stage gridded reflectron. The linear flight length is $\sim$4.2 m, and the effective path length for the reflected ions is $\sim$6.6 m. The reflected ions are detected by a coaxial dual-microchannel plate assembly with the transient signal digitized using a 500-MHz data acquisition card sampling at 500 MS/s. The mass spectrum of myoglobin shown in Fig. 1 was acquired in the DE-linear TOF mode with a total acceleration of 25 kV, an extraction pulse of 1.75 kV, and a delay time of 300 ns. The mass spectra of ACBP shown in Fig. 5 were acquired in the DE-linear TOF mode with a total acceleration of 25 kV, an extraction pulse of 1.75 kV, and a delay time of 150 ns. The high-resolution MALDI-DE/RTOF mass spectra contained in Figs. 3, 4, 7, and 10 were acquired with a total acceleration of 25 kV, an extraction pulse of 7.5 kV, and a delay time of 150 ns.

The sample spots were prepared using an overlayer method [23], which produces homogeneous surfaces and minimizes so-called "hot spots." A matrix solution was prepared by dissolving 35 mg of ACCA in 1 ml methanol. A matrix bed was prepared by applying 0.5 μl of this solution to the sample plate and allowing it to dry. The matrix/analyte solutions were prepared with 2:1 water:methanol, having final matrix and analyte concentrations of 10 mM and 1–3 μM, respectively. A small amount ($\sim$0.3 μl) of the matrix/analyte solution was applied to the matrix bed and was allowed to air dry. The small amount of solution does not redissolve the matrix bed. The dried sample was then washed with 10 μl deionized water.

ACKNOWLEDGMENTS

This mass spectrometry development research is supported by a grant from the U.S. Department of Energy, Division of Chemical Sciences, Office of Basic

Energy Sciences. The Laboratory for Biological Mass Spectrometry (LBMS) is funded by the Texas A&M University Research Infrastructure Program.

REFERENCES

1. M Karas, D Bachmann, U Bahr, F Hillenkamp. Int J Mass Spectrom Ion Processes 78:53, 1987.
2. JB Fenn, M Mann, CK Meng, SF Wong, CM Whitehouse. Science 246:64, 1989.
3. RD Macfarlane, DF Torgerson. Science 191:920, 1976.
4. A Vertes, R Gijbels. In: A Vertes, R Gijbels, F Adams, eds. Chemical analysis series vol 124, Laser ionization mass analysis. New York: John Wiley and Sons, 1993, p. 127.
5. M Barber, RS Bordoli, G Elliott, RD Sedgwick, AN Tyler. Anal Chem 54:645A, 1982.
6. FH Strobel, T Solouki, MA White, DH Russell. J Am Soc Mass Spectrom 2:91, 1991.
7. AL Burlingame, RK Boyd, SJ Gaskell. Anal Chem 66:634R–638R, 1994.
8. AL Burlingame, RK Boyd, SJ Gaskell. Anal Chem 68:599R–651R, 1996.
9. J Zhou, W Ens, KG Standing, A Verentchikov. Rapid Commun Mass Spectrom 6:671–678, 1992.
10. RC Beavis, BT Chait. Chem Phys Lett 5:479–484, 1991.
11. Y Pan, RJ Cotter. Org Mass Spectrom 27:3–8, 1992.
12. RA Zubarev, PA Demirev, P Hakansson, BUR Sundqvist. Anal Chem 67:3793–3798, 1995.
13. J Yergey, D Heller, G Hansen, RJ Cotter, C Fenselau. Anal Chem 55:353–356, 1983.
14. B Spengler, D Kirsch, R Kaufmann. Rapid Commun Mass Spectrom 5:198, 1991.
15. BH Wang, K Dreisewerd, U Bahr, M Karas, F Hillenkamp. J Am Soc Mass Spectrom 4:393, 1993.
16. RD Edmondson, DC Barbacci, KJ Gillig, DH Russell. Proceedings of the 44th ASMS Conference on Mass Spectrometry and Allied Topics, Portland, OR, May 12–16, 1996, p 736.
17. GR Kinsel, RD Edmondson, DH Russell. J Mass Spectrom 32:714–722, 1997.
18. A Ingendoh, M Karas, F Hillenkamp, V Giessmann. Int J Mass Spectrom Ion Processes 131:345, 1994.
19. GR Kinsel, LM Preston, DH Russell. Biol Mass Spectrom 23:205, 1994.
20. F Xiang, RC Beavis. Rapid Commun Mass Spectrom 8:199, 1994.
21. O Vorm, M Mann. J Am Soc Mass Spectrom 5:955, 1994.
22. O Vorm, P Roepstorff, M Mann. Anal Chem 66:3281, 1994.
23. RD Edmondson, DH Russell. J Am Soc Mass Spectrom 7:995, 1996.
24. BA Mamyrin, VJ Karatajev, DV Shmikk, VA Zagulin. Sov Phys JTEP 37:45, 1973.
25. WC Wiley, IH McLaren. Rev Sci Instrum 26:1150, 1955.
26. SM Colby, TB King, JP Reilly. Rapid Commun Mass Spectrom 8:865, 1994.
27. ML Vestal, P Juhasz, SA Martin. Rapid Commun Mass Spectrom 9:1044, 1995.
28. RS Brown, JJ Lennon. Anal Chem 67:1998, 1995.

29. RM Whittal, L Li. Anal Chem 67:1950, 1995.
30. E Murphy, RD Edmondson, DH Russell, F Schroeder. Biochim Biophys Acta 1997, in press.
31. RC Beavis. Anal Chem 65:496, 1993.
32. K Beimann. Mass spectrometry. In: JA McCloskey, ed. Methods in enzymology, vol 193, New York: Academic Press, 1990, p 295.
33. Instruction to authors. J Am Chem Soc 118:7A, 1996.
34. G Gadda, RD Edmondson, DH Russell, PF Fitzpatrick. J Biol Chem 272:5563–5570, 1997.
35. RD Edmondson, G Gadda, PF Fitzpatrick, DH Russell. Anal Chem 69:2862–2865, 1997.
36. Appendix. Mass spectrometry. In: JA McCloskey, ed. Methods in enzymology, vol 193, New York: Academic Press, 1990, p 888.
37. ON Jensen, A Podtelejnikov, M Mann. Rapid Commun Mass Spectrom 10:1371, 1996.
38. TD Lee, JE Shively. In: JA McCloskey, ed. Methods in enzymology, vol 193, Mass spectrometry. New York: Academic Press, 1990, p 361.
39. GJ Feistner, KF Faull, DF Barofsky, P Roepstorff. J Mass Spectrom 30:519, 1995.
40. Z-Y Park, S Shields, D Barbacci, RD Edmondson, DH Russell. Results presented at the 45th ASMS Conference on Mass Spectrometry and Allied Topics. Palm Springs, CA, 1997.
41. Y Wang. PhD dissertation, Texas A&M University, 1995.
42. SC Cohen, BT Chait. Anal Chem 68:31, 1996.
43. MO Dayhoff, ed. Atlas of protein sequence and structure. Silver Spring, MD: National Biomedical Research Foundation, 1968.
44. AE Romero-Herrera, H Lehmann. Biochim Biophys Acta 336:318, 1974.
45. CN McEwen, BS Larsen. Rapid Commun Mass Spectrom 6:173–176, 1992.
46. B Spengler, D Kirsch, R Kaufmann. J Phys Chem 96:9678, 1992.

3

A Novel Geometry Mass Spectrometer, the Quadrupole Orthogonal Acceleration Time-of-Flight Instrument, for Low Femtomole/Attomole Range Biopolymer Sequencing

Howard R. Morris, Thanai Paxton, Maria Panico, Roy McDowell, and Anne Dell
Imperial College of Science, Technology and Medicine, London, England

I. INTRODUCTION

Ultra-high-sensitivity biopolymer sequencing is a goal in many fields of molecular biology, and collisionally activated decomposition electrospray mass spectrometry (CAD ES MS/MS) using a triple-quadrupole mass spectrometer has become a method of choice for work in the high to mid femtomole range. However, when the detection of ions becomes statistical, as it may in that range, the mass assignment of fragment ions is inaccurate and either sequencing becomes impossible or ambiguities may result, due, for example, to the small differences in amino acid residue masses (I/L,N or K/Q,E). Some ambiguities may be resolved by synthesizing possible sequences, but this is time-consuming and costly. In considering the limitations of triple-quadrupole MS/MS with respect to scanning ion detection, resolution, transmission, and mass accuracy, we reasoned that a novel geometry quadrupole orthogonal acceleration time-of-flight (Q-TOF) instrument would have special merit for ultra-high-sensitivity MS/MS sequencing, and suggested its construction for this purpose in 1994. In the first research on this new instrument, including major histocompatibility complex (MHC) antigen and filarial nematode glycoprotein studies, low femtomole and attomole range sequencing has been demonstrated [1] with mass accuracy of better than 0.1 Da throughout the daughter ion spectrum, thus removing sequencing ambiguities in some of the most challenging work demanding the highest sensitivity.

Mass spectrometry (MS) has played a key role in protein sequencing and posttranslational modification analysis for almost three decades. Our group first solved the special problems associated with the MS analysis of protein-derived peptides in 1968 in work on silk fibroin [2,3], which concomitantly demonstrated the unique ability of mass spectrometry to sequence peptide mixtures. With the perfection of the permethylation procedure [4], a volatilizing and fragmentation-directing derivative first introduced into this field by Edgar Lederer [5], a new mixture mapping and sequencing strategy was applied to numerous peptide and protein sequencing problems in this laboratory over the following decade [6–19].

During this period, 1970–1980, it became clear that an area of major biological importance, but also of major structural difficulty, lay not so much in the protein sequence, but in the manner in which proteins were co- or posttranslationally modified and processed in order to activate or deactivate them biologically. Accordingly, our research emphasis diverged from protein sequencing per se to analysis of the structure and functional implications of posttranslational events. Key early examples of the development of mass spectrometric approaches to such problems led to the solution of the special calcium-binding properties of the blood coagulation zymogen prothrombin with the discovery of γ-carboxylation of glutamic acids in the N-terminal domain

[11,14], and the structure elucidation of the nature and multiple sites of glyco-sylation of the blood plasma antifreeze glycoproteins in Antarctic fish [20]. For posttranslational analyses such as these, mass spectrometry has always been the method of choice, and with the development of soft ionization techniques [field desorption (FD), fast atom bombardment (FAB), electrospray (ESI), and matrix-assisted laser desorption (MALDI)] the sensitivity now available with mass spectrometry has led to numerous biologically important structural solutions over the past decade. Examples from the work of this laboratory include the structure elucidation of cardioactive peptides [21], leukotriene D_4 [22], the in-sulin secretagogue [23], the glycosylation of interleukin 2 [24,25], calcitonin gene-related peptide [26,27], and pathological antithrombin variants [28], and characterization of glycans on neutrophils including the identification of sialyl Lewis[x] [29], disulphide assignments in tissue inhibitor of metalloproteinases [30], glycosylation of erythropoietin [31] and tissue plasminogen activator [32], discovery of O-GlcNAc in serum response transcription factor [33], character-ization of highly fucosylated schistosomal glycoproteins [34], and identification of glycodelin A, a human endometrial glycoprotein with immunosuppressive and contraceptive activities [35], and glycodelin S, the seminal plasma variant of this glycoprotein, showing the first example of gender-specific glycosylation in human biology [36].

Our attention was returned to basic protein sequencing in 1993, because of requests from immunologist colleagues to assist with the sequencing of pep-tides bound to MHC molecules, using electrospray triple-quadrupole tandem mass spectrometry [37,38]. One of the special problems associated with the MHC antigen field is the often tiny amount of sample available for sequencing, in the 100 fmol to 100 amol range, and while triple quadrupoles have clearly been successful in many analyses in the picomole and subpicomole range [37,38], we were initially unwilling to collaborate with the immunologists be-cause of our knowledge of the limitations of triple-quadrupole instrumentation when working in the sub 100 fmol range. These limitations relate to a number of factors including the need to lower resolving power in the observing quadru-pole of the MS/MS collision experiment (Q3) in order to improve transmission (sensitivity), together with the scanning nature of Q3 data acquisition in which much time is spent accessing areas of the mass range that turn out to be devoid of data. The net effect of nonisotopic resolution and low ion count is poor mass accuracy when signal strength becomes statistical. The consequences of these factors are limited real sensitivity together with the possibility of ambiguity in interpretation, for example, in distinguishing adjacent mass differences between fragment ions as corresponding to I/L,N,D or K/Q,E.

In contrast, the simultaneous/parallel detection of ions on a wide-angle array double-focusing mass spectrometer (ZAB2SE2FPD) has demonstrated the theoretically expected gains of one to two orders of magnitude in detection sensitivity over scanning experiments [39,40]. This work demonstrated genuine ultra-high-sensitivity data (10 fmol level) together with good mass accuracy (± 0.12 Da) on the high-resolution array, enabling unambiguous mass assignments. Wide-angle array detection with high-energy CAD MS/MS on the four-sector ZAB-T is particularly useful for carbohydrate studies where the high-energy input into the parent ions allows ring fragmentation, which can be very informative about substitution/branching positions [34,41]. However, high-energy fragmentation, in our view, does not offer significant advantages in peptide analysis, apart from the relatively unimportant distinction between Leu and Ile, and this is particularly true at the highest levels of sensitivity where it becomes disadvantageous to spread the daughter ion current between too many species. We have therefore mainly used the ZAB-T for carbohydrate ring substitution/branching studies described earlier [34,41], rather than for peptide analysis.

II. DESIGN CONCEPT

In considering which type of instrumentation design would allow us to reach the expectations of our immunological colleagues for genuine unambiguous low femtomole and subfemtomole sequencing of unknowns, our perceived requirements were that it should achieve post-CAD simultaneous/parallel (nonscanning) detection of fragment ions at good isotopic resolution with high transmission within the usable MS/MS mass range (~3000 Da), thus leading to good mass accuracy, while maintaining facile liquid interface ES coupling and, importantly, easily interpretable low-energy CAD spectra. The latter two points mitigated against a sector design at the time. An instrument achieving these requirements would then give accurate data capable of unambiguous interpretation, removing many of the problems associated with sub 100 fmol triple-quadrupole analysis.

Our solution to the problem of instrument design for ultra-high-sensitivity sequencing was to suggest the building of a novel geometry quadrupole orthogonal acceleration time of flight mass spectrometer, a Q-TOF, and we requested Fisons to build this instrument for Imperial College in 1994. A schematic of the Q-TOF, whose engineering design and construction was carried out by Fisons/Micromass, is shown in Fig. 1. The TOF mass spectrometer is used to acquire the data in both MS and MS/MS modes of operation. For normal mass spectra, the quadrupole is used in the rf-only mode as a wide-bandpass filter to transmit a wide mass range. The collision cell is not pressurized, and ions are transmitted to the TOF for mass analysis. In the MS/MS mode the quadrupole operates in the normal resolving mode and is able to select precursor ions up to m/z 4000 for collisional activation in the hexapole gas cell. The mass resolution is variable but, for the experiments reported here, was set to between two and

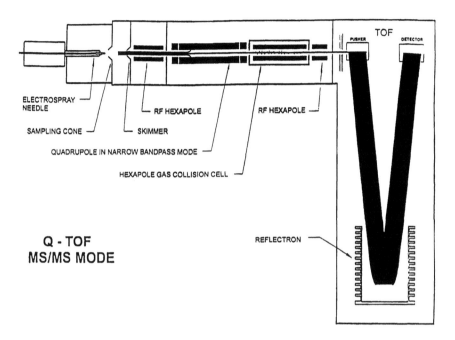

Figure 1 Schematic illustration of the Q-TOF.

four mass units to admit the isotope peaks of the parent precursor ion. The collisional energy may be varied up to 100 eV and the fragmentation patterns observed show low-energy decompositions, similar to a triple-quadrupole type mass spectrometer. Following decomposition, the daughter ions are transmitted to the TOF for mass analysis. In both MS and MS/MS modes, the ions are pulsed into the TOF with an accelerating voltage of about 7.8 kV. The time to mass relationship for the time-of-flight mass spectrometer is the same in MS and MS/MS modes, and switching is a simple operation. In both modes of operation, the effective mass resolution is determined by the time-of-flight mass analyzer, and is approximately 3000 (full width at half maximum [FWHM]). The pulse rate in this prototype instrument is set to either 16 kHZ, for which the mass range is 1500 amu, or 8 kHz for which the mass range is 6000 amu. The detector is a dual-stage microchannel plate, and acquisition is made through a time-to-digital convertor (TDC) operating at 1 GHz or 500 MHz (Precision Instruments, Inc., Knoxville, TN). The orthogonal geometry and parallel rather than sequential detection of ions lead to a significant improvement in sensitivity over scanning instruments when used to acquire full spectra. Advantages of the

Table 1   Summary of Q-TOF Performance Characteristics

Good signal/noise ratios in MS/MS spectra corresponding to low
   femtomole/attomole sample consumption
Daughter ion mass accuracies of 0.05 Da
Daughter ion resolution >3000 (FWHM)
Easy charge state differentiation
Simple low-energy MS/MS fragmentation
Definitive unambiguous sequence assignment

instrument include post-CAD nonscanning detection and resolution with good transmission, leading to ultra-high-sensitivity data with good mass accuracy even for very weak MS/MS spectra, thus meeting our requirements, as summarized in Table 1.

III. RESULTS

An immediate consideration was to establish the resolution, mass accuracy, and therefore usable sensitivity of the new instrument. This was achieved in (1) experiments involving the CAD MS/MS analysis of minor liquid chromatography (LC) peaks from a digest of a new unsequenced glycoprotein glycodelin S, (2) experiments with online LC-ESMS separation of a neuroendocrine peptide mixture with concomitant real-time CAD MS/MS of the component peptides (LC-ES MS/MS), and (3) experiments involving attomole-range sample consumption to stress the ion detection capabilities and examine resolution and mass accuracy when the data are becoming statistical [1].

Figure 2 shows the Q-TOF ESMS/MS spectrum of a fraction of a minor LC peak from a tryptic digest of human glycodelin S (GdS) run in a nanospray [42] experiment. GdS was an unsequenced seminal plasma derived glycoprotein believed to be related to glycodelin A (GdA), an endometrium-derived glycoprotein with potent contraceptive activity, whose amino acid sequence has been inferred from cDNA experiments. Our group has been studying the glycosylation of GdS and GdA by mass spectrometric strategies, and we have recently discovered that gender-specific glycosylation defines the physiological properties of these important glycoproteins including involvement in the first stages of sperm/egg interaction [35,36]. Our conclusion that differential glycosylation is the key determinant of the differential activity of GdS and GdA is, of course, based not only on differences in the glycosylation observed, but also on evidence for the identity of the protein sequences. In this regard, while peptide

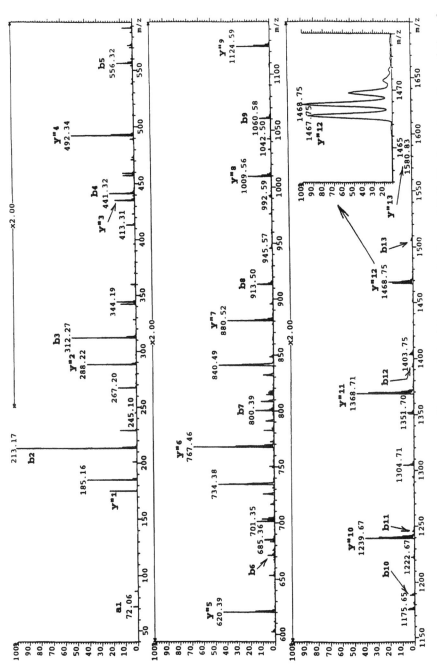

Figure 2 Daughter ion spectrum of $[M + 2H]^{2+}$ 840.4 derived from a tryptic digest of GdS. The inset shows the resolution of the daughter ion signal at m/z 1467.76.

mapping experiments on high-performance liquid chromatography (HPLC), FABMS, and ESMS suggested identity, several minor LC fractions in the GdS LC experiments that did not match earlier GdA mass data were investigated. One such fraction produced a doubly charged ion at m/z 840.4, not matching any expected GdS peptide mass. The CAD MS/MS spectrum shown in Fig. 2 illustrates the high quality of the low-energy fragmentation data achievable on the Q-TOF, giving an easily interpretable spectrum. The sequence determined from the data shown in Fig. 2 is V-L/I-V-E-D-D-E-L/I-M(O)-Q-G-F-L/I-R via the assignments shown on the spectrum. The 147 mass difference between y_5'' and y_6'' is of course isobaric for phenylalanine and methionine sulfoxide, but methionine in this position in the GdA gene-derived sequence suggested the latter assignment, which we confirmed in a MALDI magnetic sector experiment on the array ZAB2SE2FPD showing no digestion at this point in the sequence with the enzyme chymotrypsin, which would have been expected for phenylalanine.

The interpretation of these data and the MS/MS data from other projects described later was based on the knowledge that soft-ionization, low-internal-energy fragmentation processes give "N-terminal" and "C-terminal" fragmentation across the peptide bonds with charge retention on acylium (b) or ammonium (y) ion fragments [43], together with associated subfragments including the loss of neutrals such as water from hydroxyamino acid-containing fragments and the loss of CO from acylium ions to give aldimine-type ions. If necessary, derivatization experiments such as mild esterification in methanolic HCl [44] can be used to confirm assignments.

The fragment ion data shown in Fig. 2, once interpreted, were found to be well within 0.1 Da of theoretical mass throughout the mass range of the daughter ion spectrum, and the inset in Fig. 2 shows the typical resolving power observed at >3000 FWHM. This allows clear separation of isotopic clusters, which in turn contributes to preservation of mass accuracy even when the data become statistical (discussed later). Importantly, the mass spectroscopist can thus be confident of the numbers used in interpretation of the fragment ion data, an important objective in the original concept of the new instrument.

A. LC-ESMS/MS

A stringent test of MS/MS instrumentation is the ability to achieve good-quality CAD MS/MS data in online LC-MS/MS experiments where there is limited time to obtain the MS/MS data as sample peaks are rapidly eluting from the chromatography column. Figure 3 illustrates the online LC-ESMS/MS analysis of a mixture of neuroendocrine peptides on the Q-TOF, in which Fig. 3a shows the LC separation of the mixture from the composite total ion current chromatogram in the microbore LC-MS experiment. The LC-ESMS/MS spectrum shown in Fig. 3b is a single real-time 5-sec acquisition of scan 135 in Fig. 3a generated by CAD MS/MS of the $[M+2H]^{2+}$ 591.8 ion corresponding to this LC peak. This component is identified readily from the fragmentation pattern as luteinizing hormone releasing hormone (LHRH) pE-H-W-S-Y-G-L-R-P-G-NH$_2$, and the accurate high-quality data obtained in the single acquisition shown corresponded to $\sim$1 pmol sample consumption in this experiment. Clearly this does not represent the limitation of the Q-TOF in online LCMS/MS; the measured daughter ion mass accuracies for the single acquisition experiment are impressive, deviating from theoretical by <0.05 Da as seen in Table 2. Such mass accuracy has obvious and important advantages when working, for example, on database searches for matches of MS and MS/MS data when examining unknown structures either using parent ion or sequence tag fragment ion procedures.

An indication of the power of the Q-TOF in delivering the data quality we anticipated is shown in Fig. 4. This spectrum, which is derived from a nanospray experiment, shows a single 5-sec acquisition corresponding to a sample consumption of only 500 amol of the LHRH peptide. The signal/noise ratio, resolution, and mass accuracies observed are remarkable.

Table 2 Calculated and Measured Mass Values of the Daughter Ions Observed in the Online LC-ESMS/MS Experiment on $[M+2H]^{2+}$ at m/z 591.8

Daughter ion	Calculated mass	Measured mass	Δm[a]
b_1	112.04	112.04	0.00
y_2	172.11	172.07	−0.04
b_2	249.10	249.06	−0.04
y_3	328.21	328.22	+0.01
b_3	435.18	435.15	−0.03
y_4	441.29	441.31	+0.02
y_5	498.32	498.30	−0.02
b_4	522.21	522.19	−0.02
y_6	661.38	668.36	−0.02
b_5	685.27	685.28	+0.01
b_6	742.29	742.30	+0.01
y_7	748.41	748.41	0.00
y_8	934.49	934.48	−0.01
b_8	1011.48	1011.43	−0.05

[a]Error in mass units.

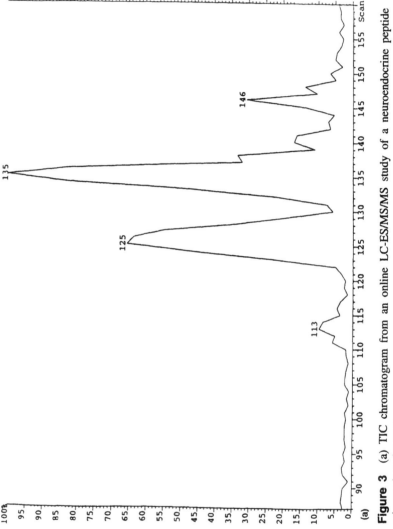

Figure 3 (a) TIC chromatogram from an online LC-ES/MS/MS study of a neuroendocrine peptide mixture (see the text).

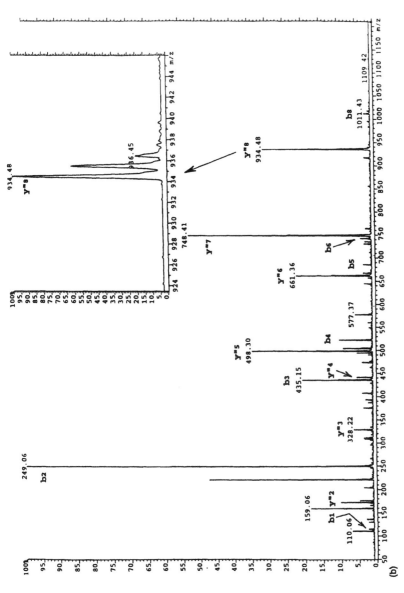

Figure 3 Continued. (b) A single acquisition (135) from the CAD MS/MS spectrum of $[M + 2H]^{2+}$ 591.8 corresponding to luteinizing hormone releasing hormone, pE-H-W-S-Y-G-L-R-P-G-NH$_2$. The work was carried out on an Applied Biosystems LC operating at a flow rate of 50 μl/min using a 1-mm microbore C$_{18}$ reverse-phase column eluted with a gradient of 0.1% trifluoroacetic acid and acetonitrile [1]. The LC effluent was stream split to give a flow of 16 μl/min into the electrospray source.

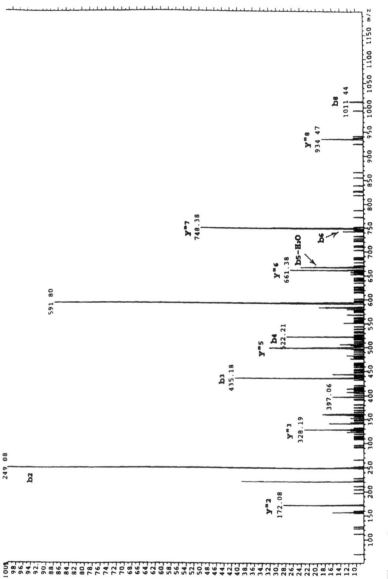

Figure 4 A single acquisition representing approximately 500 amol of peptide from CAD MS/MS analysis of $[M + 2H]^{2+}$ 591.8 in a nanospray ESMS/MS experiment. The fragment ions observed each carry the expected isotopes when the spectrum is expanded. Note that the mass accuracy is maintained.

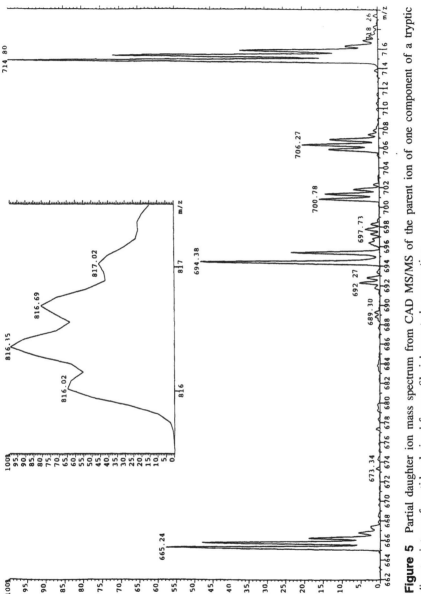

Figure 5 Partial daughter ion mass spectrum from CAD MS/MS of the parent ion of one component of a tryptic digest mixture of peptides derived from a filarial nematode preparation.

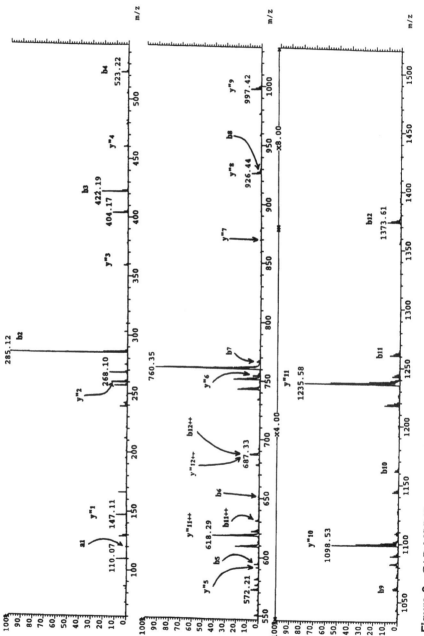

Figure 6 CAD MS/MS spectrum of a tryptic peptide from a filarial nematode preparation.

B. Sequencing Unknowns

Within the framework of this early research, the Q-TOF prototype has also been applied to problems of totally unknown structures, including work on filarial nematodes. These organisms infect more than 100 million people worldwide, causing diseases such as river blindness and elephantiasis. A major secreted glycoprotein from *Acanthocheilonema viteae*, ES-62, which infects rodents, is an important model for the study of filarial parasite diseases. This glycoprotein is of unknown protein sequence, and we have recently reported on some new phosphorylcholine decorated N-linked carbohydrate structures on ES-62, believed to confer immunomodulating activity as part of the parasitic infection mechanism [45]. Concomitantly, we are determining the protein structure of ES-62 using a digestion, mapping, and MS/MS strategy on the Q-TOF. First, Fig. 5 illustrates an important feature of the Q-TOF, derived from its resolving power, allowing easy distinction between singly, doubly, and triply charged parent or daughter ions in the MS or MS/MS spectra. Note the one-third mass difference between isotope peaks at m/z 816.02 and 816.35, for example, illustrating a triply charged ion. Doubly charged ions are assigned by the presence of one-half mass unit differences between the isotope peaks. This feature allows not only an immediate choice of collision voltage range, but also, importantly, easy distinction between fragment ion charge states, thus assisting interpretation. This is seen in Fig. 5 for the triply charged parent ion at m/z 816.02 $[M+3H]^{3+}$ and the singly and doubly charged daughters at m/z 694.38 and 665.24/714.80, respectively. An example of recent sequencing work in the nematode area on the Q-TOF instrument is shown in Fig. 6 illustrating the MS/MS spectrum of a tryptic peptide present as a digest mixture component. The data show the ease and confidence with which assignments can be made when the resolution and mass accuracy can be relied on. Note, for example, the ease of differentiation of doubly and singly charged in the m/z 550–700 region of the spectrum. Also, although some signals are very weak, such as the b_6, b_7, y_6'', and y_7'' assignments, they nevertheless were utilized in the interpretation of this unknown, by virtue of the isotopic distribution confirming the existence of a true signal as opposed to a noise spike. Overall, the sequence assigned to this spectrum is F-H-H-T-A-G-D-Y-M-T-V-L-K. Database searching of the full sequence generated from this and other peptides shows no complete match, confirming the presence of an unknown sequence, and cloning experiments (W. Harnett) for this molecule together with confirmatory studies are in progress to determine whether the Q-TOF generated sequences derive from ES-62 or from a cocontaminant.

In preliminary work on the prototype Q-TOF in the factory, we have been able to generate some MHC peptide data, although the nano-LC linkage for sample purification/introduction was not available at the time, and the micro-

Morris et al.

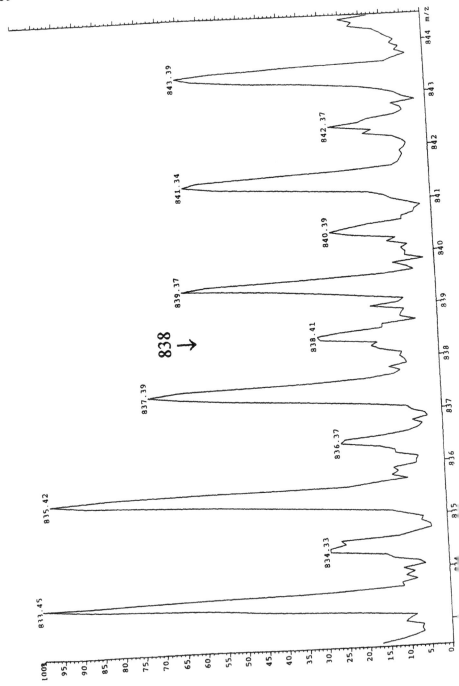

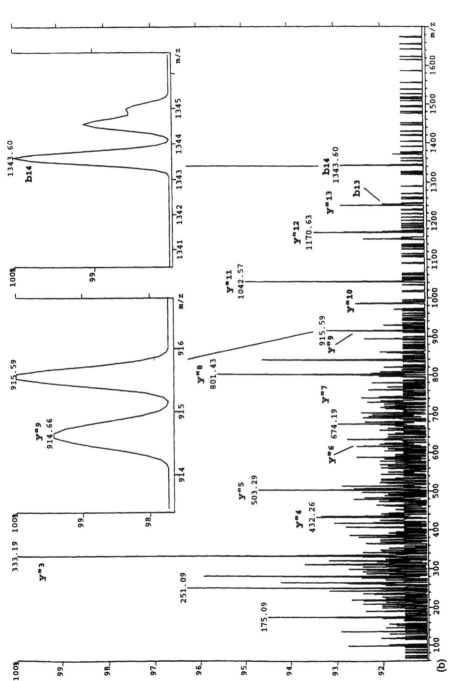

Figure 7 (a) Partial mass spectrum of class II MHC/peptide preparation in the m/z 838 region (for details see the text). (b) CAD

bore LC-MS penultimate purification step produced samples that had to be dried to remove acetonitrile for T-cell assay work. Nevertheless, an interesting insight into the application of the Q-TOF in ultra-high-sensitivity studies is shown in Fig. 7. This shows a partial mass spectrum from a nanospray experiment in which a dried fraction from the original LC-MS run was redissolved in acetic acid/acetonitrile (80/20), tuning to the area around m/z 838 where it was known that a doubly charged ion existed in the original LC-MS data from the sample. However, the fractionating and drying of the sample described earlier had unsurprisingly resulted in the loss of this peptide as an identifiable component of the preparation. Nevertheless, the m/z 838 signal seen in Fig. 7a was selected for CAD MS/MS to see whether any last vestige of sample could be identified. Remarkably, on summing acquisitions over several minutes the spectrum shown in Fig. 7b was produced. We conservatively estimated this at the time as a maximum of 500 amol total loading/consumption [1], although more recent comparisons of data suggest that this may be as low as 100 amol of sample consumed. An interesting and important principle is also illustrated here, in that the presence of the doubly charged ion hidden in the singly charged background leads to preferential decomposition of the doubly charged species in the CAD MS/MS experiment (charge repulsion in the multiply charged ion decreasing the energy barrier for decomposition), effectively amplifying its spectrum and purifying it away from the background decompositions. The spectrum produced is easily identified using only a few of the fragment ions in forward/reverse database searching as deriving from A-S-F-E-A-Q-G-A-L-A-N-I-A-V-D-K-A from the alpha chain of mouse class II MHC protein I-E^d. This sequence was found to correspond to one reported previously [37].

In more recent, as yet unpublished work with Berg and Langhorne, we have been identifying newly discovered class II MHC peptides using the Q-TOF, and an example is given in Fig. 8.

Figure 8a shows a partial MS spectrum of an antigen MHC preparation, from which a very small triply charged signal is visible as indicated. The CAD MS/MS spectrum of this signal is shown in Fig. 8b. The resolution of the Q-TOF allowed easy differentiation of the doubly and singly charged daughter ions obtained from this parent, from which the sequence . . . M-A-E-L/I-Y-K-N . . . could be readily assigned. A database search identifies the peptide V-S-N-S-V-K-E-I-M-A-E-I-Y-K-N-G-P-V, which is derived from cathepsin B, a new identification of a peptide bound to Class II MHC molecules.

C. In-Gel Digests

There is considerable interest in in-gel digestion of proteins followed by MS/MS for identification purposes, whether working with one- or two-dimensional systems [42]. The exquisite sensitivity and resolution of the Q-TOF are also of great value in this type of study. As with the very first MS mapping strategies [6,8,9,46], tryptic digestion gives the most predictable results for protein cleavage, with the majority of peptides producing doubly or triply charged ions, and terminating in Lys or Arg residues at the C-terminus, a useful feature in interpretation.

Examples of protein identification using the Q-TOF from projects requiring in-gel digestion are shown in Fig. 9. Figure 9a shows an MS spectrum taken from the product of in-gel tryptic digestion of a 19-kD protein from a polyacrylamide gel electrophoresis (PAGE) gel. The doubly charged signal at m/z 819.44 was selected for CAD MS/MS analysis on the Q-TOF and produced the spectrum shown in Fig. 9b. A virtually complete series of b and y ions was observed up to m/z 1500, as indicated, allowing identification of the sequence . . . S-V-A-E-L/I-T-Q-Q-M-F-D-A-K, which from a database search identifies the protein as tubulin beta-1 chain.

D. Glycoprotein Analysis

FP21 is a novel cytosolic protein originally identified in the slime mold *Dictyostelium*. Recently, FP21 homologues have been detected as a result of genome sequencing and expressed sequence tag (EST) projects, in plants, yeasts, nematodes, rats, and humans. In addition, FP21 has been independently identified in studies looking at the cyclin A/cdcK2 complex of human cells, yeast kinetochore function, and inner ear function. FP21 is a 162-amino-acid protein whose length and sequence are highly conserved throughout eukaryotes, which implies they share a common function, which is probably regulatory. Evidence from labeling studies with tritiated fucose suggested glycosylation of FP21 [47], and we have been using the Q-TOF to confirm glycosylation and to define the carbohydrate structure and the position and type of substitution on the peptide backbone, by examining peptides and glycopeptides obtained from FP21 by endo-LysC digestion.

A glycoform of one particular peptide, isolated by reverse-phase LC, was located initially in MALDI experiments, and the Q-TOF was then used to define the detailed structure. Nanospray experiments on the sample produced data including a doubly charged signal at m/z 1081.55 (itself a cone voltage fragment ion), which on CAD MS/MS analysis gave the spectrum shown in part in Fig. 10. The presence of C-terminal (y) fragment ions at m/z 147, 303, 416, and 544 together with N-terminal (b) fragment ions at m/z 230 and 377 allowed the peptide portion of the molecule to be locked onto region N-139 to K-151 in the gene-derived protein sequence, except that the peptide mass observed in the CAD spectrum (stripped of carbohydrate) is 16 Da higher than predicted by this sequence. The data in Fig. 10, importantly, also show some of the evidence for a novel structure in which glycosylation is assigned not to the threonine

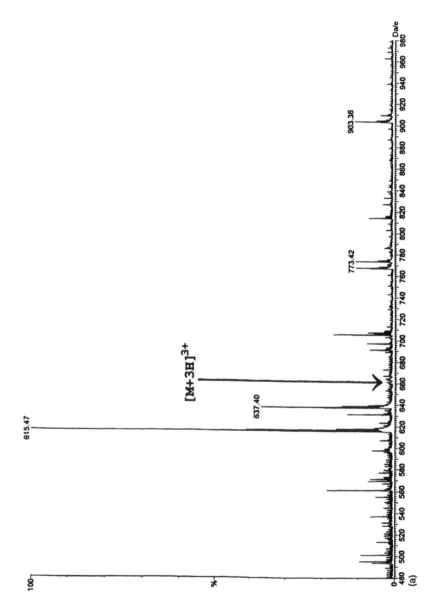

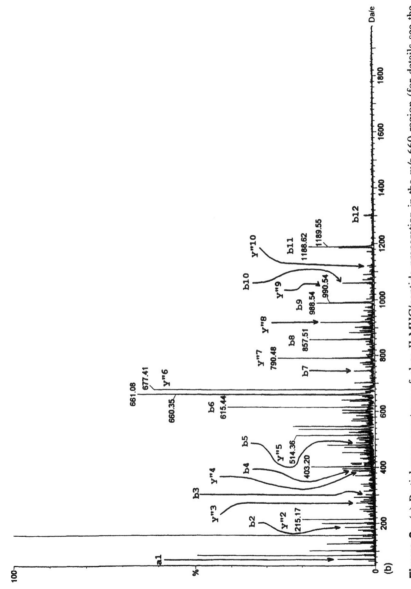

Figure 8 (a) Partial mass spectrum of class II MHC/peptide preparation in the m/z 660 region (for details see the text). (b) CAD MS/MS spectrum of the m/z 660 signal in (a) (for details see the text).

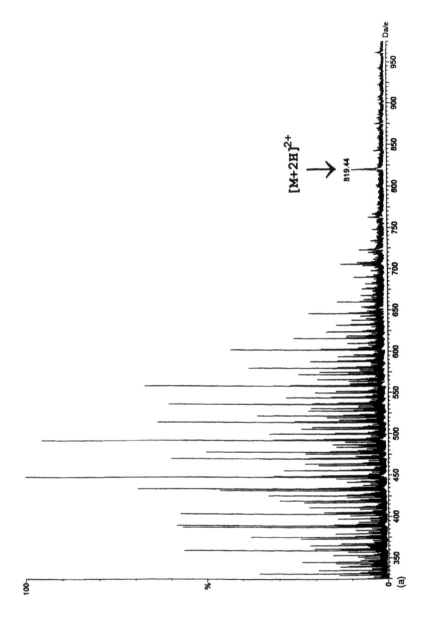

(a)

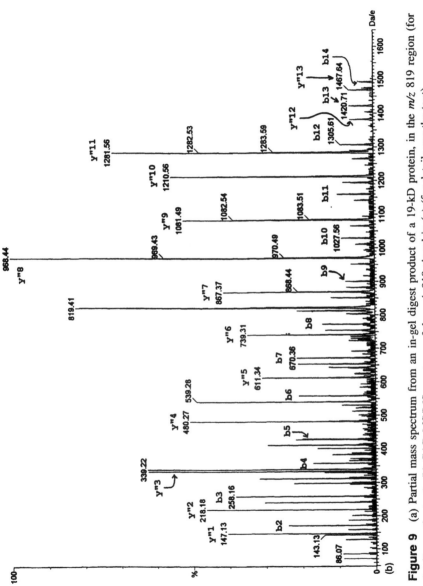

Figure 9 (a) Partial mass spectrum from an in-gel digest product of a 19-kD protein, in the m/z 819 region (for details see the text). (b) CAD MS/MS spectrum of the m/z 819 signal in (a) (for details see the text).

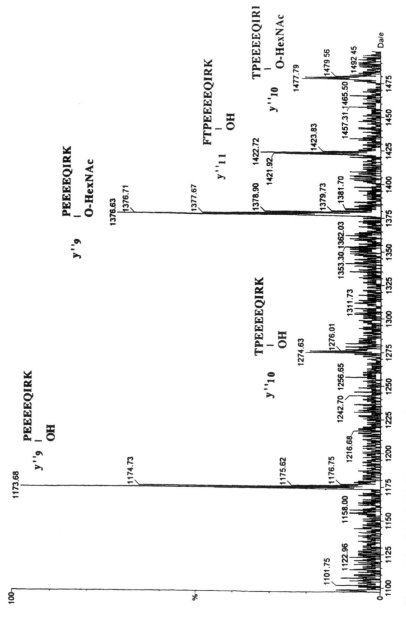

Figure 10 CAD MS/MS spectrum of the m/z 1081 signal from a FP21 Lys C digest showing the presence, positional assignment, and composition of glycosylation (for details see the text).

residue anticipated, but to a posttranslationally modified adjacent proline residue (hydroxyproline), which in the 1081.55 parent ion is substituted by Hex-NAc, Hex, Fuc (details to be reported elsewhere). Interestingly, therefore, the Q-TOF fragmentation data identify a double posttranslational modification of FP21 by hydroxylation at a specific proline residue in the sequence, followed by glycosylation.

IV. CONCLUSION

In confronting the problem of creating usable data for unambiguous ultra-high-sensitivity low femtomole/attomole range biopolymer sequencing, we suggested the building of a novel quadrupole orthogonal acceleration time-of-flight mass spectrometer. This Q-TOF instrument has been constructed by Fisons/Micromass, and the research problems discussed here illustrate the powerful advantages of the new instrument, which are summarized in Table 1.

ACKNOWLEDGMENTS

We thank the MRC, the Wellcome Trust, and the BBSRC for support for this research. We also thank Tim Alce, Debbie Smith, Richard Easton, Tony Etienne, Stuart Haslam, and Vivian Lindo (IC Biochemistry), Jean Langhorne and Matthias Berg (IC Biology), Chris West and Patana Teng-umnuay (University of Florida), William Harnett (University of Glasgow), Markku Seppala (University of Helsinki), and Gary Clark (University of West Virginia) for sample preparation; Norman Lynaugh and John Race (Micromass) for offering us exclusive research access to the prototype Q-TOF; and Robert Bordoli, Robert Bateman, and Jim Langridge for providing that access. Several of the figures in this chapter are reproduced with the permission of John Wiley and Sons Limited [1].

REFERENCES

1. HR Morris, T Paxton, A Dell, J Langhorne, M Berg, RS Bordoli, J Hoyes, RH Bateman. High sensitivity collisionally-activated decomposition tandem mass spectrometry on a novel quadrupole/orthogonal-acceleration time-of-flight mass spectrometer. Rapid Commun Mass Spectrom 10:889–896, 1996.
2. HR Morris, AJ Geddes, GN Graham. Some problems associated with the amino acid sequence analysis of proteins by mass spectrometry. Biochem J 111:38, 1969.
3. AJ Geddes, GN Graham, HR Morris, F Lucas, M Barber, WA Wolstenholme. Mass

spectrometric determination of the amino acid sequences in peptides isolated from the protein silk fibroin of Brombyx mori. Biochem J 114:695–702, 1969.

4. HR Morris. Studies towards the complete sequence determination of proteins by mass spectrometry: A rapid procedure for the successful permethylation of histidine-containing peptides. FEBS Lett 22:257–260, 1972.

5. DW Thomas, BC Das, SD Gero, E Lederer. Advantages and limitations of the mass spectrometric sequence determination of permethylated oligopeptide derivatives. Biochem Biophys Res Commun 33:199–202, 1968.

6. HR Morris, DH Williams, RP Ambler. Determination of the sequences of protein-derived peptides and peptide mixtures by mass spectrometry. Biochem J 125:189–201, 1971.

7. HR Morris, DH Williams. The identification of a mutant peptide of an abnormal haemoglobin by mass spectrometry. J C S Chem Comm 114–116, 1972.

8. HR Morris, KE Batley, NGL Harding, RA Bjur, JG Dann, RW King. Dihydrofolate reductase: Low-resolution mass-spectrometric analysis of an elastase digest as a sequencing tool. Biochem J 137:409–411, 1974.

9. HR Morris, DH Williams, GC Midwinter, BS Hartley. A mass-spectrometric sequence study of the enzyme ribitol dehydrogenase from Klebsiella aerogenes. Biochem J 141:701–713, 1974.

10. J Bridgen, HR Morris. Use of mass spectrometry and quantitative Edman degradation for the determination of repeating amino-acid sequences. Eur J Biochem 44:333–334, 1974.

11. S Magnusson, L Sottrup-Jensen, TE Petersen, HR Morris, A Dell. Primary structure of the vitamin K-dependent part of prothrombin. FEBS Lett 44:189–193, 1974.

12. HR Morris, A Dell. The sequence of the blocked N-terminal peptide from Neurospora glutamate dehydrogenase. Biochem J 149:754–755, 1975.

13. J Hughes, TW Smith, HW Kosterlitz, L Fothergill, BA Morgan, HR Morris. Identification of two related pentapeptides from the brain with potent opiate agonist activity. Nature 258:577–579, 1975.

14. HR Morris, A Dell, TE Petersen, L Sottrup-Jensen, S Magnusson. Mass-spectrometric identification and sequence location of the ten residues of the new amino acid (γ-carboxyglutamic acid) in the N-terminal region of prothrombin. Biochem J 153:663–679, 1976.

15. JV Stone, W Mordue, KE Batley, HR Morris. Structure of locust adipokinetic hormone: A neurohormone that regulates lipid utilisation during flight. Nature 263:207–211, 1976.

16. KE Batley, HR Morris. Dihydrofolate reductase from Lactobacillus casei: N-terminal sequence and comparison with the substrate binding region of other reductases. Biochem Biophys Res Commun 75:1010–1014, 1977.

17. K Batley, HR Morris. Dihydrofolate reductase: Partial sequence of the Lactobaccillus casei enzyme and homology with other dihydrofolate reductases. Biochem Soc Trans 5:1097–1100, 1977.

18. HC Thogersen, TE Petersen, L Sottrup-Jensen, S Magnusson, HR Morris. The N-terminal sequences of blood coagulation factor X_1, and X_2, light chains: Mass spectrometric identification of 12 residues of γ-carboxyglutamic acid in their vitamin-dependent domains. Biochem J 175:613–627, 1978.

19. WV Shaw, LC Packman, BD Burghleigh, A Dell, HR Morris, BS Hartley. Primary structure of a chloramphenicol acetyltransferase specified by R plasmids. Nature 282:870–872, 1979.

20. HR Morris, MR Thompson, DT Osuga, AI Ahmed, SM Chan, J Vandenheede, R Feeney. Antifreeze glycoproteins from the blood of an antarctic fish: The structure of the proline-containing glycopeptides. J Biol Chem 253:5155–5161, 1978.

21. HR Morris, M Panico, A Karplus, PE Lloyd, B Riniker. Elucidation by FAB-MS of the structure of a new cardioactive peptide from aplysia. Nature 300:643–645, 1982.

22. HR Morris, GW Taylor, PJ Piper, JR Tippins, MN Smahoun. Structure of slow-reacting substance of anaphylaxis from guinea-pig lung. Nature 285:104–106, 1980.

23. A Beloff-Chain, J Morton, S Dunmore, GW Taylor, HR Morris. Evidence that the insulin secretagogue, β-cell-tropin, is $ACTH_{22-39}$. Nature 301:255–258, 1983.

24. RJ Robb, RM Kutney, M Panico, HR Morris, WF De Grado, V Chowdry. Post-translational modification of human T-cell growth factor. Biochem Biophys Res Commun 116:1049–1055, 1983.

25. RJ Robb, RM Kutney, M Panico, HR Morris, WF De Grado, V Chowdry. Post-translational modification of human interleukin-2. Proc Natl Acad Sci USA 81:6486–6490, 1984.

26. HR Morris, M Panico, AT Etienne, I MacIntyre, JR Tippins, SI Girgis. Isolation and characterisation of human calcitonin gene-related peptide. Nature 308:746–748, 1984.

27. SD Brain, TJ Williams, JR Tippins, HR Morris, I MacIntyre. Calcitonin gene-related peptide (CGRP) is a potent vasodilator. Nature 313:54–56, 1984.

28. DA Erdjument, DA Lane, M Panico, V Di Marzo, HR Morris. Single amino acid substitutions in the reactive site of antithrombin leading to thrombosis. J Biol Chem 263:5589–5593, 1988.

29. E Spooncer, M Fukuda, JC Klock, JE Oates, A Dell. Isolation and characterisation of polyfucosylated lactosaminoglycan from human granulocytes. J Biol Chem 259:4792–4801, 1984.

30. RA Williamson, FAO Marston, S Angal, P Koklitis, M Panico, HR Morris, AF Carne, BJ Smith, TJR Harris, RB Freedman. Disulphide bond assignment in human tissue inhibitior of metalloproteinases (TIMP). Biochem J 268:267–274, 1990.

31. H Sasaki, B Bothner, A Dell, M Fukuda. Carbohydrate structure of erythropoietin expressed in CHO cells by a human erythropoietin cDNA. J Biol Chem 262:12059–12076, 1987.

32. AL Chan, HR Morris, M Panico, A Etienne, ME Rogers, P Gaffney, L Creighton-Kempsford, A Dell. A novel sialylated N-acetylgalactosamine-containing oligosaccharide is the major complex-type structure present in Bowes melanoma tissue plasminogen activator. Glycobiology 1:173–186, 1991.

33. AJ Reason, RH Treisman, R Marais, RS Haltiwanger, GW Hart, HR Morris, M Panico, A Dell. Localization of O-GlcNAc modification on the serum response transcription factor*. J Biol Chem 267:16911–16921, 1992.

34. K-H Khoo, S Sarda, X Xu, JP Caulfield, MR McNeil, SW Homans, HR Morris, A Dell. A unique multifucosylated -3GalNAcβ1-4GlcNAcβ1-3Galα 1-motif consti-

tutes the repeating unit of the complex O-glycans derived from the cercarial glycocalyx of Schistosoma mansonii. J Biol Chem 270:17114–17123, 1995b.

35. A Dell, HR Morris, RL Easton, M Panico, M Patankar, S Oehninger, R Koistinen, H Koistinen, M Seppala, GF Clark. Structural analysis of the oligosaccharides derived from glycodelin, a human glycoprotein with potent immunosuppressive and contraceptive activities. J Biol Chem 270:24116–24126, 1995.

36. HR Morris, A Dell, RL Easton, M Panico, H Koistinen, R Koistinen, S Oehninger, MS Patankar, M Seppala, and GF Clark. Gender-specific glycosylation of human glycodelin affects its contraceptive activity. J Biol Chem 271:32159–32167, 1996.

37. DF Hunt, H Michel, TA Dickinson, J Shabanowitz, AL Cox, K Sakaguchi, E Appella, HM Grey, A Sette. Peptides presented to the immune-system by the murine class-II major histocompatibility complex molecule-I-A(D). Science 256:1817–1820, 1992.

38. W Wang, LR Meadows, JMM den Haan, NE Sherman, Y Chen, E Blockland, J Shabanowitz, AI Agulnik, RC Hendrickson, CE Bishop, DF Hunt, E Goulmy, VH Engelhard. Human H-Y—A male-specific histrocompatibility antigen derived from the SMCY protein. Science 269:1588–1590, 1995.

39. HR Morris, R McDowell, A Dell, M Panico. A new range of wide angle array detector instruments for high mass high sensitivity biopolymer analysis. Proceedings of 39th American Society for Mass Spectrometry Conference on MS and Allied Topics, Nashville, TN, 1991, p 1693–1694.

40. HR Morris, A Dell, M Panico, R McDowell. Mass spectrometry in protein and glycoprotein structure elucidation Proceedings of the 43rd American Society for Mass Spectrometry Conference on MS and Allied Topics, Atlanta, GA, 1995, p 360.

41. KH Khoo, A Dell, HR Morris, PJ Brennan, D Chatterjee. Inositol phophate capping of the non-reducing termini of liporarabinomannan from rapidly growing strains of Mycobacterium: Mapping the non-reducing terminal motifs of LAMs. J Biol Chem 270:12380–12389, 1995.

42. M Wilm, A Schevchenko, T Houthaeve, S Breit, L Schweigerer, T Fotsis, M Mann. Femtomole sequencing of proteins from polyacrylamide gels by nano-electrospray mass spectrometry. Nature 379:466, 1996.

43. HR Morris, M Panico, M Barber, RS Bordoli, RD Sedgwick, AN Tyler. Fast atom bombardment: A new mass spectrometric method for peptide sequence analysis. Biochem Biophys Res Commun 101:623–631, 1981.

44. E Hunt, HR Morris. Collagen cross-links: A mass spectrometric ^{1}H- and ^{13}C-nuclear-magnetic-resonance study. Biochem J 135:833–843, 1973.

45. SM Haslam, KH Khoo, KM Houston, W Harnett, HR Morris, A Dell. Characterisation of the phosphorylcholine-containing N-linked oligosaccharides in the excretory-secretory 62 kDa glycoprotein of Acanthocheilonema viteae. Mol Biochem Parasitol 85:53–66, 1997.

46. HR Morris, M Panico, GW Taylor. FAB-MAPPING* of recombinant-DNA protein products. Biochem Biophys Res Commun 117:299–305, 1983.

47. E Kozarov, H van der Wel, M Field, M Gritzali, RD Brown, CM West. Characterization of FP21, a cytosolic glycoprotein from Dictyostelium. J Biol Chem 270:3022–3030, 1995.

4

Affinity Techniques Combined with Mass Spectrometry

Ray Wieboldt, Jerry Zweigenbaum, and Jack D. Henion
Cornell University, Ithaca, New York

I. INTRODUCTION

Affinity-based analytical methods and mass spectrometry occupy important positions in bioanalytical laboratories. Both possess extensive capabilities to detect, identify, and quantitate targeted analytes, and it is safe to say that thousands of applications have been developed involving one or the other. Many
examples demonstrate the ability of these techniques to perform highly specific
and sensitive assays. Nevertheless, the accelerated development of new pharmaceutical entities and the requirement to determine pharmacokinetic and pharma-
codynamic properties in biological matrices present important analytical challenges that reach beyond the limitations of each technique.

   Affinity chromatography and affinity extractions have the capability to
isolate and concentrate specific analytes, or classes of analytes, from complex
sample matrices. A high degree of purification of a targeted analyte is possible
in a single step from large volumes of sample. A practical example of this is
the use of affinity chromatography to selectively capture pesticides or herbicides from environmental samples for quantitation at the parts per trillion level.
Sensitive assays for these compounds have been developed that rely on the
ability of an immobilized antibody column to concentrate analytes from dilute
samples. A frequently encountered limitation of immunoassays, however, is that
cross-reactivity of the immunoaffinity reagent with analogs of the target analyte
potentially yields incorrect qualitative and quantitative results.

   The utility of liquid chromatography/mass spectrometry (LC/MS) has
been demonstrated in many applications including bioanalytical [1–5] and environmental [6–11] problems. Mass spectrometry acts as a sensitive and selective
detector that can distinguish closely related compounds on the basis of mass
differences and fragmentation behavior. Analytical methods employing mass
spectrometry provide unequivocal identification and quantitation of analytes
even when sample components coelute in a chromatographic separation. This
is also the basis for the sensitivity of MS methods, since background noise
consisting of ions from sample matrix components can be readily removed
from the detected analyte signal. Nevertheless, it is still necessary to provide
purification, concentration, and separation of analytes from biological matrices
to achieve optimal results with LC/MS. One important reason for prepurification of a sample is that the presence of excess matrix ions can suppress ionization of the analyte and greatly reduce its signal.

   Coupling immunoaffinity extraction (IAE) with mass spectrometry provides an effective approach to analyze complex samples by exploiting the advantages offered by each technique. A practical strategy is to couple IAE on-
line with LC/MS using column switching techniques. IAE minimizes sample
preparation procedures and concentrates analytes, while LC/MS provides a
rapid, specific method of detection for captured analytes eluted from the affinity

medium. This chapter provides an introduction to the development of coupled affinity–mass spectrometric methods. Examples of methods developed in our laboratory are provided with related work.

II. EXPERIMENTAL

A. Immunoaffinity Extraction

Immunoaffinity extraction concentrates and separates target analytes from unwanted sample components on the basis of their interaction with an affinity reagent such as an antibody. It is a form of sample pretreatment just as solid-phase extraction (SPE) and supercritical fluid extraction (SFE) serve as methods of sample cleanup. However, IAE is potentially more selective for an analyte than other methods because of the binding specificity of the antibody. Dissociation constants for antigen–antibody interaction typically fall in the nanomolar or subnanomolar range, so the complexation reaction has the capability to capture analytes from very dilute sample solutions.

B. Affinity Media

Robust, high-capacity immunoaffinity columns are critical to the success of the development of an extraction method. Prepared affinity media for common drugs, pesticides, and herbicides are available from commercial sources (e.g., Randox Laboratories, Ardmore, UK). However, it is more common to prepare affinity extraction columns in-house to target a specific analyte. This involves immobilization of antibodies on a column compatible with the higher pressures and flow rates required by high-performance liquid chromatography (HPLC), and it is necessary to have a source for both the antibody and the solid-phase support material. The de novo preparation of antisera is feasible when it is not possible to locate a source of a specific antibody. For small molecules, this would entail bonding the analyte to a carrier such as bovine serum albumin (BSA), injecting an animal with the modified BSA, and allowing an incubation period of about a month. The animal is injected again with the solution and bled a few days later to obtain polyclonal antibody-rich sera. To obtain monoclonal antibody, further steps are required to generate large quantities of the specific antibody. This includes selecting the specific cells responsible for generating the desired antibody, fusing them with myeloma cells, and injecting the hybridomas back into an animal or growing them in culture. It is best to get assistance from a laboratory with prior experience with the protocols for doing these procedures. Linscott's Directory [12] is helpful for locating commercial sources of antisera for specific antigens.

Immunoaffinity extraction (IAE) columns may be prepared with many different matrices, but generally the most successful are composed of uniformly sized rigid particles that are highly porous and chemically and biologically inert. Solid-phase materials are available with many coupling chemistries for bonding antibodies and proteins to the support. Some of these supports are provided with activated linkers that are used to covalently couple the affinity proteins to the support. These can be used to prepare durable affinity media dedicated to the affinity interaction provided by the antibody permanently linked to the support. Other media provide reversible immobilization of antibodies and take advantage of the strong affinity of certain bacterial cell wall proteins for antibodies. Media with covalently immobilized protein A and protein G bind specific classes of antibodies strongly, but the complexation can be disrupted by washing with low-pH buffer, leaving the protein A and protein G intact and reuseable. At neutral pH, the immobilized protein A/G–antibody complex serves as an affinity extraction media with activity defined by the antibody. Some sources of solid-phase supports are Pharmacia Biotech (Piscataway, NJ), PerSeptive Biosystems (Framingham, MA), Bio-Rad (Melville, NY), Sigma (St. Louis, MO), and Pierce (Rockford, IL). The protocols found in immunoaffinity chromatography literature [13,14] and from commercial sources provide helpful hints on column preparation.

C. Extraction Conditions

The affinity reaction is influenced by the composition of the complexation medium. It is important to equalize ionic strength and pH of standards and samples by dilution with a buffer prior to extraction. Extractions are performed by passing the sample through the IAE column in a flowing stream of binding buffer. This is often phosphate-buffered saline (PBS): 10 mM sodium phosphate buffer, pH 7.5, with 0.15 M NaCl. Capture of the analyte on the affinity medium effectively concentrates the analyte, while matrix components are removed by continued washing with PBS. Higher stringency wash buffers, typically with higher ionic strength or lower pH, aid in removal of matrix components. Chaotropic agents such as thiocyanate or urea also effectively disrupt noncovalent interactions and may be used in high-stringency wash steps.

The captured analyte is desorbed from the IAE column after washing by elution with a buffer that disrupts the analyte–antibody affinity complex. A switch to aqueous solutions of trifluoroacetic acid, formic acid, or acetic acid at pH 4 or below is usually effective. The analyte should be eluted in a minimum volume of buffer, and the purified and concentrated sample may be analyzed directly by mass spectrometry. It is usually necessary, however, to eliminate inorganic salts and to further concentrate and separate the eluted components by HPLC prior to introduction into the mass spectrometer. Inorganic salts in the effluent from the analytical HPLC column elute near the void volume, so diversion of an initial column fraction with a suitable valve eliminates most of the salt interference with MS detection. Automatically controlled column switching performed by the instrumentation hardware coordinates all sample injection, IAE steps, and HPLC analysis with the mass spectrometric data collection.

D. Coupled-Column Extraction, Trapping, and HPLC

Column-switching techniques provide a variety of possible experiments for both selective extraction and separation. Significant increases in sensitivity are achieved by focusing an analyte on one column and back-flushing to an analytical HPLC column. Both atmospheric-pressure ionization (API) mass spectrometry and ultraviolet/visible (UV/VIS) detection provide signals that are directly proportional to analyte concentration. Therefore, detection of an analyte will be enhanced if a higher concentration of analyte can be provided in a lower elution volume from smaller bore columns. Thus, the goal of column-switching methods is to provide extraction of analytes from large volumes of the sample matrix into the small volumes of salt-free analyte fractions, which are best for introduction into the mass spectrometer.

Three modes of column coupling are commonly encountered with IAE/ LC/MS. They are: (1) focusing of captured compounds on a trapping column for enhanced resolution and sensitivity, (2) selective extraction of target compounds from complex matrices, and (3) the ability to "heart-cut" a fraction from one separation mode for subsequent injection onto a column in a different mode of separation. Each of these general methods can perform the necessary exchange of an analyte fraction from a matrix incompatible with the sampling requirements of mass spectrometry into a buffer system amenable for MS detection.

1. Focusing of Target Compounds

A focusing experiment consists of placing a column in line with the sample such that the sample matrix will allow nearly complete adsorption, partition, or ion exchange onto the column. Under these conditions, it is necessary that no elution of the target compound(s) takes place. Any movement of compound through the trapping column will cause band broadening that will be evident in the analytical separation. A simple example is the column inlet focusing of polynuclear aromatic hydrocarbons on a reversed-phase trapping column in a water matrix. Large quantities of water can be passed through the column without any movement of the compounds from the head of the column. Elution of analytes from the trapping column is performed by placing it in line with the analytical column and reversing the eluent flow. The analytes elute in a tight band from the trapping column onto the analytical column as the mobile-phase solvent strength is increased.

2. Extraction of Target Compounds

Extraction of target compounds is accomplished by a medium that is selective for the type of molecule to which the target belongs. A highly deactivated or polymeric support of nonpolar nature will only extract nonpolar compounds. Weakly ionic compounds can be extracted by a weak ion exchange resin. Selection of the extraction column and the appropriate mobile phase can provide a highly selective system for specific analytes in complex matrices. Appropriate conditions for both extraction and analysis must be developed. In the examples provided next, we use an antibody generated against a specific antigen epitope for high-selectivity affinity extraction of small-molecule targets.

Figure 1 shows a column-switching configuration using automated affinity extraction, trapping, and separation. This column-switching sequence in our laboratory consists of an affinity column using immobilized protein G for noncovalent adsorption of immunoglobulin G molecules (IgGs), a reversed-phase trapping column, and a reversed-phase analytical column for analyte separation. With the switching valves in the load position, both the trapping column and analytical column are in line and connected to pump 2 for conditioning at initial analytical separation conditions. For a 1-mm analytical column, flow rates are 50 μl/min with buffers appropriate for separation of the target analyte. The affinity column is connected directly to pump 1 and is equilibrated with PBS at flow rates up to 3 ml/min and pressures less than 1000 psi. This is possible with the highly porous restricted access media used in many immunoaffinity columns. The injection loop is loaded with antibody by an autosampler. Valve 1 is then switched to place the injection loop in line with the affinity column. Up to 10 column volumes of loading buffer are allowed to pass through the affinity column to wash off nonantibody proteins that are present in the antibody solution. The elimination of these proteins in the wash step may be monitored by following absorbance at 256 nm with a UV/VIS detector on the waste line.

Valve 1 is then switched back to the load position and the selected volume of sample is loaded into the loop. The loop is again placed in line with the affinity column. Enough loading buffer must pass through the affinity column to wash to waste non-affinity-bound material in the sample. The use of a detector on the waste line is also convenient for determination of the capacity of the immunoaffinity column for its analyte. With antibody loaded on the extraction column, repeated injection of analyte standards while monitoring the

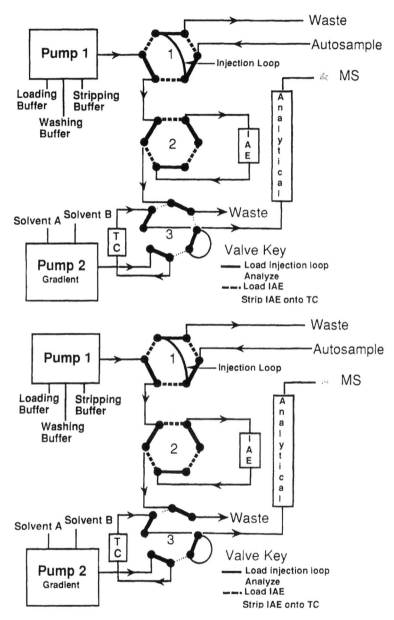

Figure 1 Schematic diagram of a general column-switching method. The diagram shows the use of an affinity capture column for specific immunoaffinity extraction of analytes, a second trapping column used to further preconcentrate, purify, and focus analytes eluted from the affinity column, and an analytical column for final separation of the target analytes. LC/MS is used online as the detector for the analytical column to provide identification and quantitative analysis of extracted sample components.

column effluent will show when the antibody-binding sites are saturated and analyte breakthrough occurs.

For trace analyses, pump 1 can be switched to a solvent reservoir containing a large volume of sample and a selected volume of sample pumped over the affinity column. In this case, the loading buffer should then be passed through the affinity column to remove any nonbound material in the sample matrix. Next, valve 3 is switched to place the trapping column in line with the affinity column. Sufficient volume of loading buffer is passed through this column to remove organics so that efficient trapping will take place. Following this, pump 1 is switched to stripping buffer, and antibody and target compounds are eluted onto the trapping column. Valve 2 is subsequently switched to take the affinity column out of line with the trapping column, and pump 1 is switched to water or the analytical aqueous buffer to remove salts used in the stripping buffer from the trapping column. Finally, all valves are returned to the original loading condition, the pump 2 gradient program is initiated, and the mass spectrometer data collection is started.

3. Heart-Cutting a Fraction from a Complex Matrix

Heart-cutting may be used both to reduce the complexity of the matrix injected onto the analytical column and to exchange buffers to ensure compatibility with the requirements of mass spectrometric detection. For example, a sample containing weakly ionic compounds of interest may be separated with a chromatographic mode that isolates a cut of the early-eluting compounds on an ion exchange column. Trapping onto a reversed-phase HPLC trapping column can be used to accomplish both matrix reduction and buffer exchange. Elution is directed onto an analytical column capable of separating the components of interest with a buffer system compatible with the mass spectrometer. Selection of appropriate column and elution chemistries can be a difficult task in developing a column-switching system, because of the many choices that are available for each dimension of the method.

III. MASS SPECTROMETRY INSTRUMENTATION

The mass spectrometer interface puts limitations on the type of sample or sample stream that can be introduced from the chromatography, and therefore careful attention should be given to selection of both chromatographic conditions and the sampling requirements of the mass spectrometric interface. For an excellent summary of LC/MS instrumentation and interfaces see the review by Niessen and Tinke [15].

The two most commonly encountered ion sources used for LC/MS are pneumatically assisted electrospray ionization (ion spray) and atmospheric pressure chemical ionization (APCI). Ion spray uses both high-pressure gas and a high electric field to produce gas-phase analyte ions from the flowing HPLC buffer. Ion spray accommodates compounds such as polar drugs, drug metabolites, peptides, oligonucleotides, and small proteins. In APCI, the HPLC effluent is evaporated with hot nebulization gas and ions are produced in a corona discharge in front of the mass spectrometer inlet. APCI performs best with lower molecular weight, reasonably volatile and less polar organic compounds. It is a robust method at conventional 1 ml/min HPLC flow rates. APCI does not usually produce multiply charged species but can degrade thermally unstable compounds. Thermally labile compounds are better suited to ion spray than to APCI.

A. Polarity

The mass spectrometer may be set for detection of positive or negative ions. Compounds such as amines and any other molecule that can readily accept a positive charge by association with a proton or other cations such as NH_4^+ or Na^+ will provide good sensitivity with positive ion detection. Organic acids are often more sensitive with negative ion detection. Buffer pH and composition affect the relative concentration of the possible ionic species and therefore can influence sensitivity. Some analytes will produce either positive or negative ions under appropriate eluant conditions, so the choice of polarity must be evaluated in the context of the requirements for each analysis. It may be possible to use the column-switching methods mentioned earlier to provide elution conditions to produce a higher concentration of a particular analyte ion.

B. HPLC Conditions Should Match the MS Interface Requirements

Volatile buffer components enhance the evaporation process. Ammonium acetate, and ammonium formate at concentrations between 5 and 25 mM, are commonly employed. Methanol, acetonitrile, and tetrahydrofuran (THF) in proportions between 5 and 95% with the aqueous buffer improve atomization efficiency in ion spray. Mobile-phase modifiers such as ion-pairing reagents and inorganic salts should be avoided since they contribute to suppression of analyte ionization and chemical background noise. Trifluoroacetic acid is a volatile ion-pairing reagent routinely used to improve HPLC separation of peptides. Its presence often causes significant ionization suppression, but the effect can be overcome by postcolumn addition of other organic acids and alcohols. If a modifier is not required, it is best to omit it. The capability of mass spectrometric detection to discriminate between closely eluting or even coeluting sample components eliminates the requirement for full chromatographic separation in most cases.

The mass spectrometer is often sensitive to the composition of the mobile phase. Impurities in the HPLC buffers are detected with great sensitivity and contribute to baseline chemical noise. All mobile phases should be evaluated for impurities. Detergents and plasticizers such as dioctyl phthalate, with a positive ion m/z of 391 Da and 149 Da for its abundant fragment, sometimes appear when buffers are allowed to age in laboratory glassware since they presumably leach out of scratches and pores in the glass surfaces. Mass spectrometric detection is not particularly responsive to the mobile-phase problems encountered with UV/VIS detection such as bubbles and temperature variation.

C. Instrument Parameters

Specific optimization procedures vary with each instrument. With quadrupole and ion-trap instruments, mass calibration and optimal lens parameters for a particular analysis are typically very reproducible and can be reused with little modification over days or weeks. Nevertheless, the performance of the spectrometer should be checked regularly with well-characterized standards (a mixture of analyte compounds at a known concentration is useful) prior to every analysis. Adjustment of sprayer position and nebulizer gas flow rates may require daily optimization, but newer designs have minimized these procedures. Optimization consists of maximizing a signal from the analyte of interest while minimizing the ion current from unwanted background noise that arises from solvent clusters and incomplete desolvation of analyte molecules. These parameter modifications are best done under actual LC/MS conditions rather than with so-called "infusion" experiments.

Most instruments automate tuning methods to aid in rapid determination of the best conditions for a particular analysis. However, higher concentrations of tuning or analyte standard solutions saturate the spectrometer signal response and contribute to chemical memory effects or carryover. Typically, infusion at 5 μl/min of a 0.1–10 μM analyte solution "teed" into an HPLC buffer stream flowing at the same rate used for the analysis provides a full-scan response for optimization of spray position, ion spray voltage, nebulizer gas flow, and ion optics voltages affecting analyte signal and background.

IV. EXAMPLES OF IAE/LC/MS METHODS

Trace analysis of drugs, pesticides, and metabolites in biological materials requires methods that are highly sensitive and specific. The analytes typically are present at very low levels and must be identified and quantitated against a high background of interfering sample components. On-line immunoaffinity extraction concentrates target analytes and eliminates most of the background constituents. The extracted sample may still contain a chemical noise background that is too high for use of UV detection for quantitative purposes, since analytical criteria require separation of analyte signal from interference of other sample components. However, the specificity of mass spectrometric detection can circumvent the problem of coelution of an analyte with matrix components.

The following examples show how different types of MS experiments can be used to improve qualitative and quantitative results of the immunoaffinity extractions. A MS scan may cover a wide mass range such as 200–2000 Da (so-called full-scan data) or it may be confined to a single m/z value (selected ion monitoring or SIM). SIM provides an enhanced signal-to-noise ratio, but is still sensitive to background noise produced by interfering species or clusters at the selected mass. This interference can be greatly reduced or entirely eliminated with tandem MS scanning modes. A typical MS/MS experiment fragments ions isolated from an analyte precursor by collision-induced dissociation (CID). The unique and reproducible precursor and product ion pairs, however, make the MS/MS experiment a molecular structure "filter" capable of providing great specificity for targeted determination of an analyte in a complex background of interfering components. This scanning method is called selected reaction monitoring (SRM).

A. Determination of LSD and Its Metabolites in Urine

Preconcentration of analytes from a sample matrix using immunoaffinity extraction combined with MS/MS detection of specific precursor/product ion pairs provides sensitivity sufficient for ultratrace quantitation. In this example [16], lysergic acid diethylamide (LSD) and its metabolites were enriched from urine sample volumes of up to 200 ml using on-line coupled immunoaffinity extraction. Figure 2 compares the results of SIM and SRM monitoring of the extracted products on a triple-quadrupole mass spectrometer. Figure 2a shows SIM traces monitoring the protonated molecules of LSD, *nor*-LSD, de-ethyl-LSD, and *nor*-allyl-LSD at m/z 324, 310, 296, and 350, respectively. Only the ion for *nor*-allyl-LSD at m/z 350 was detected at the expected retention time. It is important to note that although SIM is more selective than full-scan data acquisition, background noise at the selected m/z values is still present. The background can consist not only of residual chemical interference in the extracted sample, but also of random ion clusters from the HPLC buffer components. Figure 2b shows SRM traces derived from fragmentation of the precursor ions monitored in Fig. 2a. The levels of LSD and its metabolites spiked into the urine samples in Figs. 2a and b were identical (10 ppt). SRM provides an

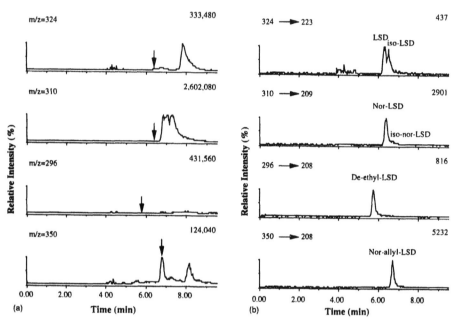

Figure 2 (a) Selected ion monitoring and (b) selected reaction monitoring detection of LSD and its metabolites preextracted with an on-line immunoaffinity column. The enhanced specificity possible with SRM lowers the detection limit to 2.5 ppt. (From Ref. 16.)

extra level of "filtering" that eliminates these sources of interference to produce significant gains in signal-to-noise ratio (S/N) and analyte specificity. Detection levels of 2.5 ppt of LSD and its metabolites were achieved using IAE with SRM detection.

A second point to note here is that the immunoaffinity media captures both LSD and its metabolites. Immunoaffinity extraction provided preconcentration but did not achieve the ultimate specificity. SRM mass spectrometric detection was needed to provide the enhanced specificity, since the metabolites elute close to each other under the conditions used for the analytical chromatography. However, this cross-reactivity of the antibody is useful in determining the presence of LSD or its metabolites, whereas if there was no cross-reactivity that determination could not be made by this method alone. In this work the immunoaffinity column used noncovalent bonding of the antibody to a support containing covalently bound protein G.

B. Determination of Carbofuran in Potato Extracts

In this example [17], extraction was accomplished directly from a preparation of whole potato using an immunoaffinity column prepared with antibodies to carbofuran. The captured analytes were then eluted onto a reversed-phase trapping column for further concentration and desalting prior to separation on an analytical column (discussed earlier). Samples of raw potato were digested by boiling in 0.25 N HCl, filtered, and diluted into PBS prior to on-line extraction and LC/MS analysis. The background components in the starchy sample matrix prevent the use of UV detection. Furthermore, the use of mass spectrometric detection without sample pretreatment often is not sufficient to provide high quality data required for analysis and quantitation. Figure 3 shows LC/MS chromatograms of the potato preparation spiked with carbofuran at 100 ng/g with (Fig. 3b) and without (Fig. 3a) immunoaffinity extraction prior to trapping and LC/MS analysis. In this work the antibody was covalently bound to the immunoaffinity column using an active aldehyde silica-based support. Successful preparation of this type of column will conserve the use of antibody. Columns where the antibody is noncovalently bound to the support have the advantage of fresh antibody for each run and the disadvantage of its continued depletion. The chromatograms were collected on an ion-trap mass spectrometer over a mass range of 100–300 Da. The full-scan plots (top traces of Fig. 3a and b) are produced by summing the signal from all ions detected in that range. Extracted ion current profiles (XICs) are produced by plotting the signal from selected ionic species at one m/z value. Full-scan data acquisition is useful for detecting unknowns and impurities and for deconvoluting coeluting peak components. An advantage of an ion-trap mass spectrometer is that sensitivity is not adversely affected by scanning over a wide mass range.

Figure 3a shows full-scan and XIC profiles for carbofuran (m/z 222) and its fragment ions (m/z 165 and 123) from an injection of sample without immunoaffinity extraction. Carbofuran is barely detectable at its retention time of 4 min. Figure 3b shows the corresponding ion current traces for a sample that was pretreated by on-line immunoaffinity extraction. The sample introduced into the spectrometer was sufficiently concentrated in the extraction step to produce a carbofuran signal in both the total ion current and the individual fragment ion currents. The use of immunoaffinity extraction prior to capture on and elution from a reversed-phase trapping column significantly enhances detection of carbofuran in this complex matrix. Detection of carbofuran down to spiked levels of 2.5 ng/g were achieved. Immunoaffinity extraction has an advantage over other forms of extraction such as SPE when the sample matrix contains an excess of interfering components.

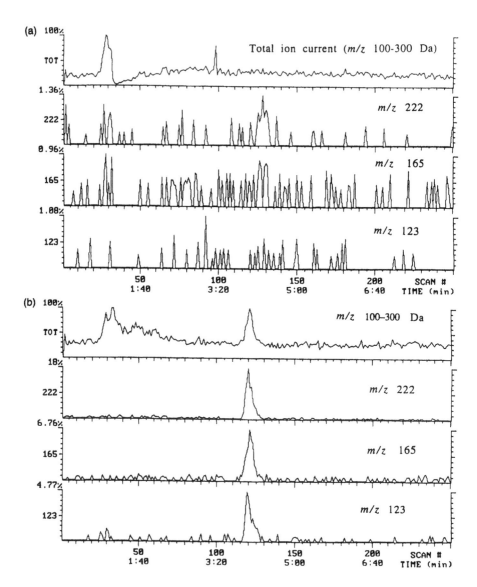

Figure 3 Full-scan and extracted ion current profiles of carbofuran spiked into a preparation of potato starch at 100 ng/g. The traces in (a) correspond to an unextracted
sample, while those in (b) were produced from a sample that was preextracted with an
on-line immunoaffinity column to concentrate the carbofuran. The XIC profiles are
[M + H]$^+$ ions at m/z 222 and fragment ions at m/z 165 and 123. (From Ref. 17.)

C. Off-Line Affinity Ultrafiltration for Combinatorial Library Screening with HPLC/Ion-Spray Mass Spectrometric Detection

Combinatorial libraries are sets of compounds that are used to rapidly explore chemical properties and pharmacological function. Small-molecule libraries consisting of mixtures of compounds may be screened for potential pharmaceutical lead compound candidates. To accomplish this, we developed an ultrafiltration procedure based on an antibody solution as the affinity reagent [18]. Ion-spray–HPLC/mass spectrometry provided analytical information about library components captured from the mixture by the antibody. A diagram of the ultrafiltration procedure is shown in Fig. 4. The method is applicable to screening libraries consisting of up to 30 closely related molecules.

Polyclonal sheep immunoglobulin G molecules (IgGs) for a series of benzodiazepines were obtained from commercial sources (Biodesign International, Kennebunk, ME, and The Binding Site, San Diego, CA) as neat sera and used without further purification. Antibody solution and a benzodiazepine library were combined in 100 μl of 10 mM ammonium acetate buffer at pH 5.0 and incubated at room temperature for 1 hr. The "library" was a pooled 10 μM mixture of 16 benzodiazepines prepared in 10 mM ammonium acetate buffer at pH 5.0. The solution was transferred to 400 μl Microcon microconcentrators having a nominal molecular weight cutoff of 50,000 Da (Amicon, Berverly MA) and centrifuged at 10,000 × g. The filters then were washed 3 times with 100 μl of pH 5.0 ammonium acetate buffer. Elution of captured benzodiaze-

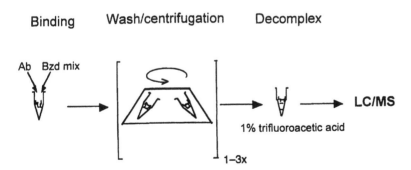

Figure 4 Schematic diagram of the affinity ultrafiltration method. The sample is allowed to interact with the antibody in solution. Unbound components are then separated from specific analyte–antibody complexes by centrifugation through an ultrafiltration membrane with a sufficiently large molecular weight cutoff limit to retain only the complexes. The complex is disrupted by addition of a low-pH buffer, and the specifically captured analytes are separated from the antibodies in a final centrifugation step.

pines was accomplished by disruption of the antibody–benzodiazepine affinity complex with 1% trifluoroacetic acid. The eluate containing the library components retained on the antibody through the wash steps was evaporated to dryness and reconstituted before LC/MS analysis.

HPLC separations were accomplished using fast-gradient elution performed with a 1 mm × 100 mm Betasil C18 analytical column (Keystone Scientific, Bellefonte, PA) and a Waters (Milford, MA) 600MS multisolvent delivery system. Typically, samples contained 1–5 ng of each component and were focused on-column by the high aqueous initial injection conditions. The HPLC effluent was analyzed directly with an API 300 tandem mass spectrometer (PE-Sciex, Concord, Ont.) equipped with a pneumatically assisted electrospray (ion spray) interface. A full-scan range of 200–500 Da was used.

Typical results of selective capture using affinity ultrafiltration are shown in Fig. 5. Selectively extracted components from a mixture of benzodiazepines by anti-nitrazepam and anti-flunitrazepam antibodies are shown in Fig. 5a and b, respectively. The selected components are identified by both retention time and the masses associated with each eluted peak. Figure 5c shows the LC/MS trace of the mixture itself and provides a reference for the retention times of

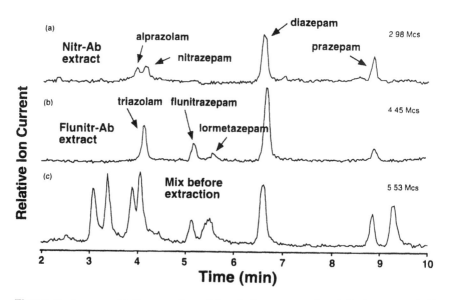

Figure 5 Immunoaffinity extraction with anti-nitrazepam and anti-flunitrazepam. (a,b) Components that are preferentially captured from the mixture by ultrafiltration extraction with antibodies for nitrazepam and flunitrazepam, respectively. (c) LC/MS reference chromatogram of the test mix.

each library component. Full-scan LC/MS data collection provides molecular-weight information for each compound detected by the method. Commercially available antibodies were shown to capture different subsets of components from libraries of known and unknown benzodiazepines. The experiment indicates that affinity ultrafiltration is feasible for screening mixtures of small molecules. The screening is performed in the solution phase, so effects of a solid support on the binding reaction are eliminated.

V. SUMMARY

We have provided some of the practical details of setting up an analytical method that combines affinity techniques with mass spectrometry. Both affinity-based methods and MS have distinct capabilities and limitations. Coupling affinity techniques with mass spectrometry provides significant improvements in sensitivity and specificity compared to either technique used by itself. In the future, robust on-line extraction methods should become available and supplant the manual sample preparation methods that are currently employed. Efficient column-switching methods using on-line extraction methods such as IAE, SPE and SFE, and API-MS for detection will become more popular as methods for handling large numbers of complex samples.

REFERENCES

1. SE Katz, M Siewierski. Drug residue analysis using immunoaffinity chromatography. J Chromatogr 624:403–409, 1992.
2. LF Statham, P Wright, CP Goddard. Immunoaffinity chromatography for the extraction of salbutamol from biofluids prior to MS analysis. Methodol Surv Biochem Anal 22:297–304, 1992.
3. E Davoli, R Fanelli, R Bagnati. Purification and analysis of drug residues in urine samples by on-line immunoaffinity chromatography/high-performance liquid chromatography/continuous-flow fast atom bombardment mass spectrometry. Anal Chem 65:2679–2685, 1993.
4. E Gelpi. Biomedical and biochemical applications of liquid chromatography-mass spectrometry. J Chromatogr A 703:59–80, 1995.
5. R Bagnati, V Ramazza, M Zucchi, A Simonella, F Leone, A Bellini, R Fanelli. Analysis of dexamethasone and betamethasone in bovine urine by purification with an "on-line" immunoafinity chromatography—high-performance liquid chromatography system and determination by gas chromatography-mass spectrometry. Anal Biochem 235:119–126, 1996.
6. GS Rule, J Henion. Determination of drugs from urine by on-line immunoaffinity

chromatography-high-performance liquid chromatography-mass spectrometry. J Chromatogr 582:103–112, 1992.

7. SJ Shahtaheri, MF Katmeh, P Kwasowski, D Stevenson. Development and optimization of an immunaffinity-based solid-phase extraction for chlortoluron. J Chromatogr A 697:131–136, 1995.

8. J Slobodnik, BLM van Baar, UAT Brinkman. Column liquid chromatography-mass spectrometry: selected techniques in environmental applications for polar pesticides and related compounds. J Chromatogr A 703:81–121, 1995.

9. EP Meulenberg, WH Mulder, PG Stoks. Immunoassays for pesticides. Environ Sci Technol 29:553–561, 1994.

10. J Slobodnik, AC Hogenboom, JJ Vreuls, JA Rontree, BLM van Baar, WMA Niessen, UAT Brinkman. Trace-level determination of pesticide residues using on-line solid-phase extraction-column liquid chromatography with atmospheric pressure ionization mass spectrometric and tandem mass spectrometric detection. J Chromatogr A 741:59–74, 1996.

11. DH Thomas, V Lopez-Avila, LD Betowski, J Van Emon. Determination of carbendazim in water by high-performance immunoaffinity chromatography on-line with high-performance liquid chromatography with diode-array or mass spectrometric detection. J Chromatogr A 724:207–217, 1996.

12. WD Linscott. Linscott's directory of immunological and biological reagents. Santa Rosa: Linscott, 1996.

13. PDG Dean, WS Johnson, FA Middle. Affinity chromatography: A practical approach. Washington, DC: IRL Press, 1987.

14. GW Jack. Immunoaffinity chromatography. Mol Biotechnol 1:59–86, 1994.

15. WMA Niessen, AP Tinke. Liquid chromatography-mass spectrometry: General principles and instrumentation. J Chromatogr A 703:37–57, 1995.

16. J Cai, JD Henion. On-line immunoaffinity extraction-coupled column capillary liquid chromatography/tandem mass spectrometry: Trace analysis of LSD analogs and metabolites in human urine. Anal Chem 68:72–78, 1996.

17. GS Rule, AV Mordehai, J Henion. Determination of carbofuran by on-line immunoaffinity chromatography with coupled-column liquid chromatography/mass spectrometry. Anal Chem 66:230–235, 1994.

18. R Wieboldt, J Zweigenbaum, J Henion. Immunoaffinity ultrafiltration with ionspray HPLC/MS for screening small-molecule libraries. Anal Chem 69:1683–1691, 1997.

5

Screening Molecular Diversity Using Pulsed Ultrafiltration Mass Spectrometry

Richard B. van Breemen, Charles P. Woodbury, and Duane L. Venton
University of Illinois at Chicago, Chicago, Illinois

I. INTRODUCTION

The classical approach to drug discovery and development has been to synthesize, purify, characterize, and test each drug candidate for a given biological activity one at a time. Once lead compounds have been identified, structural analysis of receptor active sites by x-ray crystallography combined with computer-assisted molecular modeling can provide guidance to medicinal chemists when deciding which compounds to synthesize and test. However, the process of determining structure–activity relationships as well as discovery of lead compounds remains slow, tedious, and expensive. In order to streamline this process, combinatorial chemistry has been developed as a new approach to the rapid synthesis of large collections of structurally diverse compounds for biological testing [1–3]. Although significant progress has been made in reducing the cost and increasing the speed of synthesis of large libraries of compounds, the screening of these combinatorial libraries to detect molecules with a desired biological activity or affinity for a given receptor is still a bottleneck in the process of drug discovery.

Presently, most screening approaches use classical bioassays, which usually test compounds singly in fluorescence or radiotracer competitive enzyme binding assays. High-throughput screening, a variation of this approach, typically carries out multiple assays in parallel with the assistance of automation. In order to reduce the number of assays required to screen a combinatorial library, pools of compounds may be tested for biological activity instead. Then, successively smaller pools of ligands may be screened in an iterative approach until an active compound is isolated. However, this approach is slow and labor-intensive, especially when there are several pools in a library that demonstrate activity.

Other frequently used combinatorial library screening methods are affinity-based and use either immobilized ligands or immobilized receptors [4–7]. For example, coded or indexed libraries have been developed in which the compounds are covalently linked to labeled beads, chips, pins, or other solid support for indexing the ligand [8]. These libraries are screened by identifying which bead or pin binds receptor molecules, such as by localizing the bound receptor using an enzyme-linked immunoassay that stains beads [7]. Next, the immobilized ligand is identified by using an indexing approach or by direct analysis of the ligand, such as by mass spectrometry [9]. In an analogous approach, the receptor may be immobilized on an affinity column, which is then used for the purification of lead compounds from combinatorial libraries followed by identification of lead compounds using mass spectrometry [10]. (See Chapter 16 for more details concerning this and other affinity methods of screening combinatorial libraries.) However, all of these affinity-based methods suffer from the possibility that immobilization may change the affinity characteristics of the bound species from the native, solution-phase form.

II. SOLUTION-PHASE LIBRARY SCREENING

Few methods exist that allow both combinatorial library mixtures and the target receptor to be screened in solution so as to preserve completely the native ligand–receptor binding interactions. One such approach from Smith's laboratory [11] utilizes electrospray ionization with Fourier-transform ion cyclotron resonance (FTICR) mass spectrometry to carry out screening of library receptor solutions (Fig. 1a). Using electrospray, ligand–receptor complexes are directly ionized from solution and trapped in the FTICR mass spectrometer, where dissociation and identification of the bound ligands are carried out in the gas phase. A potentially powerful technique, successful screening of combinatorial libraries using FTICR will depend upon the ability to ionize very low concentrations of ligand–receptor complexes in the presence of large amounts of non-binding library compounds, and then to trap charged ligands after dissociating them in the gas phase from the receptor molecules.

Another approach to screening combinatorial libraries and receptors in solution has been developed by Kaur and co-workers [12] and uses a combination of size exclusion chromatography, reversed phase chromatography, and mass spectrometry (see Fig. 1b). After incubating a combinatorial library with

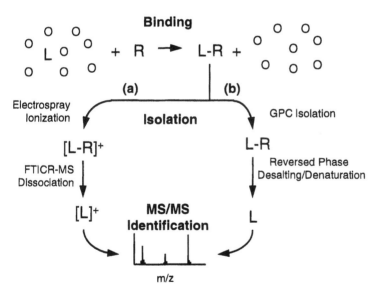

Figure 1 Schemes for solution-phase screening of combinatorial libraries using mass spectrometry. (a) Direct electrospray ionization of ligand–receptor complex from the incubation mixture followed by gas-phase dissociation of the complex and FTICR mass spectrometric identification of ligand (Adapted from Ref. 11.) (b) Following solution-phase incubation of the library with a receptor, gel permeation chromatography is used to isolate the ligand–receptor complex, which is desalted and denatured using a reversed-phase column. After denaturation to dissociate the complex, the ligand is identified using mass spectrometry. (Adapted from Reference 12.)

a receptor in solution, the ligand–receptor complex is rapidly separated from the low-molecular-weight unbound compounds by size exclusion chromatography. The speed of the separation is critically important, since the ligand–receptor complex that is being isolated will start to dissociate during this purification step. Next, the recovered complex is desalted and then denatured on a reversed-phase cartridge to liberate the bound ligands, which are eluted into an electrospray mass spectrometer for identification.

A solution-phase screening method reported by Chu et al. [13] uses electrophoresis–electrospray mass spectrometry with the receptor dissolved in the electrophoresis buffer. As the library mixture undergoes electrophoresis, the elution order of ligands is determined by their interaction with the receptor within the capillary. This method is limited by the extremely small sample quantities that may be loaded onto the capillary electrophoresis column, and electrospray ionization may be suppressed by electrophoresis buffers.

III. PULSED ULTRAFILTRATION/ELECTROSPRAY MASS SPECTROMETRY

During pulsed ultrafiltration, an aliquot of ligand or combinatorial library is injected and pumped through an ultrafiltration cell containing a solution-phase receptor that is trapped by an ultrafiltration membrane (see Fig. 2). High-affinity ligands bind and dissociate from the solution-phase receptor, causing their elution profile from the ultrafiltration cell to be altered. We have developed pulsed ultrafiltration/mass spectrometric methods that may be used either to measure the binding interactions between ligands and a macromolecular receptor or to screen mixtures of potential ligands (combinatorial libraries) for compounds with affinity for the receptor.

A. Measuring Affinity Constants

The interaction of a ligand with a macromolecular receptor may be quantitatively measured using pulsed ultrafiltration. For example, Fig. 3 shows the pulsed ultrafiltration/electrospray mass spectrometric elution profile for phenylbutazone binding to human serum albumin (see Experimental Examples section later for more details about the materials and methods). The upper curve in Fig. 3 shows the deprotonated molecule for an aliquot (or pulse) of phenylbutazone injected through the pulsed ultrafiltration chamber without any albumin present. The lower curve represents the profile for the same amount of phenylbutazone flowing through the chamber containing human serum albumin. Binding of phenylbutazone to albumin causes the elution profile of this ligand to be prolonged and the intensity of the phenylbutazone peak to be diminished. However, the

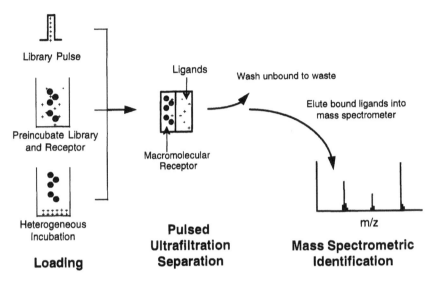

Figure 2 Scheme showing combinatorial library screening using pulsed ultrafiltration/ mass spectrometry. During the loading step (left), ligands are bound to the receptor either online (top) using a flow-through approach or offline (bottom two incubations). Next, unbound compounds and binding buffer, cofactors, etc. are washed out of the ultrafiltration chamber to waste during a separation step (middle). Bound ligands are dissociated from the receptor molecules and eluted from the chamber by introducing a destabilizing solution such as methanol, pH change, etc. Finally, released ligands are identified using mass spectrometry or tandem mass spectrometry (right).

total area under each curve is identical, since the same amount of phenylbuta-zone was injected in each case.

The ligand elution curve in Fig. 3 contains all the information necessary to produce a Scatchard plot and calculate the binding constants describing the ligand–receptor binding interactions. Our solutions to the differential equations describing the interaction of ligand and receptor during pulsed ultrafiltration coupled with the graphical method illustrated in Fig. 3 allow direct calculation of binding constants and related parameters, including K_d, K_i (competitive bind-ing), and n (number of binding sites) (see discussion in Ref. 14). At any given y value in Fig. 3, the mass spectrometer detector response is proportional to the concentration of the free ligand, phenylbutazone, and the spectrometer detector response is proportional to the concentration of the free ligand, phenylbutazone, and the area under each curve from that point on represents the ligand re-maining in the ultrafiltration chamber. However, the areas under each curve (as shown by the shaded areas A and B in Fig. 3) are different. In the control

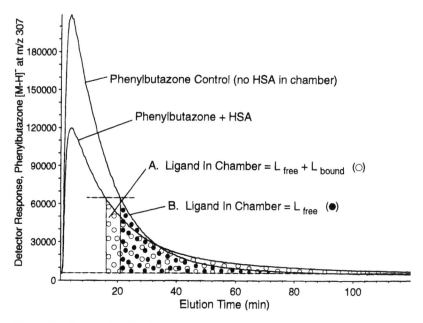

Figure 3 Pulsed ultrafiltration/electrospray mass spectrometric analysis of phenylbutazone binding to human serum albumin. The two elution profiles represent injections of
phenylbutazone through the ultrafiltration chamber either containing (lower curve) or
not containing (upper control curve) human serum albumin. Note that the peak for the
control injection is sharper and more rapidly decays to the baseline, since no binding
interactions delay phenylbutazone elution. At any given y value, the area under the curve
obtained with albumin in the binding chamber (A) represents the sum of free ligand
plus bound ligand. However, the corresponding area under the control curve (B) represents only the free ligand remaining in the ultrafiltration chamber. See text for additional
discussion.

elution profile (A), the area under the curve represents only free ligand, since
no albumin is present in the chamber, but the area under the other curve (B) is
greater and equals the sum of free plus bound ligand. Thus, the difference
between the areas under each curve at any given y value represents the bound
ligand. Since the total amount of ligand (phenylbutazone), total amount of receptor (albumin), volume of the ultrafiltration chamber, and fraction of ligand
bound at any given free ligand concentration are all known, affinity constants
and related binding data may be directly calculated.

Note that the control experiment shown in Fig. 3 automatically corrects
for any nonspecific binding to the ultrafiltration chamber or membrane. Another

convenience of the pulsed ultrafiltration method for determining affinity constants is that just two measurements (one is a control) contain all the information necessary for calculating the affinity constant for a ligand binding to a receptor. Although the example shown in Fig. 3 was carried out at room temperature, ΔH and ΔS of binding may be determined by carrying out pulsed ultrafiltration experiments at a series of different temperatures.

B. Screening Combinatorial Libraries

Based on our combination of pulsed ultrafiltration and mass spectrometry for measuring binding interactions [15], we have developed an online combination of ultrafiltration and electrospray mass spectrometry for screening combinatorial libraries so that solution-phase ligands may be identified that bind to specific solution phase receptors [16]. In addition, this screening approach permits reuse and/or recovery of the receptor molecules. Applications of this method for screening combinatorial libraries are discussed next.

The scheme outlined in Fig. 2 describes the process of pulsed ultrafiltration/electrospray mass spectrometry for screening combinatorial libraries. First, the ultrafiltration membrane installed in the chamber is selected so that its molecular weight cutoff will allow passage of ligands but not the macromolecular receptor. In the examples to follow, a 10,000 molecular weight cutoff ultrafiltration membrane was used, which retained adenosine deaminase (molecular weight approximately 40,000) and human serum albumin (molecular weight approximately 67,000) but permitted passage of ligands and other library compounds (molecular weights <1000).

Next, binding of ligand(s) to a solution-phase receptor is carried out either online using the pulsed ultrafiltration apparatus interfaced to the mass spectrometer (volatile solvents and buffers must be used) or offline in a batch-mode incubation (any solvents and buffers may be used). Online incubations require that the receptor be loaded into the ultrafiltration chamber using an HPLC injector to achieve the desired receptor concentration in the chamber (the receptor concentration is typically at least 10-fold higher than the K_d of the target ligand to be detected), followed by either (1) injection of a single pulse of concentrated library through the ultrafiltration chamber (as shown in Fig. 2, top left), or (2) continuous infusion of a dilute library solution until a steady state is reached between ligand entering the ultrafiltration chamber and ligand eluting from the chamber (used when the solubility of the library is low, which is a common combinatorial library problem). An advantage of this online approach is that the receptor protein may be used for multiple ligand injections and screening experiments, which is important when the receptor protein is expensive or in short supply.

The offline incubation approach also offers distinct advantages. For example, multiple offline incubations may be carried out simultaneously without the need for multiple ultrafiltration chambers (this allows for higher throughput screening and maximizes utilization of the ultrafiltration apparatus and electrospray mass spectrometer in the subsequent screening steps). Also, offline incubation provides the flexibility of using any buffer or cofactor necessary for optimum binding without concern for mass spectrometric compatibility, since the next step utilizes the ultrafiltration chamber for affinity purification. After the incubation mixture is injected into the ultrafiltration chamber, the ligand/receptor complexes are separated from buffer salts, excess cofactors, and unbound library compounds, which may be washed away to waste, before the bound ligands are released from their receptors into the mass spectrometer for identification. Another advantage of offline incubation is the option of using a heterogeneous incubation system consisting of solution-phase receptor and solid-phase library (see Fig. 2, lower left). The heterogeneous incubation method overcomes solubility limitations of the library by allowing the receptor to effectively extract ligands out of the solid phase until the enzyme is saturated by ligand. Finally, offline incubation allows unlimited time for conformational changes to occur in the ligand–receptor complex or other slow binding processes that might be too slow for an online pulsed ultrafiltration binding assay.

Once ligand–receptor binding has taken place using any of the approaches just discussed, ligands may be released from the receptor molecules by abruptly changing the pH, temperature, ionic strength, or organic solvent content of the binding buffer flowing through the pulsed ultrafiltration chamber. After release, these high-affinity ligands are identified using electrospray mass spectrometry and/or tandem mass spectrometry. Since mass spectrometry is used to identify "hits" within a combinatorial library based on molecular weight and structural features evident from the fragmentation patterns, indexing or tagging of library compounds for later identification is not required. If "hits" are detected during pulsed ultrafiltration/mass spectrometry screening that are isomeric with other compounds within the library and indistinguishable using mass spectrometry, then each isomer may be tested individually for binding to the target receptor in order to determine unique hits within the library.

C. Experimental Examples

A Hewlett-Packard (Palo Alto, CA) 5989B MS engine quadrupole mass spectrometer equipped with a 1090L gradient high-performance liquid chromatography (HPLC) system was used for pulsed ultrafiltration/mass spectrometric screening, except that an ultrafiltration chamber was substituted for the HPLC column. The ultrafiltration chamber was either a 1.0-in-diameter in-line solvent filtration unit (Upchurch Scientific, Oak Harbor, WA; catalog no. A-333) in which the filter disk had been replaced by a 10,000 molecular weight cutoff ultrafiltration membrane (Amicon; Beverly, MA; catalog no. YM10) or a comparable home-built chamber that was mechanically stirred. The volume of the ultrafiltration chambers used for the screening experiments was approximately 100 μl, and the volume of the stirred chamber used for measuring affinity constants was 780 μl.

The electrospray mass spectrometer operating parameters (see also Ref. 16) included a dwell time of 5 sec/ion for selected ion monitoring experiments, or a scan rate of 2 sec over the range of m/z 50–400 at unit resolution. Nitrogen gas at 80 psi was used for electrospray nebulization, and nitrogen at 10 L/min and 300°C was used as a drying gas for solvent evaporation. The formation of protonated molecules during positive ion electrospray was enhanced by acidification of the mobile phase as it eluted from the ultrafiltration chamber using methanol/water/acetic acid (49:49:2; v/v/v) at a flow rate of 5 μl/min. Alternatively, postchamber addition of 5% aqueous ammonia in methanol/water (1:1; v/v) at 10 μl/min was used during negative-ion electrospray to enhance the formation of deprotonated molecules.

Binding of phenylbutazone to human serum albumin was investigated using a 780-μl mechanically stirred pulsed ultrafiltration chamber [14] fitted with a 10,000 molecular weight cutoff ultrafiltration membrane and interfaced to the electrospray mass spectrometer. Aliquots containing 91.2 nmol phenylbutazone were injected through the chamber using a Rheodyne (Cotati, CA) 8525 injector. The chamber was loaded with 150 nmol of human serum albumin as the receptor protein. The binding buffer consisted of 33 mM ammonium acetate, pH 7.0, at a flow rate of 100 μl/min and room temperature.

Pulsed ultrafiltration/mass spectrometric screening was carried out using two different macromolecular receptors, human serum albumin and calf intestine adenosine deaminase. Serum albumin functions as an important binding and carrier protein for drugs in human blood, and adenosine deaminase is a key enzyme in purine metabolism, a target for certain antiviral and anticancer agents [17], and a potential target for new drugs that reduce ischemic heart damage [18].

A six compound library (see Table 1) was prepared in 33 mM ammonium acetate, pH 7.5, and used in a screen for binding to human serum albumin. This dilute solution (0.4 μM each, except for thyroxine, which was nearly insoluble under these conditions) was pumped for 30 min at 100 μl/min (1.2 pmol total of each compound) through a 100-μl ultrafiltration chamber containing 15 nmol human serum albumin (150 μM in the ultrafiltration chamber). After a 5-min wash with deionized water at 100 μl/min to remove unbound compounds and buffer from the ultrafiltration chamber, bound ligands were dissociated from albumin and eluted into the mass spectrometer with 60% methanol at 80 μl/ min.

Table 1 Library of Ligands

Human serum albumin ligand	$[M-H]^-$ m/z	Human serum albumin ligand	$[M-H]^-$ m/z
Ascorbic acid	175	Furosemide	329
Tryptophan	203	Salicylate	137
Warfarin	307	Thyroxine	775

Adenosine deaminase ligand	$[M+H]^+$ m/z	Adenosine deaminase ligand	$[M+H]^+$ m/z
EHNA	278	Hypoxanthine	137
Adenine	136	(−)-Inosine	269
Adenosine	268	5'-IMP	349
Adenine 9-α-D-arabinofuranoside	268	(−)-2'-Deoxyinosine	253
2'-AMP	348	2'-Deoxyinosine 5'-monophosphate	333
3'-AMP	348	Guanosine	284
5'-AMP	348	Guanosine 5'-monophosphate	364
Adenosine 5'-carboxylic acid	282	Purine	121
ADP	428	Purine riboside	253
cAMP	330	Adenine N^1-oxide	152

Source: Adapted from Ref. 16.

The resulting selected ion chromatograms for the deprotonated molecules of each library compound are shown in Fig. 4. The control chromatogram obtained without serum albumin (Fig. 4; top) shows signals for ligands that were nonspecifically adsorbed to the ultrafiltration membrane or other parts of the ultrafiltration apparatus and then eluted by methanol. The differences between the peak areas of the control and experiment containing the receptor indicate the extent of formation of the ligand–receptor complex. Specific binding to serum albumin was observed for four of the six compounds in the library including warfarin, salicylate, furosemide, and thyroxine (binding of thyroxine was lower than the other ligands due to its very low concentration in the library). These compounds have affinity constants, K_a, for serum albumin of approximately 10^5 M; for example, the K_a of warfarin binding to serum albumin is 1.5×10^5 M [19]. As expected, no binding was observed for ascorbic acid, which is not a ligand for serum albumin. In addition, no albumin binding was observed for tryptophan, which has approximately 7- to 70-fold lower affinity for albumin than any of the competing ligands warfarin, salicylate, furosemide, and thyroxine [20]. This experiment demonstrates how ultrafiltration/mass spectrometry may be used to screen for ligands that bind to a receptor with affinity constants, K_a, on the order of 1×1^5 M. (See Ref. 16 for more details on screening for ligands of albumin.)

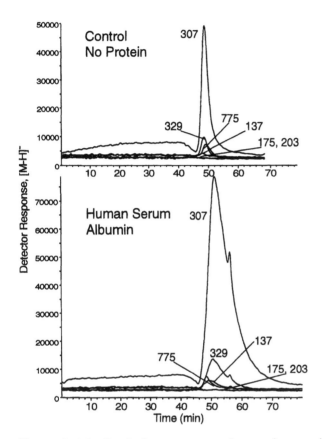

Figure 4 Ultrafiltration/mass spectrometric screening experiment showing release of drugs bound to human serum albumin. A mixture of ascorbic acid, furosemide, salicylate, thyroxine, tryptophan, and warfarin was pumped through the ultrafiltration chamber containing human serum albumin. After a 5-min wash with deionized water, bound ligands were eluted into the mass spectrometer for identification using methanol. Deprotonated molecules of each ligand were monitored using negative ion electrospray mass spectrometry. See Table 1 for list of ligands and their corresponding $[M-H]^-$ ions. (From Ref. 16.)

As another example, adenosine and 19 analogs (see list in Table 1) were used for pulsed ultrafiltration/mass spectrometric screening for compounds that bind to adenosine deaminase. Included in this library was the potent adenosine deaminase inhibitor erythro-9-(2-hydroxy-3-nonyl)-adenine (EHNA), which has a K_i of 1.7 nM and a K_d of 1.9 nM [21,22]. Dissolved in 50 nM potassium phosphate buffer, pH 7.5, all library compounds were 17.5 μM except for EHNA, which was included at a 10-fold lower concentration in order to emphasize the sensitivity and specificity of the screening method. After incubating the adenosine analog library with 2.1 μM adenosine deaminase for at least 15 min offline, an aliquot of the incubation mixture containing 420 pmol adenosine deaminase, 350 pmol EHNA, and 3.5 nmol of each library compound was injected into the ultrafiltration chamber.

The ultrafiltration chamber was washed with water to waste for 8 min at 50 μl/min (equal to 4 chamber volumes) to remove more than 98% of the nonvolatile binding buffer and unbound or weakly binding library compounds. Methanol (50%) was introduced into the mobile phase to dissociate the enzyme–ligand complex and release bound ligands for identification by electrospray mass spectrometry. During either positive- or negative-ion electrospray mass spectrometric analysis of the methanol eluate of the ultrafiltration chamber, only EHNA, $[M+H]^+$ of m/z 278, was detected (see mass spectrum in Fig. 5), because EHNA has the highest affinity for adenosine deaminase among the library compounds. In control experiments using the library without enzyme, no library compounds were detected during methanol elution (Fig. 5,

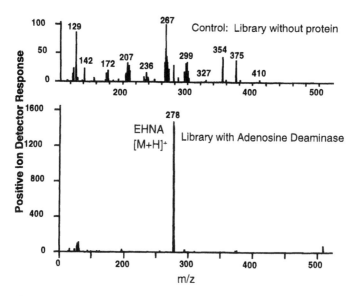

Figure 5 Identification of EHNA as the highest affinity ligand for adenosine deaminase in a combinatorial library of 20 compounds using ultrafiltration/electrospray mass spectrometry. After flushing the nonbinding and weakly bound ligands out of the ultrafiltration chamber with water, EHNA was eluted by disrupting the EHNA/adenosine deaminase complex using methanol. The abundance of protonated EHNA at m/z 278 was 16 times higher than any background ions detected in an identical control experiment using the library but no enzyme. See list of ligands and their corresponding $[M+H]^+$ ions in Table 1. (From Ref. 16.)

control). Thus, pulsed ultrafiltration/mass spectrometry was shown to provide a simple and powerful method for the screening of libraries for compounds with affinity for adenosine deaminase.

After methanolic dissociation of the EHNA/adenosine deaminase or the warfarin/serum albumin complex, the affinity of the receptor for its ligand could be restored by returning to the original aqueous binding conditions [16]. In this manner, the receptor may be reused for additional screening experiments. If necessary to prevent irreversible receptor denaturation, alternate approaches may be used to more gently disrupt ligand–receptor binding, such as changes in pH, temperature increase, exposure to other organic solvents, or a competing ligand.

IV. CONCLUSIONS

Pulsed ultrafiltration/mass spectrometry is being developed as a solution-phase method for both screening combinatorial libraries during new drug discovery and measurement of ligand–receptor interactions. The examples of pulsed ultrafiltration/mass spectrometric screening just given illustrate the range of ligand affinities that have been successfully screened (e.g., K_d values of micromolar to nanomolar), the flexibility of the method to accommodate volatile or nonvolatile buffers (i.e., ammonium acetate or potassium phosphate), the ability to reuse the receptor, and the option of loading methods that may include online or offline incubations. Additional applications of pulsed ultrafiltration/mass spectrometry are under investigation [23], and studies are in progress to establish the maximum size of a combinatorial library that may be screened, whether there are limits to the types of ligands or receptors that may be screened, and what the maximum and minimum affinity constants are that may be measured. Because of its high sensitivity, speed of analysis, and specificity, mass spectrometry has become essential to combinatorial chemistry, and pulsed ultrafiltration/mass spectrometry extends the utility of mass spectrometry even further in the area of combinatorial library screening.

ACKNOWLEDGMENTS

Use of the electrospray mass spectrometer was generously provided by the Hewlett-Packard Company. The assistance of Alexander Schilling, Chao-Ran Huang, and Dejan Nikolic with LC/MS is greatly appreciated.

REFERENCES

1. EM Gordon, MA Gallop, DV Patel. Strategy and tactics in combinatorial organic synthesis. Applications to drug discovery. Acc Chem Res 29:144–154, 1996.
2. JA Ellman. Design, synthesis, and evaluation of small-molecule libraries. Acc Chem Res 29:132–143, 1996.
3. LA Thompson, JA Ellman. Synthesis and applications of small molecule libraries. Chem Rev 96:555–600, 1996.
4. BA Bunin, MJ Plunkett, JA Ellman. The combinatorial synthesis and chemical and biological evaluation of a 1,4-benzodiazepine library. Proc Natl Acad Sci USA 91:4708–4712, 1994.
5. MA Kelly, H Liang, II Sytwu, I Vlattas, NL Lyons, BR Bowen, LP Wennogle. Characterization of SH_2-ligand interactions via library affinity selection with mass spectrometric detection. Biochemistry 35:11747–11755, 1996.
6. RS Youngquist, GR Fuentes, MP Lacey, T Keough. Generation and screening of combinatorial peptide libraries designed for rapid sequencing by mass spectrometry. J Am Chem Soc 117:3900–3906, 1995.
7. R Liang, L Yan, J Loebach, M Ge, Y Uozumi, K Sekanina, N Horan, J Gildersleeve, C Thompson, A Smith, K Biswas, WC Still, D Kahne. Parallel synthesis and screening of a solid phase carbohydrate library. Science 274:1520–1522, 1996.
8. EM Gordon, RW Barrett, WJ Dower, SPA Fodor, MA Gallop. Applications of combinatorial technologies to drug discovery. 2. Combinatorial organic synthesis, library screening strategies, and future directions. J Med Chem 37:1385–1401, 1994.
9. CL Brummel, JC Vickerman, SA Carr, ME Hemling, GD Roberts, W Johnson, J Weinstock, D Gaitanopoulos, SJ Benkovic, N Winograd. Evaluation of mass spectrometric methods applicable to the direct analysis of non-peptide bead-bound combinatorial libraries. Anal Chem 68:237–242, 1996.
10. ML Nedved, S Habibi-Goudarzi, B Ganem, JD Henion. Characterization of benzodiazepine "combinatorial" chemical libraries by on-line immunoaffinity extraction, coupled column HPLC-ion spray mass spectrometry-tandem mass spectrometry. Anal Chem 68:4228–4236, 1996.
11. JE Bruce, GA Anderson, R Chen, X Cheng, DC Gale, SA Hofstadler, BL Schwartz, RD Smith. Bio-affinity characterization mass spectrometry. Rapid Commun Mass Spectrom 9:644–650, 1995.
12. S Kaur, V Huebner, D Tang, L McGuire, R Drummond, J Csetjey, J Stratton-Thomas, S Rosenberg, G Figliozzi, S Banville, R Zuckermann, G Dollinger. Identification of specific receptor peptoid ligands in combinatorial mixtures by affinity selection and mass spectrometry. Proceedings of the 43rd ASMS Conference on Mass spectrometry and Allied Topics, Atlanta, GA, 1995, p 30.
13. YH Chu, DP Kirby, BL Karger. Free solution identification of candidate peptides from combinatorial libraries by affinity capillary electrophoresis/mass spectrometry. J Am Chem Soc 117:5419–5420, 1995.
14. S Chen. Further development of pulsed ultrafiltration analysis of ligand-macromol-

ecule interactions. PhD dissertation, University of Illinois at Chicago, Chicago, 1995.

15. RB van Breemen, CR Huang, D Nikolic, CP Woodbury, YZ Zhao, DL Venton. Quantification of ligand/receptor interactions using electrospray LC/MS. 44th ASMS Conference on Mass Spectrometry and Allied Topics, Portland, OR, 1996, p 1032.
16. RB van Breemen, CR Huang, D Nikolic, CP Woodbury, YZ Zhao, DL Venton. Pulsed ultrafiltration-mass spectrometry: A new method for screening combinatorial libraries. Anal Chem 69:2159–2164, 1997.
17. RP Agarwal. Inhibitors of adenosine deaminase. Pharmacol Ther 176:399–429, 1982.
18. GS Sandhu, AC Burrier, DR Janero. Adenosine deaminase inhibitors attenuate ischemic injury and preserve energy balance in isolated guinea pig heart. Am J Physiol 265:H1249–H1256, 1993.
19. KF Brown, MJ Crooks. Displacement of tolbutamide, glibencalmide and chorpropamide from serum albumin by anionic drugs. Biochem Pharmacol 25:1175–1178, 1976.
20. SF Sun, F Wong. Determination of L-tryptophan–serum albumin binding by high-performance liquid chromatography. Chromatographia 20:495–499, 1985.
21. DJT Porter, E Abushanab. Kinetics of inhibition of calf intestinal adenosine deaminase by (+)- and (−)-erythro-9-(2-hydroxy-3-nonyl)adenine. Biochemistry 31: 8216–8220, 1992.
22. C Frieden, LC Kurz, HR Gilbert. Adenosine deaminase and adenylate deaminase: Comparative kinetic studies with transition state and ground state analogue inhibitors. Biochemistry 19:5303–5309, 1980.
23. YZ Zhao, RB van Breemen, D Nikolic, CR Huang, CP Woodbury, A Schilling, DL Venton. Screening solution-phase combinatorial libraries using pulsed ultrafiltration (PUF)/electrospray mass spectrometry. J Med Chem, in press.

6

Mass Spectrometry in Immunology: Identification of a Minor Histocompatibility Antigen

Carthene R. Bazemore Walker, Nicholas E. Sherman, Jeffrey Shabanowitz, and Donald F. Hunt
University of Virginia, Charlottesville, Virginia

I. INTRODUCTION

A complex mixture of 10,000 fragments derived from cellular proteins is displayed on the surface of cells in association with class I glycoproteins of the major histocompatibility complex (MHC). Cytotoxic T-lymphocytes (CTL) bind to the class I molecules, sample the peptides being presented, and lyse those cells that display new antigens as a result of viral infection, tissue transplantation, autoimmune disorders, or disease states such as cancer. One copy of a foreign antigen is thought to be sufficient to stimulate an effective immune response. From 3×10^{10} cells (10 l of cells), the quantity of peptide present at one per copy per cell would be approximately 65 fmol. Identification of a single peptide antigen present at this level in a mixture with 10,000 other peptides represents a daunting analytical challenge. Here we describe the use of microcapillary high-performance liquid chromatography (HPLC) electrospray ionization (ESI) mass spectrometry (MS) in conjunction with a sensitive immunological assay to characterize a minor histocompatibility antigen involved in graft-versus-host disease   (GVHD) associated with bone marrow transplantation (BMT) [1,2].

The rapid and specific recognition of potential antigens and the production of an appropriate immune response are hallmarks of adaptive immunity. T lymphocytes are of central importance because they can kill unhealthy cells and modulate the function of other immune cells. Unlike antibodies that recognize soluble antigens (carbohydrate, lipid, or protein) in the extracellular medium, T-cell receptors recognize antigenic peptides bound to MHC class I molecules on the surface of host cells. The MHC class I molecule consists of a glycosylated 45-kDa heavy chain noncovalently associated with a 12-kDa light chain, β_2-microglobulin. At the top of the class I molecule, two alpha helices and a beta sheet form the sides and the base of the peptide-binding groove [3,4]. Peptides of 8–12 residues that associate with class I molecules are derived from cytosolic proteins and are generated by a multicatalytic protease, known as the proteosome [5]. The resulting fragments are actively translocated from the cytosol into the endoplasmic reticulum, where they bind to nascent class I molecules [6,7]. Class I MHC–peptide complexes are subsequently transported to the cell surface for perusal by CTL.

Up to six different class I molecules are expressed on the surface of nucleated cells in a fully heterozygous individual. These highly polymorphic proteins (over 150 different isoforms in the human population) vary mostly in those residues lining the antigen binding cleft [8–10]. This allows each allele to bind and present a different set of structurally diverse peptides that share a common motif [11]. MHC class I diversity is advantageous because it decreases the likelihood of a pathogen evading the protective immune responses of a population.

As alluded to earlier, the peptides presented by class I molecules are representative of the cell's expressed genes. CTL normally do not respond to autologous peptides due to their ability to discriminate self from nonself. This quality is crucial because CTL recognition of peptides derived from endogenous viral or mutated protein products leads to destruction of the diseased cell. If a self-peptide is incorrectly identified as foreign, the resulting tissue and organ damage can lead to autoimmune disorders such as diabetes, arthritis, or multiple sclerosis.

Although the need to protect the body from invading vertebrate cells is not readily apparent, the vertebrate immune system also elicits an effective immune response against foreign grafts. In fact, the deleterious immune responses to transplanted organs and tissues are the primary obstacle to successful transplantation. Graft rejections are caused primarily by differences in histocompatibility (H) antigens between a donor and a recipient. Indeed, the MHC loci, named for their significant role in graft rejections, are responsible for the most rapid and severe graft rejection reactions.

Although less vigorous, transplant rejection can be induced by antigens encoded by non-MHC loci as well. These so-called minor H antigens are peptides recognized in a manner analogous to those from viruses or tumors [12]. Minor antigens were originally identified by their ability to cause rejection of tissue transplants between MHC-matched inbred strains of mice [13]. The recipient mounted an effective immune response against donor tissue because the non-MHC genes encoded antigens that were recognizable by recipient CTL. Genetic loci coding for minor H antigens is widely, and unpredictably, distributed throughout the murine genome [14], suggesting great complexity in the outbred human species. More than 50 murine minor H genes have been mapped; however, up to several hundred loci may actually encode H antigens [15,16]. The overall number and complexity in humans remain unclear.

Organ and tissue transplants are generally performed to restore a functional deficit resulting from a disease or treatment for a disease. Although some types of transplants, such as kidney and corneal transplants, are being performed with high success rates, others, such as allogeneic BMT, are most often characterized by their morbidity and mortality rates following the onset of graft failure or GVHD. In BMT, minor H disparities cause about 80% of the cases of graft failure or GVHD although the donor and recipient are HLA-matched

[17,18]. The mature immunocompetent T cells found in the donated bone marrow recognize host tissue as foreign and react against it.

CTL reactive to host minor H antigens (Ags) have been detected in HLA-identical BMT patients suffering from GVHD [19]. Five non-sex-linked minor H Ags, HA-1 through HA-5, have been identified using these reactive CTL clones. Each minor H antigen exhibits a unique phenotype frequency [20] and tissue distribution profile [21]. All are recognized within the context of the human MHC class I molecule HLA-A2.1 except HA-3, which is restricted by HLA-A1 [20]. Only cells of hematopoietic origin (including leukemic cells) express HA-1 and HA-2, whereas HA-3, -4, and -5 are expressed ubiquitously on all tissues [21].

Many hematologic diseases, such as severe aplastic anemia and leukemia, have been treated using BMT [22,23]. However, the prophylactic and therapeutic protocols currently in place are disappointingly inadequate, warranting continued investigation of the immunologic basis of GVHD. Identification of the antigens responsible for GVHD would make it possible to screen potential patients and to predict the outcome of a BMT between two HLA-matched individuals. Identified antigens could also be administered in advance of BMT to tolerize donor T cells to recipient proteins and thus to avoid GVHD altogether [24]. Here we describe the identification and structural characterization of the first human minor histocompatibility antigen, HA-2. Approximately 95% of the humans that express HLA-A2.1 (40% of the Caucasian population) also express this antigen.

II. EXPERIMENTAL

A. MHC Purification and Peptide Isolation

The naturally processed HA-2 epitope [2] was obtained from a B-cell lymphoma by using immunoaffinity chromatography followed by acid extraction and size filtration. HA-2 was expressed in these cells in association with the MHC class I molecule HLA-A2.1. The B cells, taken from a female allogeneic BMT patient, were immortalized by the Epstein-Barr virus (EBV) and cultivated in 1-l roller-bottle flasks with media and antibiotics. A cocktail of detergent and protease inhibitors was employed to lyse about 7×10^{10} cells, and the lysate was centrifuged to separate the supernatant from the cellular debris. Supernatant was then passed through an HLA-A2.1-specific monoclonal antibody column that retained the targeted peptide–MHC complexes while nonspecific material washed through the column. Acid, which denatures MHC molecules, was used to liberate the peptides from the affinity column. Low-molecular-weight material was separated from high-molecular-weight contaminants (antibody fragments and β_2-microglobulin) by ultrafiltration using a

10-kDa cutoff filter (Millipore Corp., Marlborough, MA), and the extract was concentrated by vacuum centrifugation.

B. Reversed-Phase HPLC Fractionation

Over 10,000 different endogenous peptides are found in material extracted from the class I MHC molecule HLA-A2.1 [1]. Prior to analysis by mass spectrometry, this mixture can be simplified by multiple stages of HPLC purification. The peptide extract containing HLA-A2.1-bound peptides was fractionated by reversed-phase HPLC using a LiChrospher 60 RP-Select B column (5 μm, 250 mm×4 mm, E. Merck, Darmstadt, FRG). The two elution buffers were 0.1% trifluoroacetic acid (TFA) (buffer A) and 0.1% TFA in acetonitrile (buffer B). Peptides were eluted using an 80-min gradient of 0% buffer B (20 min), 0–12% buffer B (5 min), and 12–50% buffer B (55 min) at a flow rate of 100 μl/min with fractions collected each minute. Each fraction was analyzed for HA-2-sensitizing activity (bioassay described in the following section), and sample in the biologically active fraction from this first dimension separation was refractionated using a shallower gradient. This 108-min gradient consisted of 0% buffer B (29 min), 0–22% buffer B (5 min), 22% buffer B (5 min), 22–27.9% buffer B (59 min), 27.9% buffer B (5 min), and 27.9–100% buffer B (5 min). The flow rate was 100 μl/min. Fractions were collected each minute and monitored at 214 nm for both separations. It should be noted that the gradient profile used in the second dimension fractionation was developed specifically for the elution of the HA-2 antigenic peptide. Usually, the first dimension of chromatography will consist of the already described 80-min gradient using heptafluorobutyric acid (HFBA), followed by a second separation of the biologically active fractions with the same acetonitrile gradient but using TFA as the ionic modifier instead. This protocol has been used in our laboratory to successfully isolate other minor H antigens and tumor-specific epitopes [25–27].

C. Cytotoxicity Assay

An in vitro chromium release assay, used to determine the amount of cell-mediated killing, has been described previously [1]. In general, the experiment exploits the phenomenon of effector recognition of antigenic peptides presented on the surface of target cells. When the epitope of interest is encountered by the T cell, the target cell is lysed by the cytolytic activity of the CTL.

An HA-2-reactive CTL clone 5H17, used as the effector cell line, was derived from a female patient suffering from chronic GVHD after undergoing BMT for severe aplastic anemia. It recognizes the HA-2 epitope within the context of HLA-A2.1 HLA-A2.1-positive, HA-2-negative, T2 cells served as the target cell line. These cells are impaired in antigen processing because they lack the genes encoding the TAP (transporter associated with antigen processing) proteins that translocate peptides from the cytosol into the endoplasmic reticulum (ER). T2 cells do express MHC molecules on the cell surface, and many of them are devoid of peptide. These empty MHC molecules readily take up peptide that is provided exogeneously [28].

   T2 cells were incubated with chromium. After labeling, the target cells were washed to remove free chromium and then incubated for 30 min with aliquots taken from individual fractions. Aliquots (25 μl) taken from each HPLC fraction were diluted 1:50 and 1:30 for first and second dimension HPLC fractions, respectively, in Hanks balanced salt solution (HBSS) containing 50 mM HEPES. CTL were then cocultured with radiolabeled target cells at an effector:target (E:T) ratio of 18:1 (first dimension fractions) and 30:1 (second dimension fractions). After 4 hr, the supernatant was removed and measured for chromium radioactivity. The cytotoxic activity (% specific lysis) was determined as

$$\frac{\text{experimental release} - \text{spontaneous release}}{\text{maximum release} - \text{spontaneous release}} \times 100$$

Target cells cultured in media alone or with detergent served as the spontaneous chromium release control and maximum release control, respectively. All reported values are the average of triplicate experiments.

   Each well of a 96-well microtiter plate was also tested for CTL sensitizing activity following a splitting experiment (described later). In this case, 50 μl of media was removed from each well and incubated with T2. CTL were added to give an E:T ratio of 40:1. The experiment proceeded as described earlier. For the testing of synthetic peptides, a 1-μM solution was added to the targets instead of an aliquot from an HPLC fraction.

D. Microcapillary HPLC-Electrospray Ionization/Tandem Mass Spectrometry

Our laboratory uses a modified version of the procedure described by Kennedy and Jorgenson to prepare reversed-phase microcapillary columns [29,30]. A 0.5-mm plug of 5-μm spherical silica particles (60 Å pores, LiChrosorb Si 60, EM Science, Gibbstown, NJ) was created at one end of a piece of fused-silica capillary (100 μm ID $\times$ 25 cm length, Polymicro Technologies, Inc., Phoenix, AZ). The security of the glass frit was tested at $\sim$1000 psi using a helium pressure vessel (bomb). Constant pressure (500 psi) supplied by the helium bomb was then used to force a continuous stream of C_{18} (10 μm, 120 Å pores, YMC-Gel, Morris Plain, NJ) beads in isopropanol into the capillary until a bed length of 12 cm had been reached. Aliquots of HPLC fractions were loaded onto the column using the same pressurization technique. The amount of sample loaded was determined indirectly by measuring the amount of displaced solvent using a 1–5-μl disposable glass pipet. Sample was eluted directly into the ESI source from the microcapillary column (flow rate of 0.5 μl/min) using a 34-min linear gradient of 0–60% acetonitrile in 0.1 M acetic acid for splitting experiments, and 0–80% acetonitrile in 0.1 M acetic acid for MS/MS experiments.

Data for the splitting experiments (described in the next section) were acquired on an upgraded Finnigan MAT TSQ-70 (San Jose, CA) triple-quadrupole mass spectrometer equipped with a Finnigan APCI electrospray ionization (ESI) source. Acidic eluant was sprayed from the tip of the ESI stainless steel needle maintained at +4.6 kV relative to the heated inlet capillary. This electric field produces a fine mist of multiply protonated droplets, which are attracted to the capillary held at a lower potential. The electrospray was stabilized by a coaxial flow of sheath liquid (0.125% acetic acid in 70% methanol) and ambient nitrogen gas (used as both sheath and auxiliary gases). Charged droplets were desolvated as they traveled through the heated capillary (150°C), producing positively charged gaseous peptide ions. The peptides were subjected to mass analysis by placing a combination of rf and dc potential on Q1 and only rf potential on Q2 and Q3. In this mode of operation, Q2 and Q3 function as ion-focusing devices and transmit all ions from Q1 to the detector. Q1 functions as a mass filter by separating the ions according to their mass-to-charge (m/z) ratios. Ions are detected by a high-voltage conversion dynode (15 kV) electron multiplier (10^8 gain). The result is a main-beam mass spectrum containing singly or multiply charged peptide ions, $[M+nH]^{n+}$, characteristic of the molecular weight of each peptide in the mixture. Q1 was operated at less than unit resolution (2 amu wide near the baseline) to maximize ion transmission, or sensitivity. Tandem mass spectrometry experiments were performed on a Finnigan MAT TSQ-7000 equipped with a Finnigan APCI/ESI source. A mass range from 50 Da to the molecular mass of the selected peptide was normally scanned at a rate of 500 Da/sec. Methodology for determining the amino acid sequence of class I peptides by collision-activated dissociation (CAD) is described in a later section of the chapter.

E. On-Line Postcolumn Effluent Splitting

Mass spectrometric analysis of the second dimension HPLC fraction found to be biologically active revealed that the fraction still contained between 50 and 100 different HLA-A2.1 associated peptides. Since the remaining material was still too complex and insufficient in quantity to sequence each peptide, the HA-2 active fraction was analyzed by an on-line microcapillary column effluent splitter [1,2]. The column effluent was split so that material was simultaneously directed into the mass spectrometer and into a 96-well microtiter plate for subsequent analysis using the cytotoxicity assay described earlier. A C$_{18}$ (YMC-

Gel) microcapillary HPLC column (100 μm by 25 cm) was butt-connected with a zero dead volume union (Valco, Houston, TX) to two capillaries of different lengths and interior diameters (25 μm and 40 μm; Polymicro Technologies, Phoenix, AZ) and eluted with a 34-min gradient of 0–60% acetonitrile. The 25-μm capillary deposited one sixth of the material into microtiter plate wells containing 150 μl of RPMI 1640 (GIBCO-BRL, Grand Island, NY) with 10% human serum culture media. The 40-μm capillary directed the remaining five sixths of the material into the electrospray ionization source, and mass spectra of the peptides deposited in each well were recorded on the mass spectrometer. This allows for an accurate mass spectrometric account of the peptide molecular masses that are present within each well. This mass spectrometric record is then correlated to the results obtained from the CTL assay.

F. Synthesis of Synthetic Peptide(s)

Peptides corresponding to the sequence of interest were synthesized using standard N-α-fluorenylmethoxycarbonyl (FMOC) chemistry with a Gilson AMS 422 peptide synthesizer (Middletown, WI) with target yields of 10 μmol. Py-BOP/NMM (benzotriazole-1-yl-oxy-tris-(pyrrolidino)-phosphonium hexafluorophosphate/N-methylmorpholine) was used as the activator for the in situ activation of the FMOC amino acids [31]. Peptides were cleaved from the Wang type resins (Calbiochem, La Jolla, CA) using a mixture of TFA/EDT/anisole/thioanisole/water (89:3:2:3:3) for 3 hr [31]. This cleavage solution also removed and scavenged the protecting groups used to safeguard the functional groups of the amino acid residues. The synthetic peptides were precipitated using cold diethyl ether, which was decanted after centrifugation, and dried under argon gas. All peptides were resuspended in 2 ml of 10% acetic acid and eluted using an 8-min linear gradient from 0 to 80% acetonitrile (0.085% TFA) in 0.1% TFA at 3 ml/min. An ABI 140A solvent delivery system (Foster City, CA) coupled to an ABI 759A ultraviolet (UV) detector was used to purify the synthetic peptides. In addition, a model 505 column oven (SSI, State College, PA) was used to house the reversed-phase column (POROS II/RH or R2/H column, 4.6 mm × 10 cm; PerSeptive BioSystems, Cambridge, MA). The column effluent was monitored at 214 nm, and peptides were collected by peak. Sequences were confirmed by tandem mass spectrometry performed on a triple-quadrupole mass spectrometer.

G. HLA-A2.1 Binding Assay

A quantitative assay was used to determine the ability of synthetic test peptides to compete for binding to the HLA-A2.1 molecule. The dose of the test peptides giving 50% inhibition of binding of the standard, an iodinated HLA-A2.1 restricted epitope from the hepatitis B virus core protein (residues 18–27), was calculated using radioactivity. Briefly, a predetermined concentration of purified HLA-A2.1 molecules and radiolabeled peptide were incubated for 48 hr at room temperature with varying amounts of test peptides. Gel filtration was used to separate the free peptide from the bound peptide–HLA-A2.1 complexes, and both were measured for radioactivity.

H. Peptide Dose-Response Curve

Synthetic peptide solutions of greater than micromolar concentrations were made using HBSS and 50 mM HEPES. Serial dilutions were made in individual wells of a 96-well microtiter plate until final concentrations of 1 μM to 100 fM were reached. A standard chromium release assay was then performed on each well to determine the concentration of synthetic peptide necessary for half-maximal lysis of T2.

III. SEQUENCE ANALYSIS ON A TRIPLE-QUADRUPOLE MASS SPECTROMETER

A. General Method

Sequence information on an individual peptide is obtained by conducting a collision-activated dissociation (CAD) experiment, which is carried out in three stages. The first quadrupole (Q1) is set to transport all ions within a 2 mass unit window centered around the ion of interest. All other ions introduced into Q1 from the ion source are ejected due to the oscillating electric field. Q2, also called the collision chamber, is a bent-octapole mass filter that has only rf potential applied to the rods. When the peptide ion enters Q2, it encounters an inert gas, such as argon, which is present at a total pressure of 3 mtorr in our system. It suffers 10–100 low-energy (15–40 eV) inelastic collisions with the gas, and this process converts translational kinetic energy into vibrational kinetic energy, resulting in random fragmentation along the amide backbone of the peptide chain. A collection of neutrals and positively charged fragments is produced by this process. The neutrals are pumped away by the vacuum system, while charged ions are transmitted to Q3, which then separates them by mass. Q3 is also operated at less than unit resolution (2 mass units wide) in order to increase ion transmission. Computer control of critical parameters allows for facile execution of this entire procedure, which lasts only a few seconds.

B. Interpretation of CAD Spectra

As stated earlier, the peptides cleave preferentially along the polyamide backbone during the CAD process. Using the terminology of Biemann [32], ions of type b and y predominate the CAD spectra obtained using low-energy conditions. All y-type ions contain the carboxyl terminus of the peptide plus one or more additional residues. In like manner, all b-type ions contain the amino terminus of the peptide plus one or more amino acid residues. The ions of a particular series are used to establish the sequence of the peptide by calculating the mass differences between fragment ions and noting those that are separated by the residue masses of the lightest (57 amu) and heaviest (186 amu) amino acids. Due to the multiple collisions a precursor ion suffers during the CAD process, it may undergo other fragmentation pathways that provide additional information. As a result, there may be ions resulting from losses of water, ammonia, carbon monoxide, or low-mass ions indicative of specific amino acids present in the spectrum. These extra ions often facilitate recognition of the next ion in a certain series [33]. Since b and y ions offer complementary or even redundant information, sequence deduction is often possible from both the C- and N-termini of the peptide.

C. Modifications of Candidate Antigen

Two sets of amino acids (leucine/isoleucine and glutamine/lysine) cannot be distinguished from one another using a triple-quadrupole mass spectrometer because they are of identical mass (isobaric). Under low-energy conditions, there is no way to differentiate leucine and isoleucine isomers, and this uncertainty is denoted by the single-letter code X, or the three-letter code Lxx.* In the case of the lysine/glutamine uncertainty, discrimination between the two is accomplished by performing an acetylation reaction on the peptide of interest. The free amine group of lysine is vulnerable to modification, whereas the terminal amide group of glutamine is not. Repeating the CAD experiment on the modified peptide will determine which of the two residues are present in the peptide. Esterification of peptides is also used extensively in our laboratory to resolve ambiguities, since we operate at less than unit resolution to maximize sensitivity. Acidic residues are modified thereby differentiating them from amino acids differing by only 1 amu (asparagine/aspartic acid and lysine or glutamine/glutamic acid).

D. Database Searches

As implied in the preceding discussion, the manual interpretation of CAD mass spectra is in many cases the rate-limiting step in the deduction of a peptide sequence. Therefore, in some instances the partially interpreted or totally unin-

---

*Hulst and Kientz recently described the use of high-cone-voltage fragmentation and low-energy tandem mass spectrometry to distinguish leucine from isoleucine [34].

terpreted peptide MS/MS spectra is compared to amino acid sequences from protein and nucleotide databases using the SEQUEST program [35–37]. The program correlates tandem mass spectra to a primary sequence by using the nonredundant OWL database, which is a compilation of protein sequences taken from the GenBank (National Center for Biotechnology Information, Washington, DC), National Research Laboratory (Brookhaven National Laboratory, Brookhaven, NY), Protein Information Resources (National Biology Resource Foundation, Georgetown University, Washington, DC), and SWISS-PROT (University of Geneva, Geneva, Switzerland) databases. This strategy is advantageous because it combines computer-assisted interpretation of tandem mass spectra with protein and DNA database searching. However, this approach does not prove expedient in all cases. A sequence may be determined without the aid of computer algorithms, and then all protein and nucleotide databases are searched using the basic alignment search tool (BLAST).

IV. RESULTS AND DISCUSSION

Peptides extracted from HLA-A2.1 were purified by reversed-phase HPLC. One fraction (Fr33) was recognized by the HA-2-specific clone 5H17 (Fig. 1a). This fraction was subjected to a second dimension of HPLC using a shallower gradient. HA-2-sensitizing activity was observed in fraction 37 and, to a lesser extent, in fraction 38 (Fig. 1b). More than 100 different peptide species were contained within the active fractions as assessed by microcapillary HPLC electrospray tandem mass spectrometry. This number of peptides could not be sequenced with the available material. Therefore, the strongly active fraction 37 was rechromatographed using microcapillary HPLC to identify the peptide responsible for HA-2-sensitizing activity. An on-line splitter was used to split five sixths of the effluent into the mass spectrometer, while one sixth was directed into a 96-well microtiter plate for a subsequent epitope reconstitution assay [1]. In this way, individual peptide abundances observed in the mass spectrum could be correlated to the epitope reconstituting activity. HA-2 activity was found in wells 50–53 as shown in Fig. 2a. Although many peptides are still present in this third dimension separation (Fig. 2b), the relative abundance profiles of only five ions [with mass-to-charge ratios (m/z) of 651, 869, 965, 979, and 1000] matched HA-2 epitope reconstitution activity. At this point, it is important to stress that from a extract of over 10,000 different MHC class I peptides (simplified to 50–100 by two HPLC separations), five candidate peptides have been identified by the use of the splitter apparatus.

Collision-activated dissociation (CAD) analysis performed on the species at m/z 979 revealed the existence of two different peptides, YXGEVXVSV and SXDFGTXQV (Fig. 3a and b). The candidate ions were also converted to their corresponding methyl ester derivatives in order to fully establish the cor-

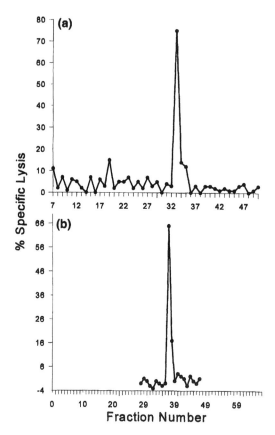

Figure 1 Reconstitution of HA-2-specific epitopes using reversed-phase HPLC fractions of peptides extracted from HLA-A2.1. Individual HPLC fractions were added to 2500 ^{51}Cr-labeled target cells and then incubated in the presence of HA-2-specific CTL. (a) First dimension fractionation. (b) Second dimension fractionation of fraction 33 from (a).

rect sequences. The two peptide sequences, made with an equimolar mixture of L and I in place of X, were assayed for HA-2-specific CTL-sensitizing activity. Only incubation with peptide mixture YXGEVXVSV resulted in lysis of T2. The SXDFGTXQV peptide mixture was negative at all concentrations tested. To determine the sequence of the naturally processed HA-2 epitope, four peptides with I or L at positions 2 and 6 were synthesized. The elution time of the four synthetic peptides was investigated using microcapillary HPLC. The co-elution experiments revealed that three of the peptides (YIGEVLVSV, YLGEVLVSV, YLGEVIVSV) coeluted with the naturally processed epitope. This experiment demonstrated that peptide YIGEVIVSV was not the peptide of

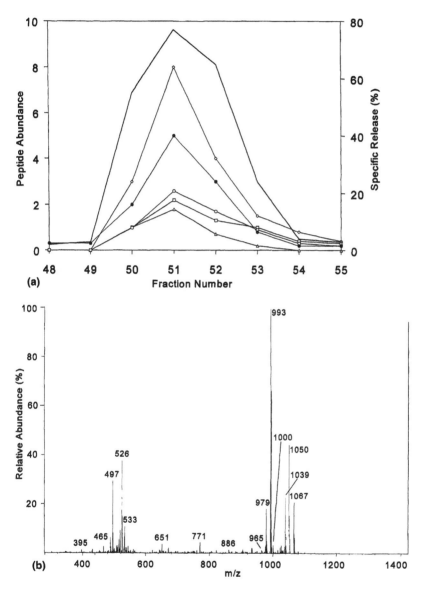

Figure 2   (a) Identification of candidate peptides by mass spectrometry. Epitope reconstituting activity is measured as percent specific release of ^{51}Cr (solid line). Peptide ion
abundance for candidate peptides with *m/z* 651 (diamonds), 869 (triangles), 965
(squares), 979 (open circles), and 1000 (solid circles). Full scale is 10^5 counts except
for 979, which is 10^6 counts. (b) The summed mass spectral data for the most active
well (51) of the split shown in (a). Full scale is $\sim 10^7$ counts.

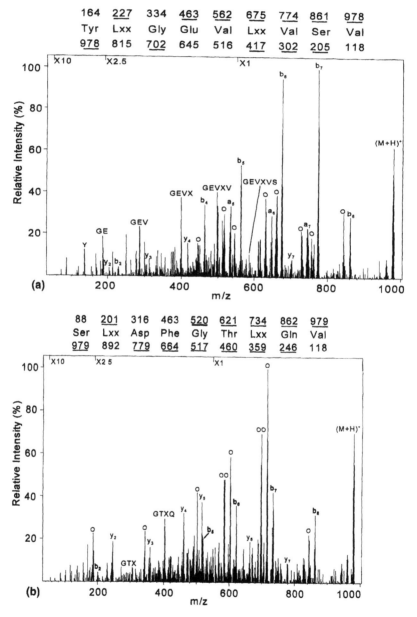

Figure 3 (a) CAD of peptide [M + H]⁺ 978 using ~25 fmol of material from HA-2-positive cells after two dimensions of HPLC. (b) CAD of peptide [M + H]⁺ 979 using ~25 fmol of material from HA-2-positive cells after two dimensions of HPLC. Predicted masses for fragments of b and y type are written above and below the sequence, respectively. Losses of water or ammonia are indicated by open circles. Observed ions are underlined.

interest. Further attempts to chromatographically resolve the remaining three peptides from one another or from the naturally processed epitope failed.

Each of the three peptides was recognized by clone 5H17 when incubated with the T2 cell line (Fig. 4a). Peptides YLGEVLVSV and YLGEVIVSV sensitized targets for half-maximal lysis at nanomolar concentrations (1.5 and 2.25 nM, respectively), whereas the sensitizing concentration for peptide YIGEVLVSV was substantially lower at 40 pM. All of these concentrations were within the range of 10 pM to 50 nM established for other naturally processed epitopes [1,38–41]. Additionally, an independently derived CTL clone (5H13) that recognizes HA-2 but differs slightly from clone 5H17 in its fine specificity of antigen recognition also recognized all three peptide variants (Fig. 4b). Peptide YIGEVLVSV still sensitized at a lower concentration in this experiment. Thus, both HA-2-specific CTL clones recognized the same peptide epitope despite their fine specificity differences. Binding studies were performed to determine the relative affinity of the three peptides for HLA-A-2.1, and peptide YIGEVLVSV exhibited the strongest binding to the class I molecule. The concentration necessary to inhibit the binding of the iodinated standard peptide to purified HLA-A2.1 by 50% (IC$_{50}$) was 6.7 nM for YIGEVLVSV (Fig. 5). The IC$_{50}$ values for peptides YLGEVLVSV and YLGEVIVSV were 17 and 27 nM, respectively (Fig. 5). Although the differences in binding affinities are less than a factor of 4 among these three peptides, the lower concentration of YIGEVLVSV needed to reconstitute activity cannot be explained in terms of binding. These data suggest that peptide YIGEVLVSV is the actual HA-2 epitope because it is recognized with the highest affinity by the T-cell receptors of both CTL clones.

In order to determine the amount of HA-2 expressed on the B cells from which it was extracted, a biologically active HPLC fraction was compared to a known amount of synthetic peptide on the mass spectrometer. Assuming an overall purification yield of 12% [1], HA-2 is present at approximately 260 peptide–HLA-A2.1 complexes per cell. This value is similar to values obtained for other sequenced T-cell epitopes [1,38–41], indicating that HA-2 is moderately abundant as compared to other naturally processed peptides [1,42].

The HA-2 epitope, YIGEVLVSV, was searched against all DNA and protein sequence databases. No exact match of all nine amino acids was found to any human or nonhuman protein sequence. However, six coding sequences were found that agreed with eight of nine residues, and an additional seven proteins matched to seven out of nine residues. The homologous sequences are all from the same family of myosin proteins, non-filament-forming class I. These proteins participate in organelle transport and cell locomotion [43–45]. The class I myosin family has not been completely characterized in humans. In fact, the only completely sequenced class I myosin gene in humans is the unconventional myosin IC gene [46]. However, all of the known class I myosin

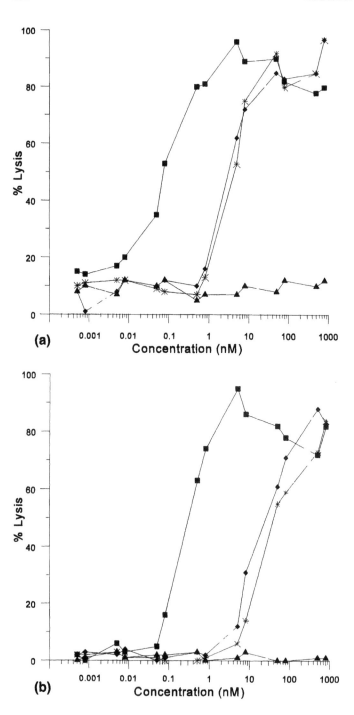

(a)

(b)

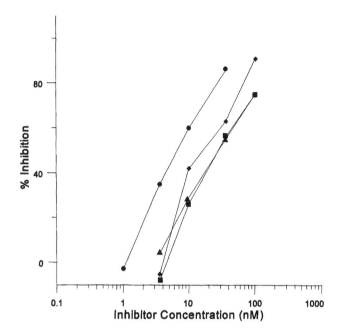

Figure 5 Binding of synthetic peptides to purified HLA-A2.1. HPLC-purified peptides were assayed for their ability to inhibit binding of iodinated hepatitis B core antigen peptide, FLPSDYFPSV. Peptides: YIGEVLVSV (circles), YLGEVLVSV (squares), YLGEVIVSV (triangles), and influenza M1 protein antigen GILGFVFTL (diamonds—standard). All data points are the average of at least two experiments.

sequences show identity at six or more of the nine positions within the sequence that corresponds to the HA-2 epitope. Therefore, the HA-2 peptide is thought to be derived from a human class I myosin protein yet to be characterized.

Until the recent identification of human minor H Ags [2,25] using our technique, information on this subject was extremely limited. They play a pivotal role in organ transplantation in general, and in BMT in particular. However,

Figure 4 (a) Dose-response curves for peptides: YIGEVLVSV (squares), YLGEVLVSV (diamonds), YLGEVIVSV (stars), and SXDFGTXQV (triangles—control) with CTL 5H17. Synthetic (25 μl) peptide was incubated with 2500 target cells and 5H17 at an effector-to-target ratio of 26:1. (b) Dose-response curves for peptides: YIGEVLVSV (squares), YLGEVLVSV (diamonds), YLGEVIVSV (stars), and YXDFGTXQV (triangles—control), with CTL 5H13. Synthetic peptide (25 μl) was incubated with 2500 target cells and 5H13 at an effector-to-target ratio of 7:1.

their normal physiological function is still unclear. In vivo modification of GVHD-related T cell responses is now possible since the sequence of HA-2 is now available. HA-2 is found in 95% of the human population expressing HLA-A2.1. Presentation of HA-2 is restricted to cells of hematopoietic origin, including leukemic cells, making its potential use in immunotherapy for leukemia prior to BMT promising.

V. CONCLUSIONS

We have developed a technique to identify an individual peptide epitope found in a mixture estimated to contain at least 10,000 species [1,11]. Reversed-phase HPLC alone could not completely resolve the minor H Ag from other endogenous peptides found within the biologically active fraction of 50–100 peptides. Simultaneous immunological and mass spectrometric analysis was the most efficient means to accomplish this objective. Moreover, the direct analysis of HLA-A2.1-associated peptides using tandem mass spectrometry allowed the selection of a human minor H Ag without prior knowledge of its source protein. This has important implications because prediction of potential antigens using the genetic approach has been difficult and the DNA may not correctly identify the naturally occurring peptide sequence [47]. Our approach has proven to be generally useful for the identification and complete sequencing of individual peptides associated with melanoma and colon cancers, with malaria, and with transplantation rejection [1,2,25–27].

REFERENCES

1. AL Cox, J Skipper, Y Chen, RA Henderson, TL Darrow, J Shabanowitz, VH Engelhard, DF Hunt, CL Slingluff. Identification of a peptide recognized by five melanoma-specific human cytotoxic T cell lines. Science 264:716–719, 1994.
2. JMM den Haan, NE Sherman, E Blokland, E Huczko, F Koning, JW Drijfhout, J Skipper, J Shabanowitz, DF Hunt, VH Engelhard, E Goulmy. Identification of a graft versus host disease-associated human minor histocompatibility antigen. Science 268:1476–1480, 1995.
3. PJ Bjorkmann, MA Saper, B Samraoui, WS Bennet, JL Strominger, DC Wiley. The foreign antigen binding site and T cell recognition regions of class I histocompatibility antigens. Nature 329:506–512, 1987.
4. DH Fremont, M Matsumura, EA Stura, PA Peterson, IA Wilson. Crystal structures of two viral peptides in complex with murine MHC class I H-2 Kb. Science 257:919–927, 1992.
5. KL Rock, C Gramm, L Rothstein, K Clarke, R Stein, L Dick, D Hwang, AL Goldberg. Inhibitors of the proteosome block the degradation of most cell proteins

and the generation of peptides presented on MHC class I molecules. Cell 78:761–771, 1994.

6. M-T Heemels, H Ploegh. Generation, translocation, and presentation of MHC class I-restricted peptides. Annu Rev Biochem 64:463–491, 1995.

7. W-K Suh, MF Cohen-Doyle, K Fruh, K Wang, PA Peterson, DB Williams. Interaction of MHC class I molecules with the transporter associated with antigen processing. Science 264:1322–1326, 1994.

8. JG Bodmer, SG Marsh, ED Albert, WF Bodmer, B Dupont, HA Erlich, B Mach, WR Mayr, P Parham, T Sasazuki. Nomenclature for factors of the HLA system. Hum Immunol 34:4–18, 1991.

9. PJ Bjorkman, MA Saper, B Samraoui, WS Bennett, JL Strominger, DC Wiley. Structure of the human class I histocompatibility antigen, HLA-A2. Nature 329:506–512, 1987.

10. TPJ Garrett, MA Saper, PJ Bjorkman, JL Strominger, DC Wiley. Specificity pockets for the side chains of peptide antigens in HLA-Aw68. Nature 342:692–696, 1989.

11. VH Engelhard. Structure of peptides associated with class I and class II MHC molecules. Annu Rev Immunol 12:181–207, 1994.

12. O Rotzschke, K Falk, H-J Wallny, S Faath, H-G Rammensee. Characterization of naturally occuring minor histocompatibility peptides including H-4 and H-Y. Science 249:283–287, 1990.

13. DC Roopenian, GJ Christianson, AP Davis, AR Zuberi, LE Mobraaten. The genetic origin of minor histocompatibility antigens. Immunogenetics 38:131–140, 1993.

14. BE Loveland, K Fischer Lindahl. The definition and expression of minor histocompatibility antigens. In: J McCluskey, ed. Antigen processing and recognition. Boca Raton, FL: CRC Press, 1991, pp 173–192.

15. DW Bailey. Sources of subline divergence and their relative importance for sublines of six major inbred strains of mice. In: H Morse, ed. Origins of inbred mice. New York: Academic Press, 1978, pp 197–215.

16. L Johnson. At how many histocompatibility loci do congenic mouse strains differ? J Hered 72:27–31, 1981.

17. PG Beatty, P Hervé. Immunogenetic factors relevant to acute graft-versus-host disease. In: SJ Burakoff, DHJ Deeg, S Ferrara, K Atkinson, eds. Graft-versus-host disease: Immunology, pathophysiology, and treatment. New York: Marcel Dekker, 1989, pp 415–423.

18. PG Beatty, JA Hansen, GM Longton, ED Thomas, JE Sanders, PJ Martin, SI Bearman, C Anasetti, EW Petersdorf, EM Mickelson. Marrow transplantation from HLA-matched unrelated donors for treatment of hematologic malignancies. Transplantation 51:443–447, 1991.

19. MM Bortin, MM Horowitz, M Mrsic, AA Rimm, KA Sobocinski. Progress in bone marrow transplantation for leukemia: A preliminary report from the Advisory Committee of the International Bone Marrow Transplant Registry. Transplant Proc 23:61–62, 1991.

20. MM Bortin. A compendium of reported human bone marrow transplants. Transplantation 9:571–587, 1970.

21. GMT Schreuder, J Pool, E Blokland, C van Els, A Bakker, JJ van Rood, E Goulmy. A genetic analysis of human minor histocompatibility antigens demonstrates Mendelian segregation independent of HLA. Immunogenetics 38:98–105, 1993.

22. C van Els, JD Amaro, J Pool, E Blokland, A Bakker, PJ van Elson, JJ van Rood, E Goulmy. Immunogenetics of human minor histocompatibility antigens: Their polymorphism and immunodominance. Immunogenetics 35:161–165, 1992.

23. M de Bueger, A Bakker, JJ Van Rood, F Van der Woude, E Goulmy. Tissue distribution of human minor histocompatibility antigens: Ubiquitous versus restricted tissue distribution indicates heterogeneity among human cytotoxic T lymphocyte-defined non-MHC antigens. J Immunol 149:1788–1794, 1992.

24. REM Toes, R Offringa, RJJ Blom, CJM Melief, WM Kast. Peptide vaccination can lead to enhanced tumor growth through specific T-cell tolerance induction. Proc Natl Acad Sci USA 93:7855–7860, 1996.

25. W Wang, LR Meadows, JMM den Haan, NE Sherman, Y Chen, E Blokland, J Shabanowitz, AI Agulnik, RC Hendrickson, CE Bishop, DF Hunt, E Goulmy, VH Engelhard. Human H-Y: A male-specific histocompatibility antigen derived from the SMCY protein. Science 269:1588–1590, 1995.

26. AYC Huang, PH Gulden, AS Woods, MC Thomas, CD Tong, W Wang, VH Engelhar, G Pasternanck, R Cotter, D Hunt, DM Pardoll, EM Jaffee. The immunodominant major histocompatibility complex class I-restricted antigen of a murine colon tumor derives from an endogenous retoviral gene product. Proc Natl Acad Sci USA 93:9730–9735, 1996.

27. L Meadows, W Wang, JMM den Haan, E Blokland, C Reinhardus, JW Drijfhout, J Shabanowitz, R Pierce, AI Agulnik, CE Bishop, DF Hunt, E Goulmy, VH Engelhard. The HLA-A*0201-restricted H-Y antigen contains a posttranslationally modified cysteine that significantly affects T cell recognition. Immunity 6:273–281, 1997.

28. RA Henderson, H Michel, K Sakaguchi, J Shabanowitz, E Appella, DF Hunt, VH Engelhard. HLA-A2.1-associated peptides from a mutant cell line: A second pathway of antigen presentation. Science 255:1264–1266, 1992.

29. R Kennedy, JW Jorgenson. Preparation and evaluation of packed capillary liquid chromatography columns with inner diameters from 20 to 50 μm. Anal Chem 61:1128–1135, 1989.

30. MA Moseley, LJ Deterding, KB Tomer, JW Jorgenson. Nanoscale packed-capillary liquid chromatography coupled with mass spectrometry using a coaxial continuous-flow fast atom bombardment interface. Anal Chem 63:1467–1473, 1991.

31. B Fields, RL Noble. Solid phase peptide synthesis utilizing 9-fluorenylmethoxycarbonyl amino acids. Int J Peptide Protein Res 35:161–214, 1990.

32. K Biemann. Sequencing of peptides by tandem mass spectrometry and high-energy collision-induced dissociation. Methods Enzymol 193:445–479, 1990.

33. DF Hunt, JE Alexander, AL McCormack, PA Martino, H Michel, J Shabanowitz, NE Sherman, MA Moseley, JW Jorgenson, KB Tomer. Mass spectrometric methods for protein and peptide sequence analysis. In: JJ Villafranca, ed. Techniques in protein chemistry II. San Diego: Academic Press, 1991, pp 441–454.

34. AG Hulst, CE Kientz. Differentiation between the isomeric amino acids leucine and isoleucine using low-energy collision-induced dissociation tandem mass spectrometry. J Mass Spectrom 31:1188–1190, 1996.

35. J Eng, AL McCormack, JR Yates III. An approach to correlate tandem mass spectral data of peptides with amino acid sequences in a protein database. J Am Soc Mass Spectrom 5:976–989, 1994.

36. JR Yates III, J Eng, AL McCormack, D Schieltz. Method to correlate tandem mass spectra of modified peptides to amino acid sequences in the protein database. Anal Chem 67:1426–1426, 1995.

37. JR Yates III, J Eng, KR Clauser, AL Burlingame. Search of sequence databases with uninterpreted high-energy collision-induced dissociation spectra of peptides. J Am Soc Mass Spectrom 7:1089–1098, 1996.

38. K Udaka, TJ Tsomides, HN Eisen. A naturally occuring peptide recognized by alloreactive CD8+ cytotoxic T lymphocytes in association with a class I MHC protein. Cell 69:989–998, 1992.

39. RA Henderson, AL Cox, K Sakaguchi, E Appella, J Shabanowitz, DF Hunt, VH Engelhard. Direct identification of an endogenous peptide recognized by multiple HLA-A2.1-specific cytotoxic T cells. Proc Natl Acad Sci USA 90:10275–10279, 1993.

40. O Mandelboim, G Berke, M Fridkin, M Feldman, M Eisenstein, L Eisenbach. CTL induction by a tumour-associated antigen octapeptide derived from a murine lung carcinoma. Nature 369:67–71, 1994.

41. A Uenaka, T Ono, T Akisawa, H Wada, T Yasuda, E Nakayama. Identification of a unique antigen peptide pRL1 on BALB/c RL male 1 leukemia recognized by cytotoxic T lymphocytes and its relation to the AKT oncogene. J Exp Med 180:1599–1607, 1994.

42. EL Huczko, WM Bodnar, D Benjamin, K Sakaguchi, NZ Zhu, J Shabanowitz, RA Henderson, E Appella, DF Hunt, VH Engelhard. Characterization of endogenous peptides eluted from the class I MHC molecule HLA-B7 determined by mass spectrometry and computer modeling. J Immunol 151:2572–2587, 1993.

43. MA Titus. Myosins. Curr Opin Cell Biol 5:77–81, 1993.

44. E Coudrier, A Durrbach, D Louvard. Do unconventional myosins exert functions in dynamics of membrane compartments? FEBS Lett 307:87–92, 1992.

45. M Mooseker. A multitude of myosins. Curr Biol 3:245–248, 1993.

46. WM Bement, JA Wirth, MS Mooseker. Cloning and mRNA expression of human unconventional myosin-IC: A homologue of amoeboid myosins-I with a single IQ motif and an SH3 domain. J Mol Biol 243:356–363, 1994.

47. JCA Skipper, RC Hendrickson, PH Gulden, V Brichard, A Van Pel, J Shabanowitz, T Wolfel, CL Slinghoff, T Boon, DF Hunt, VH Engelhard. An HLA-A2 restricted tyrosinase antigen on melanoma cells results from post-translational modification and suggests a novel processing pathway for membrane proteins. J Exp Med 183:527–534, 1996.

7
Strategies for the Rapid Characterization of Recombinant and Native Proteins by Mass Spectrometry

Daniel B. Kassel
CombiChem, Inc., San Diego, California

R. Kevin Blackburn
Glaxo Wellcome, Inc., Research Triangle Park, North Carolina

Bruno Antonsson
Geneva Biomedical Research Institute, Glaxo Wellcome Research and Development S.A., Geneva, Switzerland

I. INTRODUCTION

High-speed, high-sensitivity peptide mapping and protein sequencing are of paramount importance to both the biotechnology and pharmaceutical industries. The speed by which aberrant forms of overexpressed, recombinant proteins are identified impacts the timeline on which the therapeutic protein makes it to market. Important to the drug discovery process in the pharmaceutical industry is the time frame for which novel therapeutic targets (e.g., receptors or enzymes) are isolated and identified. Equally important is to understand the structure/function of the therapeutic target, and this often requires knowledge and role of any posttranslational modifications. Bioanalytical chemistry has played an increasingly important role in these areas, and this is attributed principally to developments in electrospray ionization (ESI) [1,2] and matrix-assisted laser desorption (MALDI) [3,4] mass spectrometry. A major reason for the success of these two techniques has been their ability to reach levels of sensitivity not achievable by well-established protein chemistry methods, such as gas-phase Edman sequencing. Most recently, tryptic mapping by nanospray ESI-MS [5,6] and MALDI [7] has been achieved at the subpicomole level. ESI and, more recently, delayed extraction (DE)-MALDI [8] have enabled peptide mass determinations to be made very precisely and accurately, thereby permitting rapid protein identification from protein, gene, and expressed sequence tag (EST) databases [9,10].

Another major instrumentation advance that has impacted protein chemistry and is now routinely employed in many bioanalytical laboratories is capillary high-performance liquid chromatography (HPLC). Ultrahigh sensitivities have been achieved for the analysis of biomolecules using submillimeter ID (e.g., 100 μm, 300 μm) columns and capillary (z-shaped) flow-cells [11] and, more recently, using perfusion chromatography resins for enhanced recoveries of hydrophobic peptides [12,13]. The miniaturization of HPLC columns has enabled direct coupling with ESI-MS [13,14] and gas-phase Edman sequencing instruments [15,16], thus enabling current detection limits to be improved upon significantly. Consequently, the advent of new ionization methods and the miniaturization of column chromatography have permitted much higher sensitivity protein molecular weight and peptide map determinations to be made routinely in many laboratories.

One of the current challenges is in the isolation, handling, and chemical and enzymatic manipulation of proteins prior to bioanalytical analysis. These steps are often time-consuming and when performed manually are subject to sample losses. The challenge has been to effectively handle small amounts of protein while minimizing the losses during the chemical and enzymatic manipulation steps and subsequent liquid chromatography/mass spectroscopy (LC/MS) analysis. Several groups have addressed these issues effectively [17,18]. Burkhart demonstrated that dilute protein solutions can be preconcentrated rapidly on octadecylsilyl (ODS) cartridges and reduced/alkylated and digested with a variety of endoproteases on column [19]. For these experiments, microgram quantities of endoprotease are introduced on-column in the presence of organic buffer (acetonitrile) to aid in the digestion process. The advantages of the method are that (1) sample handling steps can be virtually eliminated following isolation of the protein and (2) products of digestion can be analyzed directly by Edman sequencing. The limitations to the method are (1) the total time for reduction, alkylation, and enzymatic digestion approaches 24–36 hr; (2) the digests are frequently incomplete; and (3) the digests are contaminated by a large proportion of autocatalytic cleavage products (due to the relatively large amounts of enzyme applied to the column).

Recently, these steps have been implemented on a fully automated multi-dimensional HPLC-based workstation directly coupled to ESI-MS. The system described provides the basis for more complete protein digestions and higher protein recoveries, all of which facilitate the more complete characterization of recombinant proteins. The power of the technique for rapid characterization of recombinant proteins is demonstrated for the Src homology 2 (SH2) domain of $pp60^{c\text{-}src}$, stathmin and its neuronal homologue, SCG10. Dilute protein solutions are trapped, reduced, and alkylated on-column (when necessary), proteolytically digested (i.e., with trypsin) and analyzed by LC/MS. The steps are fully automated, extremely rapid, and involve minimal sample handling. This process permits rapid identification of posttranslational modifications in recombinantly expressed proteins and, because of the speed of the technique, can be applied to enzyme kinetics as well.

The sensitivity of nanospray ESI (nanoES) has made this technique particularly attractive for the identification and sequence determination of poorly expressed recombinant proteins and native proteins. The power and versatility of the technique is demonstrated for the characterization of the site of phosphorylation of stathmin. Digestions are performed rapidly on immobilized trypsin columns and the products are trapped transiently onto a reversed-phase cartridge. Peptide fragments are eluted in minimal volume directly into a nanoES capillary. The results are shown later.

II. EXPERIMENTAL

A. Column Packing Procedures

POROS R2/H reversed-phase and Poroszyme trypsin columns (PerSeptive Biosystems, Framingham, MA) were packed into 500-μm ID PEEK tubing. For packing of reversed-phase columns, the amount of dry resin (mg) used was equivalent to the PEEK column volume (ml). The appropriate amount of resin was weighed out and slurried in 1 ml ethanol and placed into a 4.6-mm ID open tubular "packing bomb." The inlet of the packing bomb was attached to an HP1090 HPLC and the outlet of the packing bomb was connected to a 10 cm × 500 μm ID piece of PEEK tubing (Fig. 1). The outlet of the PEEK tubing was connected to a zero dead volume (ZDV) union (UpChurch Scientific) fitted with a 0.062 × 0.028 in stainless steel frit (Mott Metallurgical, CT). The packing bomb was pressurized to roughly 150 bar with 95/5 acetonitrile/water for 5–10 min. Following pressurization of the packing material, the PEEK packed perfusion column was disconnected from the packing bomb and fitted with a second ZDV union containing a stainless steel frit at the head of the column. The column was washed extensively by running several blank gradients of standard HPLC buffers prior to use. Preparation of PEEK packed immobilized enzyme cartridges was achieved in an identical manner except that Tris buffer, pH 8.5, was used as the column loading and wash buffer.

B. Automated Reduction/Alkylation and Digestion Procedures

In situ alkylation/tryptic digestion experiments were carried out using an INTE-GRAL Microanalytical Workstation (Persepetive Biosystems, Framingham, MA) (Fig. 2a). Column 1 was a Poros R2/H desalting column, and was equilibrated in 15% acetonitrile (ACN) containing 0.04% tritfluoroacetic acid (TFA). A 20-pimol sample of SH2 was loaded onto this column and desalted for 3 mins at a flow rate of 80 μl/min. Next, 40 μl of 4-vinylpyridine in reduction/alkylation buffer [6 M guanidine-HCl; 0.5 M Tris-HCl; 5 mM ethylene diamine tetraacetic acid, (EDTA), pH 8.6] was injected on-column and allowed to flow until column 1 was saturated with the alkylation solution (approximately 0.4 min). After saturation, column 1 was programmed to switch offline and incubate for 15 min at 37°C. Column 1 was then switched inline and washed extensively with 15% ACN solution to remove excess alkylation solution. Column 1 was connected in line with column 2 (the Poroszyme immobilized trypsin column). Alkylated protein bound on column 1 was eluted with 0.025% TFA in 45/55 ACN/water, through a postcolumn mixing tee onto column 2, which was held at ambient temperature. Coupling of a high-resolution C_{18} column directly to the outlet of the immobilized trypsin column was not possible because of high back pressure at the flow rates used for digestions. For this reason, a 300 μm × 1 cm C_{18} reversed-phase trapping cartridge (LC Packings, San Francisco, CA) was attached to the outlet of column 2 in order to trap tryptic peptides as they eluted from the Poroszyme column. Two postcolumn pH adjusts were required. First, a postcolumn 1 mix with 20 mM Tris, pH 8.5, was made to adjust the pH of column 1 eluate to make the buffer pH optimal for immobilized trypsin digestion. Secondly, a postimmobilized enzyme column (column 2) pH adjust with 0.5% TFA was required, to adjust the ion pairing capacity of the

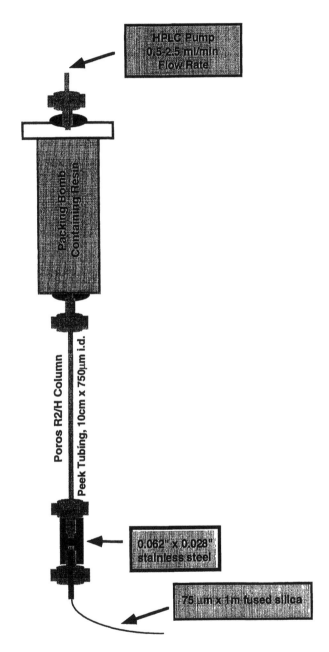

Figure 1   Procedures for self-packing perfusion chromatography resins into PEEK tubing.

reversed-phase column and to lower the ACN concentration to <5% (v/v) to facilitate binding of hydrophilic tryptic peptides. During elution/digestion, the flow rate of 45% ACN through column 1 was held at 10 μl/min. The Tris buffer was introduced into the postcolumn 1 mixing tee at 10 μl/min, for an overall flow rate of 20 μl/min through the trypsin column. This flow rate corresponded to a residence time on the trypsin column of 1 min. A third pump was used to deliver 0.5% TFA in H_2O at 80 μl/min at the outlet of column 2 and prior to column 3. This facilitated the binding of the tryptic peptides to the C_{18} trapping cartridge by providing a source of ion pairing.

After digestion/trapping, the C_{18} trapping cartridge was removed from the workstation and connected to the Rheodyne 8125 injector of an HP1090 HPLC system (Hewlett Packard, Palo Alto, CA). A high-resolution 300 μm × 15 cm C_{18} packed capillary column (LC Packings, USA, San Francisco, CA), was connected to the outlet of the reversed-phase trapping cartridge to facilitate high-resolution separations of the tryptic peptides. Tryptic SH2 peptides were then eluted at a flow rate of 5 μl/min using the following gradient: 1–41%B in 10 min; 41–61%B in 10 min; buffer A = 0.05% TFA in water; buffer B = 0.045% TFA in 90/10 ACN/H_2O. Mass spectra were acquired over the range m/z 450–1950 using a step size of 0.5 Da and a dwell time of 1 msec, corresponding to a scan time of 3 sec/scan.

C. Automated On-Column Digestion of Stathmin and SCG10

For stathmin and SCG10, it was not necessary to perform on-column reduction/alkylation to facilitate full protein digestion. Therefore, human and chicken stathmin (20 pmol) was injected directly (unmodified) onto a 30 cm × 500 μm Poroszyme trypsin column preequilibrated in digestion buffer as shown in Figure 2b. Digestion was carried out at a flow rate of 30 μl/min, corresponding to an on-column residence (i.e., digestion) time of 2 min. A 20-pmol sample of SCG10 was treated in an identical manner.

Figure 2 (a) Schematic representation of the automated two-dimensional chromatography/mass spectrometry workstation used for preconcentration and reduction alkylation on the desalting cartridge, elution of reduced and alkylated protein onto an immobilized trypsin column, and trapping of the resultant tryptic peptides at the head of a C_{18} "trapping" cartridge. (b) Configuration for direct injection of recombinant proteins onto an immobilized enzyme column and trapping of the resultant tryptic peptides at the head of a C_{18} trapping cartridge.

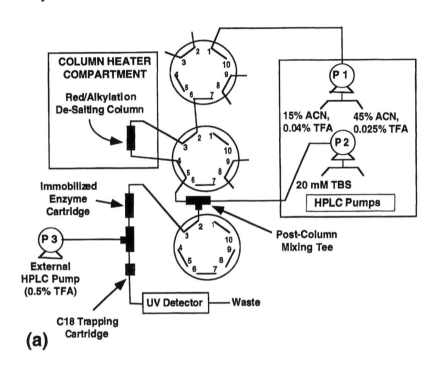

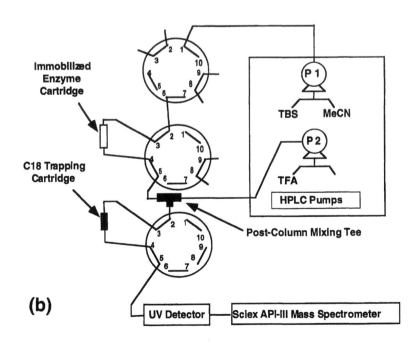

D. Protein Identification by NanoES and MALDI-TOF

An aliquot of chicken stathmin and SCG10, corresponding to 1–5 pmol, was desalted and eluted into a NanoES capillary following the procedures of Wilm et al. [20]. From 1 μL of sample it was possible to acquire the full-scan mass spectrum for the tryptic digest, m/z 79 (PO_3^-) phosphate ion scan, a precursor ion scan of m/z 79, and the MS/MS spectra of the precursors for PO_3^-.

III. RESULTS AND DISCUSSION

The ability to chemically modify and digest a protein in situ has a number of advantages over conventional, solution-based digest procedures for the characterization of recombinant proteins. Excess reagents can be removed readily and, perhaps more importantly, sample handling steps can be reduced, thereby enhancing recoveries of peptides. The method of Burkhart [19] was shown to have the advantages of (1) enabling dilute protein solutions to be rapidly desalted and concentrated onto the ODS support, (2) enabling them to be digested in the presence of organic buffers (up to 40% acetonitrile), and (3) the capacity to be coupled directly to HPLC or a gas-phase Edman sequencer.

As a model, we evaluated a column-based strategy for characterizing the model protein Src SH2 domain. The polypeptide is a 107-amino acid, 12.2-kD domain of the nonreceptor tyrosine kinase, pp60[c-src]. It has been shown to play a role in the binding of endogenous, phosphorylated protein substrates in colon and breast carcinoma cell lines [21,22]. The domain contains three cysteine residues, though none are involved in disulfide bonding. Although the protein is highly expressed in *Escherichia coli* and is soluble at high millimolar concentrations, a dilute solution (5 μM) of the protein in "typical" purification buffer [i.e., 350 mM NaCl, 20 mM HEPES, 5 mM dithiothreitol (DTT), 5 mM EDTA] was used for these model studies.

Maleknia et al. [23], Kassel et al. [24], and, more recently, Blackburn et al. [25] have shown that high concentrations of endoproteases immobilized to large through-pore resins enable digests to occur extremely rapidly and with very high efficiency. We evaluated this technique for the rapid digestion of Src SH2. The microanalytical workstation was configured, as shown in Fig. 2b. An aliquot corresponding to 20 pmol of SH2 was injected directly onto a PEEK packed 500 μm $\times$ 10 cm (20 μl total volume) immobilized trypsin column (prepared in-house) at a flow rate of 20 μl/min. This flow rate corresponded to an on-column residence digestion time of 1 min. The digestion products were trapped onto a C_{18} trapping cartridge and analyzed by LC/ESI/MS, as shown in Fig. 3a. The immobilized enzyme digest occurred with very high efficiency. Importantly, there was very little qualitative difference between the immobilized enzyme digest relative to the more commonly used solution-based digest, which was performed overnight (12 h at 37°C) (see Fig. 3b). Further, with a total digestion time of 1 min, there was no evidence of trypsin autocatalytic activity (none of the observed molecular weights corresponded to tryptic fragments of trypsin itself). On-column residence times could be reduced to less than 20 sec and still produce complete enzyme digestions [26].

It was found that trapping the enzyme digest products directly onto a C_{18} reversed-phase trapping cartridge enables peptides to be recovered with much higher efficiencies than with direct collection into an Eppendorf tube and injecting the collected digest onto a reversed-phase column. Samples of SH2 (20-pmol aliquots) were digested using the immobilized enzyme column and either (1) transiently collected into an Eppendorf tube and then reapplied to a reversed-phase column or (2) trapped directly onto a reversed-phase trapping cartridge. In both cases, the digests were then eluted into the ion source of the mass spectrometer, as described previously. The ultraviolet (UV) and mass spectrometric responses of a number of tryptic peptides were compared for these two experiments and are summarized in Fig. 4. Clearly, recoveries were

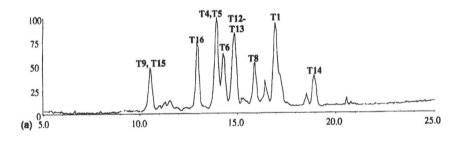

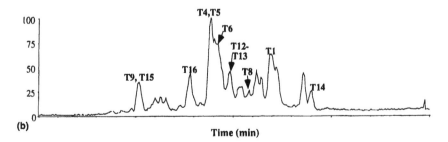

Figure 3 Total ion current (TIC) chromatogram for 20-pmol SH2 tryptic digest separated on a Hypersil 300 μm × 15 cm C_{18} column using the gradient described in Materials and Methods. (a) Tryptic digestion on the immobilized trypsin column. (b) Tryptic digest in solution using an enzyme:substrate ration of 1:50.

far greater when the enzyme digest was trapped directly onto a reversed-phase trapping cartridge. In some cases, the differences in recoveries were as great as 70–80%. Further, it was found that these differences were accentuated at lower protein concentrations.

During the course of studies on other proteins, it was found that a post-enzyme column pH adjust was necessary for retaining several of the more hydrophilic tryptic peptides on the trapping cartridge. This pH adjust was achieved by addition of an equal volume of 0.5% aqueous TFA through a mixing tee following the digestion. To illustrate this effect, the total ion current (TIC) chromatograms for the immobilized enzyme digest of 20 pmol of human stathmin [27] trapped onto the reversed-phase C_{18} trapping cartridge in the presence and absence of a pH adjust solution are shown in Fig. 5a and b.

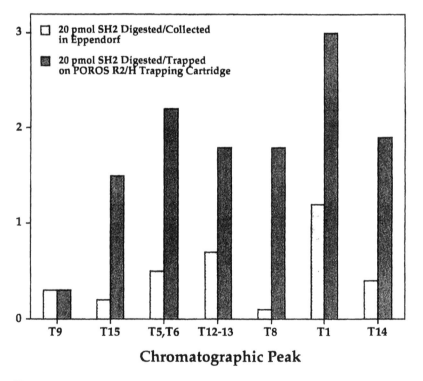

Figure 4 Comparison of recoveries of tryptic peptides from a 20-pmol immobilized enzyme digest of SH2 collected directly into an Eppendorf tube and the entire contents reapplied to the C_{18} column for LC/MS analysis (open boxes), and the same 20-pmol sample injected onto the immobilized enzyme directly coupled to a C_{18} 300 $\mu m \times 1$ cm trapping cartridge as shown in Fig. 2b (shaded boxes).

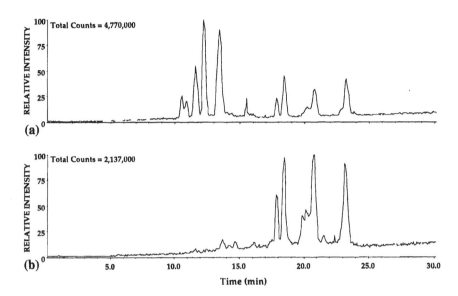

Figure 5 Binding to and recovery of tryptic peptides from the C_{18} trapping cartridge is enhanced when the pH of the enzyme digest buffer is adjusted with a 0.5% aqueous TFA solution (postenzyme column). (a) LC/MS analysis of tryptic peptides trapped onto the C_{18} trapping cartridge using the postenzyme column pH adjust. (b) LC/MS analysis of tryptic peptides trapped onto the C_{18} trapping cartridge in the absence of the postenzyme column pH-adjusted solution.

Strikingly, only the most hydrophobic peptides were retained in the absence of this postcolumn pH adjustment, whereas the hydrophilic peptides were not retained at all. When the pH-adjusted solution was used to aid in the trapping of the peptide digest, dramatically enhanced recovery of both the hydrophilic and hydrophobic peptides was achieved.

Because a makeup flow is required to adjust the pH of the solution to facilitate trapping on the cartridge, there was some concern as to the effect of increased flow rate on the binding capacity of the short 1-cm C_{18} columns used for trapping peptide fragments. Figure 6 shows the results of a study assessing the effect of increasing the trapping cartridge flow rate on the efficiency of trapping enzyme digest products. At moderate flow rates (40–100 μl/min), binding capacity on-column appeared to be relatively unaffected for the hydrophobic peptides (those compounds eluting after 10 min) and appeared to have a modest effect on the retention of the more hydrophilic peptides. At significantly higher flow rates through the trapping cartridge (i.e., 200 μl/min), the binding capacity of the column was significantly compromised. These re-

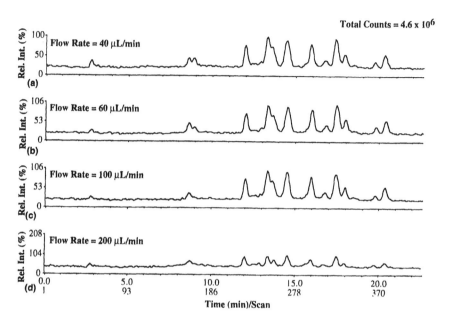

Figure 6 Effect of flow rate through the C_{18} 300 μm $\times$ 1 cm trapping cartridge on peptide digest trapping efficiency. The flow rate through the enzyme column was held fixed at 20 μl/min. The flow rate of the pH adjust solution was varied: (a) 20 μl/min, (b) 40 μl/min, (c) 80 μl/min, and (d) 180 μl/min. The effect of increasing flow rate on trapping efficiency is shown.

sults suggest that the optimal conditions for performing on-line digestion/trapping experiments are when total flow rates are less than or equal to 100 μl/min (i.e., maximum enzyme column flow rate = 50 μl/min, maximum postenzyme column pH-adjusted solution flow rate = 50 μl/min).

Often, reduction/alkylation procedures are required to aid in the digestion of a protein. Therefore, a multistep on-column reduction/alkylation/digestion procedure was developed. A reduction/alkylation cartridge was placed in the first dimension of the multidimensional chromatography workstation as shown in Fig. 2a. A 20-pmol solution of the SH2 protein was concentrated at the head of this reversed-phase reduction/alkylation cartridge and desalted. The protein was reduced and alkylated on-column as described in the Materials and Methods section. Excess alkylating agent was removed from the reduction/alkylation cartridge by extensive washing with 15% ACN. The efficiency of the reduction/alkylation step on the microanalytical workstation was monitored by eluting the protein from the reversed-phase support into the ion source of the mass spectrometer. Shown in Fig. 7 are the reconstructed mass spectra for 20 pmol of SH2 eluted from the desalting column before and after reduction/alkylation. The reduction/alkylation reaction proceeded to completion as evidenced by the shift in molecular weight of the protein by 315 Da, consistent with the addition of three pyridylethyl moieties to the three cysteine residues contained within the polypeptide chain of SH2.

It remained to be demonstrated that reduced and alkylated protein could be step eluted directly onto the immobilized enzyme column, thereby enabling the protein to be modified and digested under conditions of virtually no sample handling. The methodology was validated initially by injecting a 20-pmol sample of SH2 onto the desalting column, as previously described. Reduction/alkylation buffer was introduced on-column and allowed to incubate for 10 min at 37°C. Excess alkylating reagent was removed by extensive washing of the desalting column with a solution of 15% ACN in H_2O containing 0.02% TFA. Column 2 (the immobilized enzyme column) was then placed in line. The protein was step eluted from the desalting cartridge with a 45/55 ACN/H_2O solution containing 0.025% TFA at a flow rate of 20 μl/min. The eluate from column 1 was adjusted to pH 8.5 by addition of an equal volume solution of Tris

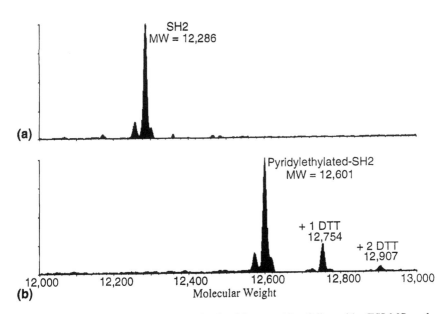

Figure 7 Step elution of SH2 from the desalting cartridge followed by ESI-MS analysis. Shown are the reconstructed molecular weight spectra for SH2 (a) before and (b) after oncolumn reduction/alkylation on the desalting cartridge.

buffer through a mixing tee placed between the outlet of column 1 and the inlet of column 2 (as shown in Fig. 2). The protein was digested in real time on the immobilized enzyme column and trapped onto the reversed-phase trapping cartridge, as described earlier. The TIC chromatogram for this multistep automated analysis is shown in Fig. 8. The three tryptic peptides containing cysteine residues were observed and each was fully pyridylethylated.

The methods described earlier have been applied routinely for identifying sites of posttranslational modifications in proteins [28]. The high-speed digestions possible on immobilized enzyme cartridges permits time-course kinetics of reactions (e.g., phosphorylation kinetics) to be monitored as well [29,30]. As an example, we have used the automated immobilized enzyme digest procedure with online peptide trapping to identify the differences in phosphorylation between two stathmin homologues (chicken and human) and a human neuronal homologue of stathmin, SCG10. Stathmin is a highly conserved 17-kD cytosolic phosphoprotein, ubiquitously expressed in the cell. It has been found to be a substrate for mitogen activating protein (MAP) kinase, cyclic AMP (cAMP)-dependent kinase, DNA-dependent kinase, cGMP-dependent kinase, and the cell-cycle-dependent kinase p34cdc2 [31]. Chicken stathmin shares 93% identity to human stathmin, the principle difference being that the known MAP kinase site for human stathmin (Ser-25) is replaced with Gly-25. The sequence of chicken stathmin is shown in Fig. 9. As isolated from baculovirus Sf9 cells, chicken stathmin has been shown to be partially phosphorylated (Antonsson et al., Glaxo Institute of Molecular Biology, unpublished results). Further, from in

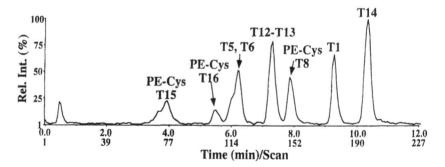

Figure 8   Following reduction/alkylation of 20 pmol of SH2 on the desalting cartridge, the alkylated protein was step eluted from the reversed-phase C_{18} support onto an immobilized enzyme cartridge. To facilitate digestion, a solution of Tris, pH 8.5, was plumbed in line (as described in Materials and Methods). Digest products were trapped onto the C_{18} trapping cartridge. Tryptic peptides were eluted from the column into the ESI ion source. The TIC chromatogram is shown. All three of the cysteine-containing tryptic peptides were found to be pyridylethylated.

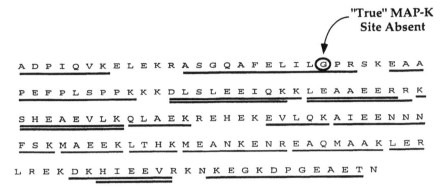

Figure 9 Sequence of chicken stathmin. The true MAP kinase site, Ser-25 in human stathmin, is replaced by glycine in chicken stathmin. The underlined regions are those peptides identified by LC/MS following digestion of chicken stathmin on the immobilized enzyme column.

vitro phosphorylation assays, this protein has been found to be a substrate for MAP kinase, although the enzyme is 10-fold more active against human stathmin than chicken stathmin [32]. The protein molecular weight spectrum for chicken stathmin incubated with and without MAP kinase is shown in Fig. 10. The deconvoluted mass spectrum shown in Fig. 10b shows that the protein is partially monophosphorylated ($\sim$20%).

The microanalytical workstation was configured for online enzyme digestion and trapping of tryptic peptides, as shown in Fig. 2b. A 20-pmol sample of chicken stathmin incubated with MAP kinase was injected onto the immobilized enzyme column at a flow rate of 20 μl/min, corresponding to an on-column residence time of 1 min. The digest products were trapped onto a C_{18} cartridge as described previously. In order to detect the low level of phosphorylation in this sample (estimated amount of the phosphorylated protein isoform was 4 pmol), the stepped orifice potential technique was used. This technique allows for selective detection of phosphopeptides based on their ability to form prominent PO_3^- marker anions when a high orifice voltage is applied [33,34]. The negative ion TIC chromatogram and reconstructed ion chromatogram for m/z 79 (PO_3^-) of the chicken stathmin + MAP kinase tryptic digest is shown in Fig. 11. The MS/MS spectrum of the phosphopeptide (shown eluting at 20.2 min in Fig. 11b) was acquired in a subsequent run and is shown in Fig. 12. From the MS/MS spectrum, it was possible to identify Ser-38 conclusively as the MAP kinase phosphorylation site for this stathmin variant.

In order to identify the exact site of phosphorylation by this method, two analyses were required: one to acquire the negative ion LC/MS data to identify

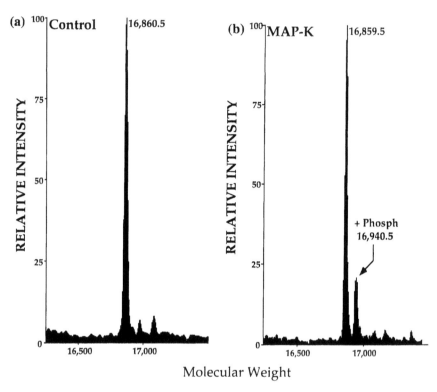

Figure 10 Comparison of the reconstructed molecular weight electrospray mass spectra for stathmin (a) before and (b) after incubation with MAP kinase. The protein, following MAP kinase activation, was found to be approximately 20% phosphorylated.

the mass and retention time of the phosphopeptide, and a second run, in the positive ion mode, to acquire the LC/MS/MS spectrum of the phosphopeptide to conclusively identify the site of phosphorylation. Unfortunately, when protein is in limited quantity, it may not be possible to perform more than one digestion and analysis. For this reason, we used nanoES, which enables multiple experiments to be performed from a single sample preparation. An identical aliquot of chicken stathmin incubated with MAP kinase (20 pmol) was digested on the immobilized enzyme column and the products were trapped onto the C_{18} cartridge. The peptides were step eluted from the cartridge and a 2-μl aliquot, corresponding to 1–2 pmol total protein, was transferred to a nanoES tip for analysis, as described previously by Mann et al. [35]. From this 2-μl sample, it was possible to acquire the full-scan positive- and negative-ion spectra, precur-

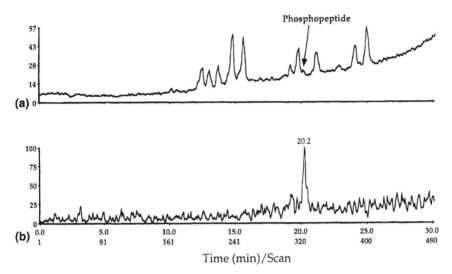

Figure 11 To facilitate phosphopeptide identification, the stepped orifice potential technique was used. (a) The negative-ion TIC chromatogram for chicken stathmin + MAP kinase digested on-column with trypsin. (b) Mass chromatogram for the m/z 79, PO_3^- phosphate ion, showing a single phosphorylated tryptic peptide.

sor ion scans of m/z 79, and the MS/MS spectra of the phosphopeptides identified from precursor ion scans. The negative-ion nanoES full-scan spectrum of the tryptic digest is shown in Fig. 13. A smaller number of tryptic peptides was observed relative to the LC/MS analysis. The transfer step into the nanoES capillary involves the use of a desalting column containing a small quantity (≤ 100 nl) of POROS R2/H resin. We found that some of the more hydrophilic peptides were lost during this desalting step, and this could be explained by the poorer retention of low-molecular-weight species on the perfusion material relative to C_{18} reversed-phase supports. Nonetheless, a precursor ion scan for m/z 79 (PO_3^-) showed a prominent doubly charged anion at m/z 680.4 (MW = 1362.8) (Fig. 14). The MS/MS spectrum of this peptide was recorded and was found to match the LC/MS/MS spectrum shown in Fig. 11 (data not shown).

A similar approach was used to characterize SCG10, a neuronal homologue of stathmin. SCG10 shares greater than 74% sequence identity with stathmin within the overlapping regions [27]. SCG10 is characterized by an amino-terminal stretch of 34 amino acids (not present in stathmin), which aids in its membrane anchoring. Further, it has been found to be both an in vivo and in vitro substrate for several serine/threonine kinases that are activated by a vari-

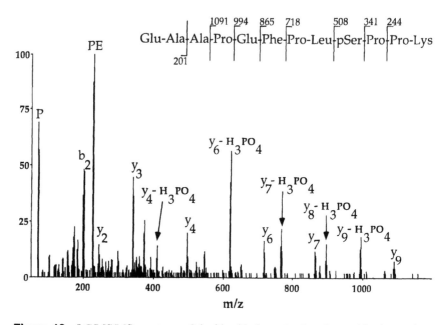

Figure 12 LC/MS/MS spectrum of the identified tryptic phosphopeptide. A prominent $y_n - H_3PO_4$ ion series was observed that aided in the identification of Ser-38 (based on human stathmin nomenclature) as the MAP kinase phosphorylation site in chicken stathmin.

ety of extracellular signals. SCG10 contains multiple potential serine phosphorylation sites and has been demonstrated recently to be a substrate for several signal transduction kinases [36]. Using the same rapid methodology, the sites of in vitro phosphorylation of SCG10 by several serine/threonine kinases were identified. The results are summarized in Fig. 15.

Upon stimulation by neuronal growth factor, in vivo expression and phosphorylation levels of SCG10 are enhanced. Attempts to identify the sites of in vivo phosphorylation of SCG10 using both MALDI and NanoES were unsuccessful and are the focus of our current work.

IV. CONCLUSIONS

The methods presented in this chapter show that protein samples can be manipulated both chemically and enzymatically in a fully automated, column environ-

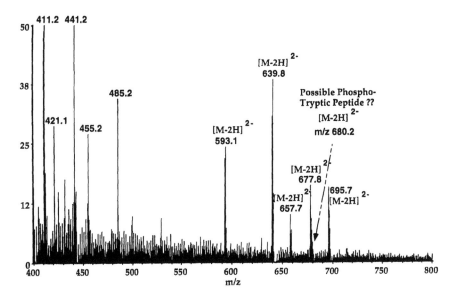

Figure 13 NanoES negative-ion spectrum of stathmin tryptic digest. The digest was loaded onto a Poros R2H nanospray needle, desalted, and eluted from the column into a NanoES tip using 2 μl of 50% MeOH containing 0.1% formic acid. Both 2+ and 3+ charge state ions were observed for a number of tryptic peptides.

ment. After purification of the protein, almost all sample handling steps can be eliminated. This permits much better recoveries of peptides following reduction/alkylation and digestion procedures. On-column digestions were shown useful for the rapid characterization and sequence identification of recombinant proteins. Using this technique, it was possible to readily map the sites of protein phosphorylation for stathmin and SCG10. A postenzyme column addition of aqueous 0.5% TFA was shown to have a dramatic effect on the ability to trap hydrophilic peptides. This pH adjustment was particularly important for mapping the sites of protein phosphorylation. Phosphopeptide identification was made either using the stepped orifice potential technique or nanoES. The advantage of nanoES was that positive- and negative-ion full-scan mass spectra, precursor ion scans, and MS/MS spectra could all be recorded from the same sample (consuming no more than a total of 2 μl of sample). Further, the technique has the advantage of being more sensitive than the LC/MS-based strategy presented, permitting sequence identifications readily at the subpicomole level, as has been shown by others.

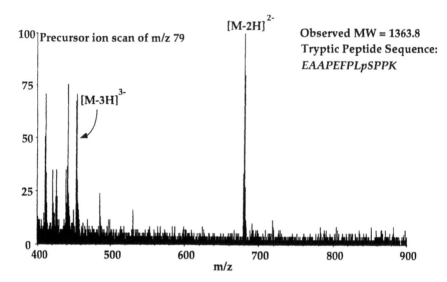

Figure 14 Parent ion scan of m/z 79 (PO_3^-) from the same sample by nanoES. A prominent ion was observed at m/z 680, corresponding to $[M-2H]^{2-}$ ion of the chicken stathmin tryptic phosphopeptide. The MS/MS spectrum of this peptide was recorded from the same sample in the positive ion mode, confirming the site of phosphorylation, previously identified by LC/MS/MS (Fig. 12).

```
◯ = cGMP, PKA Sites                          ◇ = MAP-K Site
▢ = p34cdc2, DNA-Dep Kinase Sites

SCG10    A K T A M A Y K E K M K E L S M L S L I C S C F Y P E P R N I N Y

                                       50                         62
SCG10    T Y D D M E V K Q I N K R A Ⓢ G Q A F E L I L K P P◇Ⓢ P I S

                                  73
SCG10    E A P R T L A Ⓢ P K K K D L S L E E I Q K K L E A A E G R R

SCG10    K Ⓢ Q E A Q V L K Q L A E K R E H E R E V L Q K A L E E N N

                  97
SCG10    N F S K M A E E K L I L K M E Q I K E N R E A N L A A I I E

SCG10    R L Q E K E R H A A E V R R N K E L Q V E L S G
```

Figure 15 A combination of stepped orifice potential LC/MS and nanoES experiments were used to identify the "preferred" sites of phosphorylation of SCG10 by several serine/threonine kinases.

ACKNOWLEDGMENTS

The authors gratefully acknowledge T. G. Consler and D. H. Willard (Glaxo Wellcome, Inc., Research Triangle Park, NC) for providing purified SH2 for these studies. The authors are indebted to G. Grenningloh (Glaxo Institute for Molecular Biology, Geneva, Switzerland) for providing stathmin and SCG10 for these studies. Matthias Mann is gratefully acknowledged for providing access to nanoES and MALDI for protein analyses.

REFERENCES

1. JB Fenn, JM Mann, CK Meng, CM Whitehouse. Science 246:64–71, 1989.
2. JA Loo, HR Udseth, RD Smith. Anal Biochem 179:404–412, 1989.
3. M Karas, U Bahr, A Deppe, B Stahl, F Hillenkamp. Makromol Chem Macromol Symp 61:397–406, 1992.
4. R Wang, BT Chait, SB Kent. In: R Hogue-Angeletti, ed. Techniques in protein chemistry IV. San Diego: Academic Press, 1993, pp 471–478.
5. M Mann, M Wilm. Trends Biochem Sci 20:219–224, 1995.
6. A Shevchenko, M Wilm, O Vorm, ON Jensen, AV Podtelejnikov, G Neubauer, P Mortensen, M Mann. Biochem Soc Trans 24:893–896, 1996.
7. DJC Pappin, P Hojrup, AJ Bleasby. Curr Biol 3:327–332, 1993.
8. M Vestal, L Juhasz, SA Martin. Rapid Commun Mass Spectrom 9:1044, 1995.
9. KR Clauser, SC Hall, DM Smith, JW Webb, LE Andrews, HM Tran, LB Epstein, AL Burlingame. Proc Natl Acad Sci USA 92:5072–5076, 1995.
10. MA Adams. Science 252:1651–1656, 1991.
11. JP Chervet. High Res Chromatogr Chromatogr Commun 12:278–282, 1989.
12. FE Regnier. Nature 350:634–635, 1991.
13. NB Afeyan, NF Gordon, I Maxsaroff, L Varady, SP Fulton, YB Yang, FE Regnier. J Chromatogr 519:1–29, 1990.
14. PR Griffin, JA Cvoffman, LE Hood, JR Yates. In J Mass Spectrom Ion Processes 111:131–149, 1991.
15. DB Kassel, B Shushan, T Sakuma, JP Salzmann. Anal Chem 66:236–243, 1994.
16. P Tempst, S Geromanos, C Elicone, H Erdjument-Bromage. METHODS: A companion to methods in enzymology 6:248–261, 1994.
17. WJ Henzel, TM Bileci, JT Stults, SC Wong, C Grimley, C Watanabe. Proc Natl Acad Sci USA 90:5011–5015, 1993.
18. M Moyer, D Rose, W Burkhart. In: R Hogue Angeletti, ed. Techniques in protein chemistry V. San Diego: Academic Press, 1994, pp 195–204.
19. W Burkhart. In: R Hogue Angeletti. ed. Techniques in protein chemistry IV. San Diego: Academic Press, 1993, pp 399–406.
20. M Wilm, M Mann. Anal Chem 66:1–8, 1996.
21. DK Luttrell, A Lee, TJ Lansing, RM Crosby, KD Jung, D Willard, M Luther, M Rodriguez, J Berman, TM Gilmer. Proc Natl Acad Sci 91:83–87, 1994.

158   **Kassel et al.**

22. WJ Wasilenko, DM Payne, DL Fitzgerald, MJ Weber. Mol Cell Biol 11:309–321, 1991.
23. SD Maleknia, JP Mark, JD Dixon, CP Elicone, BF, McGuiness, SP Fulton, NB Afeyan. Proceedings of the 42nd ASMS Conference on Mass Spectrometry and Allied Topics, Chicago, 1994, p 304.
24. DB Kassel, TG Consler, M, Shalaby, P Sekhri, N Gordon, T Nadler. In: J Crabb, ed. Techniques in protein chemistry VI. San Diego: Academic Press, 1995, pp 39–46.
25. RK Blackburn, RJ Anderegg. J Am Soc Mass Spectrom 8(5):483–494, 1997.
26. D Wagner, RJ Anderegg. J Am Chem Soc 117:1374–1377, 1995.
27. A Sobel. Trends Biochem Sci 16:301–305, 1991.
28. DB Kassel, JM Lenhard, WJ Rocque, L Hamacher, WD Holes, I Patel, C Hoffman, M Luther. Biochem J 316:751–758, 1996.
29. CR Lombardo, TG Consler, DB Kassel. Biochemistry 34:16456–16466, 1995.
30. RJ Boerner, DB Kassel, SC Barker, P Ellis Delacy, WB Knight. Biochemistry 35:9519–9525, 1996.
31. PA Curmi, A. Maucuer, S. Asselin, M. Lecourtois, A. Chaffotte, J-M Schmitter, A Sobel. Biochem J 300:331–338, 1994.
32. DB Kassel, G DiPaolo, G Grenningloh, B Antonsoon. Differences in phosphorylation of human and chicken stathmin by MAP kinase. Submitted for publication.
33. MJ Huddleston, RS Annan, MF Bean, SA Carr. J Am Soc Mass Spectrom 4:710–717, 1993.
34. J Ding, W Burkhart, DB Kassel. Rapid Commun Mass Spectrom 8:94–98, 1994.
35. M Wilm, A Shevchenko, T Houthaeve, S Breit, L Schweigerer, T Fotsis, M Mann. Nature 379:466–469, 1996.
36. B Antonsson, R Lutjens, G Di Paolo, D Kassel, B Allet, A Bernard, S Catsicas, G Grenningloh. Protein Expression Purification 9:363–371, 1997.

8
Mass Spectrometry in Drug Discovery

M. Arthur Moseley III, Douglas M. Sheeley, R. Kevin Blackburn, Robert L. Johnson, and Barbara M. Merrill
Glaxo Wellcome, Inc., Research Triangle Park, North Carolina

I. INTRODUCTION

Mass spectrometry can be applied to an incredible diversity of structural problems in the biological sciences, ranging from the quantitation of organic molecules by gas chromatography/mass spectrometry (GC-MS) to the molecular
weight profiling of intact proteins by matrix-assisted laser desorption ionization
time-of-flight mass spectrometry (MALDI-TOF/MS). A cursory perusal of the
chapters in this book makes it clear that most of the work in the development
of robust mass spectrometric methods for biochemistry applications has occurred in the past 15 years. In the early 1980s the drug discovery process relied
on mass spectrometry principally for the characterization of small organic molecules and their metabolites. After moving through a period in which fast atom
bombardment allowed us to see the tremendous benefits of a straightforward
method for mass analysis of biopolymers, we have arrived at a state of affairs
in drug discovery in which we rarely use electron impact or chemical ionization, or the expensive, high-resolution sector instruments that were formerly the
workhorses of mass spectrometry.

At the same time, our role as analytical and structural chemists in the
drug discovery process has changed as well. The revolutionary changes that
have occurred in mass spectrometry have been driven in part by changes just
as fundamental in the biological sciences and in the drug discovery process.
The development of fast, reliable methods for the cloning and expression of
protein target molecules has meant a constantly increasing demand for rapid,
high-throughput techniques for biopolymer characterization. The sensitivity and
flexibility of modern mass spectrometry have allowed mass spectrometrists to
make real contributions to the process of protein characterization and rational
drug design.

Over the past 10 years the role played by mass spectrometry in the discovery of therapeutic drugs has expanded dramatically, such that it has become
an integral part of essentially each step in the drug discovery process. The goal
of this chapter is to illustrate the different roles played by mass spectrometry
in this process. For the purposes of this chapter the drug discovery process may
be roughly divided into three parts: protein chemistry, medicinal chemistry, and
drug metabolism/drug pharmacokinetics. First the biologists identify the pro-

teins in the phenotype of the disease state, and the molecular biologists then use recombinant DNA technology to cause organisms to express these nonnative proteins that are the targets of the small drug molecules. The primary structure of these proteins, and any posttranslational modifications, are characterized using MALDI-TOF/MS as well as electrospray and nanoelectrospray. Next, the medicinal chemists synthesize small molecules that serve as agonists or antagonists of these proteins, and after a hit is identified they use structure–activity studies to modify the structure of these molecules to optimize their biological activity. Analysis of these compounds is primarily accomplished using single-quadrupole mass spectrometers equipped with either atmospheric-pressure chemical ionization (APCI) or ESI sources. The relatively recent advent of combinatorial chemistry has greatly increased the throughput required to support the medicinal chemists. Here, "open-access" mass spectrometers have been successfully used to meet the sample throughput demands. After optimization of the biological activity, the metabolic stability of these molecules is evaluated by studying their pharmacokinetic behavior in vivo. The drug's bioavailability and half-life are determined using quantitative liquid chromatography/mass spectrometry (LC/MS) or, more often, LC/MS/MS. If the biological half-life is too short, the molecular sites of metabolic instability are elucidated by the structural identification of the metabolites using primarily LC/MS/MS, and the medicinal chemists modify the drug's structure to minimize this instability.

Pharmaceutical companies have come under ever-increasing pressure from both the business community and the patient population to increase the rate of the discovery of drugs, and mass spectrometry has proven to be not only effective at providing the information necessary for drug discovery but also has improved the speed of the process. Moreover, the speed of analysis provided by modern mass spectrometric techniques has proven to offset the cost of the spectrometers by the ability to bring the drugs to market more quickly. This is particularly evident in the area of quantitation of drugs and drug metabolites in biological tissues and fluids, where "fast" LC/MS/MS is now the method of choice despite the high cost of triple-quadrupole mass spectrometers compared with the ultraviolet (UV) and fluorescence detectors they have replaced. Every major pharmaceutical company now has tens of triple-quadrupole mass spectrometers dedicated to this activity.

II. CHARACTERIZATION OF PROTEINS, PEPTIDES, AND CARBOHYDRATES

Over the course of the past decade, with the development of electrospray and matrix-assisted laser desorption ionization, mass spectrometry has gone from a marginal contributor in the structural analysis of biomolecules, to being a widely used, standard technique for the characterization of proteins, peptides, and carbohydrates. There is now a vast and still expanding literature on the integration of mass spectrometry into protein chemistry laboratories. We use several different instruments in our laboratory to analyze everything from the molecular weights of intact proteins to site-specific glycosylation patterns to the sites of interaction of proteins with small molecules. Some of these topics are addressed in other chapters, so in this discussion we try to describe how these techniques are applied in our laboratory. Two areas that are addressed in somewhat greater depth are the detailed characterization of glycoproteins, and the emerging technique of nanoelectrospray.

The characterization of unknown and recombinant proteins often presents different challenges. In the case of a recombinant protein, one is generally interested in learning as much as possible about the sequence, posttranslational modifications, and other structural features that may affect its biological activity. In the case of an unknown, of course very little may be known about the sequence or sometimes even the function of the molecule. Often the primary goal of the analysis is fairly modest in terms of the type and amount of information required, which may be limited to enough amino-terminal or internal sequence for cloning, or simply identification of the protein, using mass mapping, molecular weight, or amino acid composition in combination with one or more stretch of amino acid sequence. The amount of sample available is generally quite limited, restricting the number and depth of analyses that can be performed. It is critical to choose the correct tool for each project in order to maximize the return on the investment of a few picomoles of protein. Sample preparation and separation techniques are often more critical than the mass spectrometric portion of the experiment, and mass spectrometry must be able to interface seamlessly with techniques such as in-gel digestion and amino-terminal sequencing in order to extract the maximum information from a vanishingly small sample.

A. Quick and Dirty Information from Unknown Proteins by MALDI-MS

MALDI-MS presents several advantages for protein characterization. It is a fast, simple, and sensitive technique for measuring protein molecular weight, and is very tolerant of sample contamination. Only a small amount of material is needed, typically 1–3 μl of a 0.1–1 mg/ml solution of protein. We generally apply the protein to the target, allow it to dry, and then add 0.6 μl of a 10 mg/ml solution of matrix [generally sinapinic acid for proteins or α-cyano-4-hydroxycinnamic acid for peptides, in 60% acetonitrile/0.04% trifluoroacetic acid (TFA)]. If no signal is observed, often the sample target can be rinsed with water, dried, treated with another aliquot of matrix, and rerun. This rinse removes interfering substances in biological buffers that can cause suppression. When suppression is suspected before a sample is run, we use nitrocellulose-coated targets to aid in protein binding during the water rinse. The rinse approach worked well in determining the mass of the enzyme that catalyzes the conversion of valaciclovir to acyclovir, valaciclovir hydrolase (Fig. 1) [1].

We have used MALDI-MS to obtain accurate molecular weights of unknown proteins that are only available in small quantities in biological matrixes. The information is useful for molecular biologists as it provides more accurate data than that obtained from gel electrophoresis. The molecular weight information, in conjunction with other data, can be used in determining protein open reading frames from cDNA clones, for example. It can also give a preliminary indication of whether a protein is posttranslationally modified. Of course these techniques are equally applicable to analysis of recombinant proteins as well.

In addition to using MALDI-MS to measure masses of intact proteins, we also screen peptides derived from proteolytic digests using this technique, prior to Edman sequencing. Screening the digest using MALDI-MS gives the approximate length of a peptide contained in a fraction of interest (which allows more accurate use of sequencer time), as well as indicating if more than one peptide is present. A caveat with using MALDI-MS as a purity check is that the presence of only one peptide signal for a particular sample does not guarantee that only one peptide is present. We have sequenced fractions that by MALDI-MS appeared to contain one species and found more than one peptide to be present. Conversely, at times peptides can form adducts during ionization and thus more than one species can appear to be present with only one actual peptide sequence.

MALDI-MS data also provide a component for database searching of proteins. For example, protein molecular weight information plus amino acid analysis data can be used to identify an unknown protein with a fair degree of success. MALDI-MS of a proteolytic digest can also provide a quick fingerprint of the digested protein for database searching. Although we have not seen significant advantages in performing protein analyses in reflectron mode, digest mass mapping is an area where this capability can be helpful for two reasons. The effectiveness of database searches relies on the number of peptides one can include in the search, and the accuracy of the molecular weights assigned to these peptides. The improved resolution of mass maps generated in the reflectron mode is helpful in this regard. However, more importantly, detection of metastable fragment ions that result from postsource decay is possible in reflectron-mode MALDI-MS. Identifying even a few residues of sequence information in addition to peptide masses can help to narrow a database search,

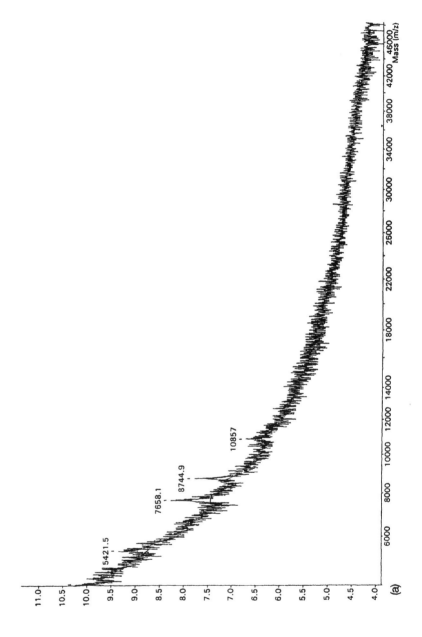

(a)

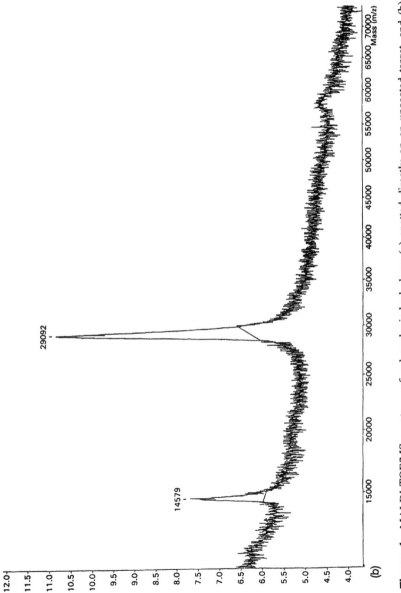

Figure 1 MALDI-TOF/MS spectrum of valacyclovir hydrolase: (a) spotted directly on an uncoated target, and (b) spotted on a nitrocellulose-coated target and washed prior to addition of matrix. The matrix was sinnapinic acid.

bringing MALDI-MS data closer to the accuracy achieved with nanoelectrospray mass mapping of digests and MS/MS of one or more peptide molecular ions.

Direct MALDI-MS of a digest also provides a check that the digest is complete. It is a much faster method and requires much less material than an analytical HPLC run. To observe good peptide signals, digests performed in 2 M urea may need to be diluted to 1 M before mixing with the matrix.

Finally, MALDI-MS has been used to characterize in vitro modifications to known proteins, such as monitoring the biotinylation of antibodies. MALDI-MS of the intact and the reduced and alkylated molecules (which resolves heavy and light chains) provides a much faster, more sensitive, and more accurate method of quantitating the extent of modification to biotinylated proteins than does the traditional dye-binding assay. When reduced and alkylated molecules are examined, distinct species can be seen for modified and biotinylated light chains (Figure 2). MALDI-MS is also the method of choice for examining proteins that have been covalently modified with polyethylene glycol (PEG) chains. For example, Watson et al. [2] used MALDI-MS to examine the extent of PEG attachment to recombinant stem cell factor, and were able to distinguish several different forms of PEGylated intact protein. The principle advantage of MALDI-MS in examining PEG-modified proteins is that the multiple charging typical of ESMS is not present, which greatly simplifies spectral interpretation. MALDI-MS of PEGylated proteins also requires little or no sample cleanup. Another important consideration is the charge-carrying capacity of the PEG. If the protein or peptide to which the PEG is attached is small, PEGylation may increase the mass to charge ratio that can be achieved in electrospray to a point above the useful mass range of most quadrupole analyzers.

B. Recombinant Proteins

By the time a recombinant protein has been expressed and submitted for characterization, the questions being asked are usually very different from the case of an unknown sample. Although many of the same techniques are used, the goal is generally to determine as much about the protein as possible, and one is not nearly as limited in the availability of material. This is, after all, why proteins are cloned and expressed. Priorities diverge yet again as one considers a recombinant protein that has been expressed for study as a drug target, or a recombinant therapeutic protein that must be scrutinized according to regulatory requirements.

Some analytical requirements are the same for any recombinant protein. In each case, a sequence is available, so one has a clear idea of what the amino- and carboxyl-terminal sequences should be, the potential sites of proteolysis, and the molecular weight. The first priority is to determine whether the protein

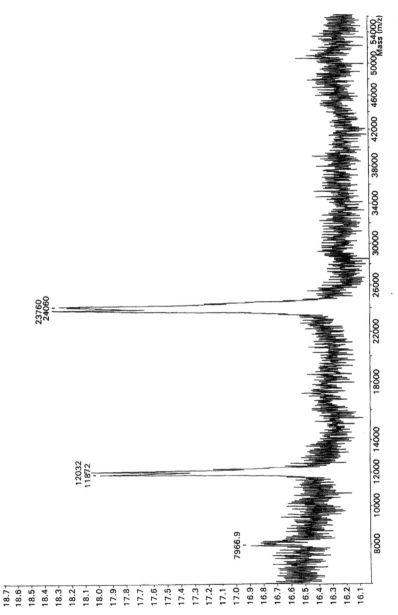

Figure 2 MALDI-TOF/MS spectrum of the biotinylated light chain of a monoclonal antibody. The matrix was sinnapinic acid.

has been expressed appropriately with respect to these parameters. Electrospray mass spectrometry is performed as a first step, usually as a rapid LC/MS experiment using a capillary column containing a perfusive reversed-phase packing, providing information on purity. The molecular weight is accurate enough to determine incomplete or inappropriate processing, many but not all mutations, and posttranslational modifications. LC/MS and amino-terminal sequencing may be the only analyses performed, if there are no surprises. However, most recombinant proteins inevitably enter a second phase of analysis in which proteolytic digests are mapped by LC/MS and peptides are sequenced by MS/MS, particularly if a mass discrepancy in LC/MS of the intact protein suggests a mutation, processing anomaly, or the presence of posttranslational modifications. Often when a protein is phosphorylated or glycosylated, if these features are functionally important, structural characterization of these features is a major focus. Similarly, if the three-dimensional structure of a target protein is being studied, the locations of disulfide bonds must often be determined in our laboratory. Because of the advantages of coupling LC with electrospray MS, LC/MS is the best technique available for many of these analyses. Although we are generally not severely limited by sample availability, we usually do not have more than a few hundred picomoles of material. This necessitates the use of capillary columns wherever possible, and makes methods such as proteolysis using immobilized enzyme columns and nanoelectrospray very attractive.

C. Nanoelectrospray Ionization Mass Spectrometry

Nanoelectrospray ionization mass spectrometry (nanoES) was recently introduced as a technique for the characterization of subpicomole amounts of proteins [3,4]. With nanoES, a protein is initially digested with a specific protease followed by infusion of the resulting 1–2 μl of sample into the mass spectrometer (no separation of the mixture) at nanoliter per minute flow rates through a finely drawn glass capillary with a metallic coating. These low flow rates result in analysis times of approximately 1 hr for a 1-μl sample. This allows time for MS/MS fragmentation/sequencing of multiple peptides from the digest, as well as permitting careful optimization of collision-induced dissociation (CID) experiments for different peptides, if necessary. As described earlier, using a database searching algorithm, proteins can then be identified from individual peptide fragmentation spectra, often by calling only two or three residues of sequence in a single spectrum [5]. This process is rather error tolerant, and proteins can often be identified even if posttranslational modifications are present. Besides being applicable to identifying proteins whose sequences are in the databases, nanoES has also been used to obtain sequence of novel proteins for cloning [6].

    NanoES is most useful in situations where sample is limited and N-terminal sequencing or capillary LC/MS lacks the sensitivity needed to solve the problem at hand. However, even in situations where capillary LC/MS might have adequate sensitivity, nanoES offers additional flexibility in performing multiple mass spectrometric experiments on a sample by essentially removing the time limitation imposed by an online separation. NanoES has been particularly useful for (1) monitoring initial expression and purification of recombinant proteins, (2) characterizing N-terminally blocked proteins and fusion proteins containing N-terminal protein sequences such as glutathione *S*-transferase (GST), and (3) characterizing posttranslational modifications and sequence errors in proteins.

The identification of an endonuclease associated with spermatocyte apoptosis in 2-methoxyethanol-treated rats [7] using nanoES is illustrated in Figs. 3 and 4. In Fig. 3, lane 1 shows a silver-stained gel separation of the nuclear extract from testes of 2-methoxyethanol-treated rats, with the band of

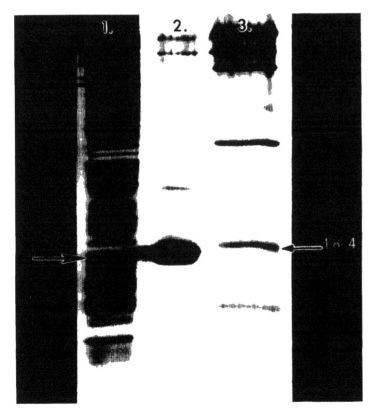

Figure 3 Lane 1: Silver-stained polyacrylamide gel electrophoresis (PAGE) separation of nuclear extract from testes of 2-methoxyethanol-treated rats with band of interest marked with arrow. Lane 2: Recombinant cyclophilin A. Lane 3: Molecular weight markers with 18.4-kD band marked.

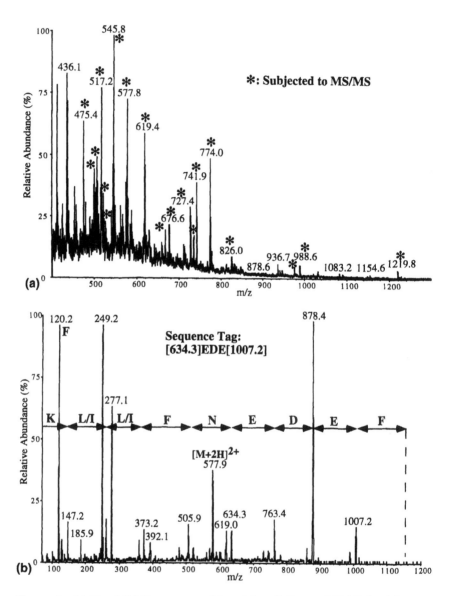

Figure 4   (a) NanoES-MS spectrum of in-gel Lys-C digest of 18-kD band from nu-clear extract of testes of 2-methoxyethanol-treated rats. (b) Product ion spectrum of Lys-C peptide at m/z 577.8.

interest marked with an arrow. This band was found to comigrate with recombinant cyclophilin A (lane 2), a known nuclease that has an approximate molecular weight of 18 kD as illustrated by the molecular weight markers in lane 3. Numerous attempts at N-terminal sequencing of this band failed to return any sequence, presumably due to the low amount of protein present in the band (estimated 5 pmol). The band marked in Fig. 3, lane 1, was excised and subjected to in-gel Lys-C digestion [8] followed by nanoES analysis. The nanoES MS spectrum of this band (Fig. 4a) contained numerous peptides, several of which matched expected masses of cyclophilin A Lys-C peptides. Several of the peptides in Fig. 4a were subjected to MS/MS analysis in order to obtain sequence data. MS/MS of the peptide at m/z 577.8 (Fig. 4b) gave a complete y-ion and partial b-ion series, and upon a search of the OWL database using the sequence tag [634.3]EDE[1007.2] (see Ref. 5 for a description of sequence tags), four matching cyclophilin protein sequences were retrieved including rat cyclophilin A. Over 73% of the rat cyclophilin sequence could be accounted for in the mass spectrometric data. Along with Western blot data, this provided the confirmation needed to conclude that cyclophilin A was the nuclease associated with the rat spermatocyte apoptosis.

Another useful technique for protein characterization by nanoES is the ability to use precursor ion scanning to detect analytes that cannot be observed in MS mode because of high chemical background [9]. Precursor ion scanning is a very selective type of tandem (or MS/MS) mass spectrometric experiment in which a spectrum is acquired of ions (precursors) that undergo fragmentation to form a characteristic fragment (or product) ion. For example, Leu- or Ile-containing peptides will often form a fragment ion at m/z 86, the Leu/Ile immonium ion. An example of using precursor ion scanning to overcome chemical noise is shown in Fig. 5. Following purification of a 100-kD recombinant GST fusion protein, the identity of the protein in this band was determined by nanoES for two reasons. N-Terminal sequencing at best would provide approximately 80 residues in a single experiment, which would be insufficient to sequence completely through the approximately 220 residues of GST sequence at the amino terminus. Also, based on the intensity of this band with copper stain, less than 20 pmol of protein was estimated to be present, requiring the sensitivity of nanoES. Figure 5a shows the MS spectrum of the in-gel Lys-C digest of what was thought to be the 100-kD recombinant protein. Because the digest was conducted using a high concentration of guanidine in the digestion buffer, the spectrum is dominated by a series of guanidine HCl cluster ions differing in mass by 95 Da. By scanning for precursors of m/z 147, the y_1 fragment ion (see Ref. 10 and Chapter 10 by Dass for explanation of peptide fragment ion nomenclature) of lysine-containing peptides (Fig. 5b), several Lys-C peptides were observed including a peptide at m/z 609.7. The product ion scan of this peptide shown in Fig. 5c allowed identification of the protein from the OWL

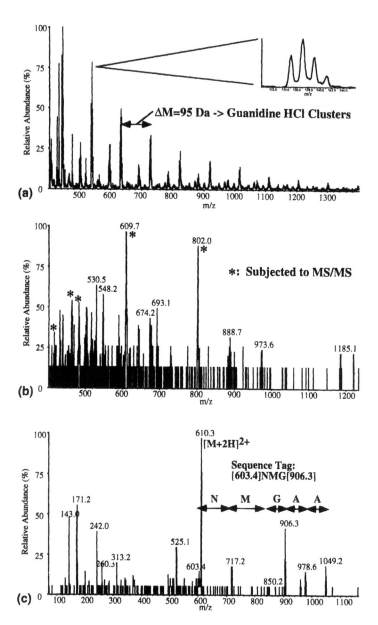

Figure 5 (a) NanoES-MS spectrum of in-gel Lys-C digest of 100-kD band from recombinant protein sample (digested in 2 *M* guanidine). (b) Spectrum of precursors of *m/z* 147. (c) Product ion spectrum of *m/z* 609.7.

database as 60-kD chaperonin from *Escherichia coli*. Instead of the expected recombinant protein, the 100-kD band is thought to be a dimer of *E. coli* chaperonin. In this situation, the sensitivity of nanoES allowed us to provide feedback to the protein biochemist as to the success of the expression and purification earlier (because of being able to work with less material) than if another technique were used.

Besides being useful for overcoming high chemical background, precursor ion scanning with nanoES is also useful for identifying peptides with post-translational modifications such as phosphorylation or glycoslyation [9,11]. For example, the nanoES MS spectrum of a tryptic digest of a recombinant glycoprotein is shown in Fig. 6a. By scanning for precursors of m/z 204 (oxonium ion of N-acetylhexosamine) as shown in Fig. 6b, doubly and triply charged ions of a single glycopeptide are detected with the heterogeneity associated with glycosylation. The product ion spectrum of the glycopeptide at m/z 868.6 (Fig. 6c) yields useful information about the composition of the glycan such as the presence of a core fucose. Precursor ion scanning has also been shown to be useful for detecting phosphopeptides from phosphoprotein digests [11]. We have found this technique useful for selectively detecting peptides from digests that contain unusual amino acids or modifications (i.e., detecting selenomethionine-containing peptides from digests of fully selenomethionated proteins).

We have found nanoES particularly useful in the characterization of proteins that are not amenable to N-terminal sequencing, because they are either blocked or expressed as fusion proteins, such as the example of the GST fusion protein described earlier. If a protein is N-terminally blocked or contains a long GST domain, direct sequencing of the intact protein to confirm its identity is either impossible or uninformative. In both these cases, a labor-intensive chromatographic separation of a proteolytic digest would be required in order to obtain internal sequence information. However, by using nanoES, a digest can be analyzed directly without separation and sequence obtained by MS/MS on multiple peptides in the mixture. An example of this is given in Fig. 7a, which shows the nanoES MS spectrum of an in-gel tryptic digest of the fusion protein GST-Tie2. Tie2 is a kinase involved in cell signaling. Peptides identified in the MS spectrum shown in Fig. 7a account for 13% of the GST-Tie2 sequence, confirming the presence of the Tie2 portion of the protein. Figure 7b shows the nanoES product ion spectrum of the ion at m/z 868.5, which did not match any expected GST-Tie2 peptide. Upon a database search, we found this peptide to indeed be a Tie2 peptide, but containing a Thr to Lys mutation as shown in Fig. 7b. By detecting this mutation, the construct could be corrected prior to scaling up purification.

NanoES will play an increasing role in drug discovery, particularly in the area of protein characterization. Its ability to successfully characterize proteins at subpicomole levels with a relatively simple apparatus allows protein charac-

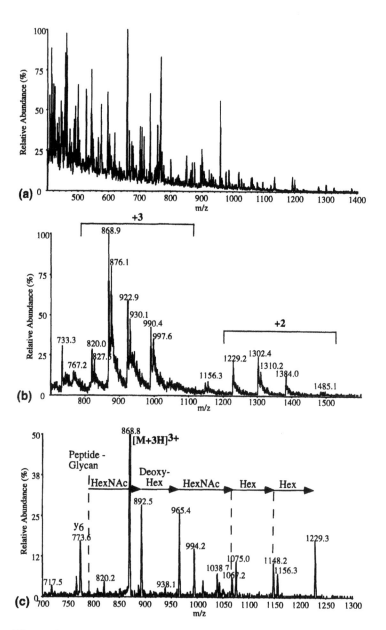

Figure 6 (a) NanoES-MS spectrum of tryptic digest of recombinant glycoprotein. (b) Spectrum of precursors of *m/z* 204. (c) Product ion spectrum of tryptic glycopeptide at *m/z* 868.6.

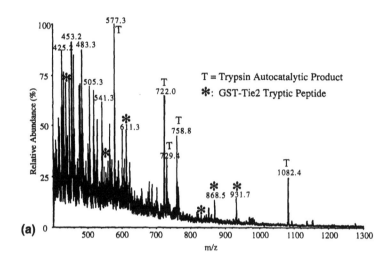

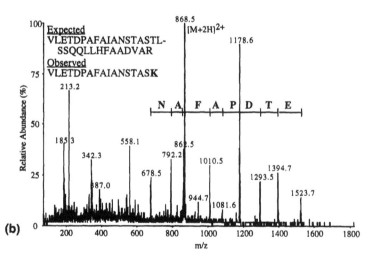

Figure 7 (a) NanoES-MS spectrum of in-gel tryptic digest of recombinant GST-Tie2. (b) Product ion spectrum of GST-Tie2 peptide at m/z 868.5.

terization problems to be solved that would otherwise be difficult or impossible to solve using other strategies. Perhaps the most exciting application of nanoES deals with the mapping of proteomes and subsequent correlation with genomic data [12,13]. Another area in which nanoES is likely to impact drug discovery is in the area of differential gene expression, monitoring the expression levels of proteins in response to disease or treatment with a potential drug. NanoES may also be found useful in the combinatorial chemistry area as a sensitive means for decoding compounds released from single beads from combinatorial libraries [14].

D. Using Mass Spectrometry to Map Interactions Between Proteins and Small Molecules

In addition to using mass spectrometry to identify posttranslational modifications, combined high-performance liquid chromatography (HPLC)/MS techniques can be used to identify sites of covalent modifications of proteins by other molecules such as nucleotides. Difference maps between modified and unmodified proteins can point to specific peptides as the site of modification. Mass spectrometry allows the modification discovered by HPLC peptide mapping to be identified as well as the site determined. We have used a combination peptide mapping/mass spectrometry approach in protein modification problems. For example, MS in combination with other techniques has been used to identify the site of suicide substrate binding to a protein [15]. The nucleoside 2,6-diamino -9-(2'-deoxy-2'-fluoro-β-D-arabinofuranosyl)-H-purine (dFDAP) is an inhibitor of nucleoside 2-deoxyribosyltransferase. The nucleoside inhibits the enzyme by forming a stable enzyme-bound intermediate. Electrospray mass spectrometry (ES-MS) of the inhibited versus the native enzyme showed a mass difference of 135 Da, or that expected for a bound 2-fluoro-2-deoxyarabinosyl group. A map of the Lys-C digested protein indicated the presence of one modified peptide. The mass of this peptide was 21 Da less than calculated for the expected Lys-C peptide, and the sequence of the peptide was abruptly terminated prematurely. This suggested that a modification such as dehydration occurred in the middle of the peptide (it also suggested that the sugar moiety was lost from the protein during the digestion procedures). The inhibited protein was found to be easily cleaved to fragments under basic conditions. The smaller of these fragments was found by Edman sequence analysis to be NH_2-terminally blocked. LC-MS of a Lys-C digest of this fragment indicated that the block was due to formation of pyroglutamate at Glu-98, or just beyond the site where sequencing was terminated in the modified protein. Thus, a combination of protein chemistry techniques, in which MS played an important part, was used to determine that deoxyribosyltransferase was modified by dFDAP on Glu-98. This particular problem points out another important aspect of the role of mass spectrometry, in that a number of complementary techniques are used to solve problems in the most effective manner possible.

E. Biopharmaceuticals

The most recent push in the analysis of biopharmaceuticals, that is, therapeutic molecules derived from or produced in biological systems, is toward an analytical strategy closer to that for small molecules—in other words, using a well-characterized molecule as the standard for release, rather than having the release be process dependent. This eliminates the need to recharacterize the molecule's biological activity when a change is made in the manufacturing, purification, or formulation process. Mass spectrometry–based techniques thus play a vital role in the analysis of biopharmaceuticals. While what actually constitutes a "well-characterized molecule" is not yet certain, and may vary from case to case, it is clear that mass spectrometry will play a role in this characterization.

LC/MS peptide mapping is a fast and accurate way of confirming the primary structure of recombinant proteins. For example, one LC/MS run of a single digest of a monoclonal antibody provided confirmation of approximately 95% of the primary structure of the molecule, including information on the glycoforms present (discussed later). This is a quick way of ensuring integrity of the DNA sequence as well as a check on possible posttranslational modifications to the primary structure. Glycosylation is especially common in biopharmaceuticals, as these molecules are often secreted proteins, in which glycosylation modulates activity, serum half-life, and other properties [16,17]. Other modifications may include processing at the amino- or carboxyl-terminal ends. The characterization of a recombinant monoclonal antibody provides an example of the characterization of many of these structural features [18].

CAMPATH-1H is a recombinant humanized murine monoclonal immunoglobulin G1 (IgG1) antibody that went through clinical trials for the treatment of non-Hodgkin's lymphoma and rheumatoid arthritis. CAMPATH contains a single consensus sequence site for N-linked glycosylation in the CH_2 domain of each heavy chain, which is a conserved site, found in every IgG, and the attached oligosaccharides are always biantennary complex structures [19]. The CAMPATH expressed in Chinese hamster ovary (CHO) cells and a murine myeloma cell line (NS0) have been studied and the IgG1 and IgG4 isotypes compared. LC-MS was used to profile the intact heavy and light chains, as well as proteolytic digests, of several variants of CAMPATH, in order to rapidly confirm protein primary sequence and create a site-specific profile of glycan heterogeneity. These data also allows processing at the COOH terminus to be characterized. The major focus of the work on this molecule, however, was characterization of glycosylation.

Characterization of the glycosylation pattern of a recombinant therapeutic protein is important because the cell type used for its expression is usually not closely related to the cells that produce the protein in nature. This can result in significant changes in the glycosylation of the recombinant protein relative to the native form [20]. An example is the variation in glycosylation of tissue plasminogen activator isolated from different cell lines [21]. The demonstrated potential for significant changes in the biological activities of recombinant glycoproteins depending on their expression systems is a real concern from a pharmacological standpoint [22].

The structural analysis of glycoconjugates presents significant challenges because of some unique features among biopolymers. These branched arrays of largely isomeric monosaccharides exist as heterogeneous populations, each component of which must be characterized. Current mass spectrometric techniques represent one of the most effective strategies available for the structural analysis of intact oligosaccharides [23,24]. Although LC-MS profiling cannot determine the sequence or branching of the oligosaccharide chains, this site specificity of glycan mapping is a major advantage in a protein that contains more than one site of glycosylation, and because N-linked glycan structures tend to follow a few well-defined patterns, often one can tell a great deal from the molecular weights of glycopeptides. To perform a detailed analysis of the oligosaccharides, they are released from the protein by hydrazinolysis or endoglycosidase digestion, profiled, and sequenced by MS/MS. Linkage position and anomeric configuration can then be determined by methylation analysis and exoglycosidase digestion.

As discussed earlier, frequently a great deal can be learned quickly about a protein by a simple LC-MS analysis of the intact molecule. In the case of this monoclonal IgG, because it is made up of two copies each of two distinct protein molecules (heavy and light chains), the antibody was reduced and alkylated prior to analysis, in order to reduce the disulfide linkages that hold the multisubunit complex together. The resulting LC-MS data provided molecular weights for the light and heavy chains of the antibody. Figure 8 shows the deconvoluted molecular weight plots derived from LC-MS of CAMPATH grown in NS0 cells. Figure 8a shows a single peak corresponding to the molecular weight of the reduced and alkylated light chain. The molecular weight determined from these experiments is within 1 Da of the predicted molecular weight for the CAMPATH light chain sequence. The deconvoluted molecular weight plot shown in Fig. 8b for the CAMPATH heavy chain shows five major species present, which differ in molecular weight by approximately 162 Da, the interval molecular weight of one hexose subunit. This glycoform heterogeneity corresponds exactly in relative abundances to the glycosylation observed at the conserved Asn-linked glycosylation site of the heavy chain in subsequent LC-MS mapping of proteolytic digests. These molecular weights are within 25 Da of the expected masses, representing an error of 0.05%. While this is not adequate to confirm sequence, it provides an accurate representation the overall glycosylation of the protein.

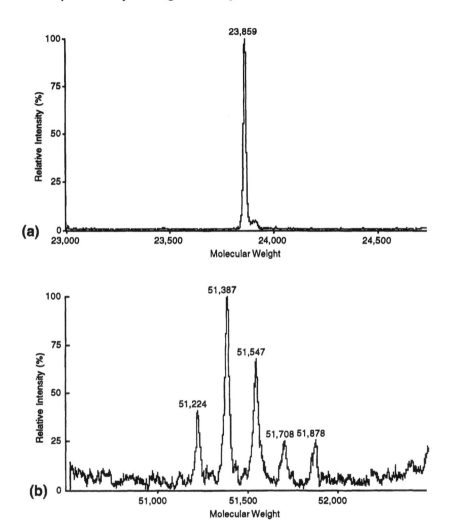

Figure 8 Deconvoluted molecular weight plots obtained from LC-MS of CAMPATH-1H (a) light and (b) heavy chains, following reduction with dithiothreitol and alkylation with iodoacetamide.

Figure 9a shows the total ion chromatograms (TIC) for Lys-C digests of CAMPATH produced in NS0 cells. All expected Lys-C peptides greater than four amino acids in length were detected. Of the 665 residues in the heavy and light chains of the IgG1 isotype, 95% were accounted for by LC-MS of the Lys-C digest. Likewise, the masses for all major HPLC peaks were consistent

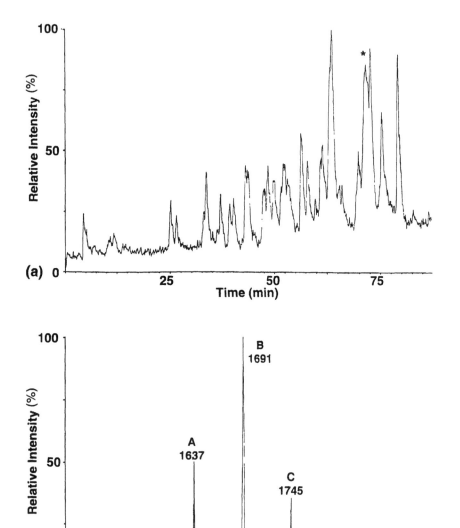

Figure 9   LC-MS of the Lys-C digest of CAMPATH expressed in NS0 cells: (a) TIC and (b) mass spectrum of the glycopeptide over the range of the 3+ charge state.

with Lys-C cleavage peptides of CAMPATH. LC/MS of CAMPATH proteolytic digests illuminated two important structural features of this molecule. The peptides corresponding to the amino and carboxyl termini of the CAMPATH heavy chain indicate posttranslational modifications. In the case of the amino-terminal peptide, the terminal glutamine residue has been converted to pyroglutamate, effectively blocking the amino terminus of the heavy chain. This modification is a common feature of immunoglobulins. Likewise, at the carboxyl terminus, a terminal lysine residue has been removed. In each case, assignment of ions observed in LC/MS rapidly identified these modifications, which could be confirmed by complimentary techniques if necessary. In addition, LC/MS provided a profile of CAMPATH glycosylation.

Figure 9b shows the electrospray spectrum of the glycosylated peptide indicated by the asterisk in Fig. 9a. Only the portion of the spectrum showing the $(3+)$ charge envelope is shown. Those peaks in the spectrum not specifically labeled as corresponding to the glycosylated peptide are due to other co-eluting peptides. Multiply charged ions that correspond to five glycoforms of the glycopeptide were observed for the NS0-derived sample shown here. Because IgG glycosylation follows specific patterns, it is reasonable to propose carbohydrate structures for the glycopeptide based only on the expected biantennary motif and molecular weight data. Possible structures for the peaks labeled A through E are shown above the corresponding peaks in the spectrum. Structures A, B, and C are biantennary core-fucosylated structures, all of which are typical of IgG glycosylation. However, two structures, D and E, were not observed in CHO-expressed samples. Peaks D and E are not as abundant as the other glycoforms, but do indicate a difference in glycosylation between expression systems, and are significant because they are due to the addition of an epitope to the glycan structure, which is immunogenic in humans [25,26]. Although the addition of either one or two hexose residues might theoretically result in a number of possible structures, the most likely possibility for the structures of the D and E glycoforms is that they result from an α-galactosyltransferase activity present in murine cells like NS0 but not CHO cells. These structural assignments were confirmed by detailed characterization of the N-linked oligosaccharides.

CAMPATH N-linked oligosaccharides were released by hydrazinolysis, permethylated, and analyzed by electrospray MS under conditions designed to generate natriated parent ions exclusively [24]. The ES-MS spectra of permethylated N-linked glycans generally show molecular ions with little or no fragmentation. This concentration of ion current in the molecular ions favors the use of MS/MS for sequence analysis. When the $[M + 2Na]^{2+}$ precursor is selected, abundant sequence-related fragment ions are produced. Figure 10 is the MS/MS spectrum of the $[M + 2Na]^{2+}$ parent of the permethylated glycan corresponding to glycoform B in Fig. 9b. The fragmentation of this ion is dia-

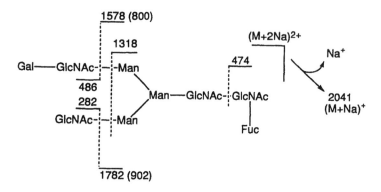

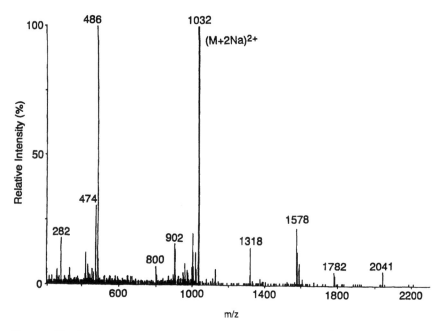

Figure 10 Product ion spectrum obtained for the $[M + 2Na]^{2+}$ parent ion of the monogalactosyl biantennary glycan. The sample was infused directly from a 10 mM solution of sodium acetate in 70/30 methanol/water; 50 scans were summed.

grammed above the spectrum. The ion at m/z 2041 is due to the loss of one Na^+ ion, and corresponds to the singly charged molecular ion. The fragment ion at m/z 474 represents the loss of the reducing-terminal core GlcNAc residue, and an attached fucose. This ion showed that in the CAMPATH glycans, a fucose residue is attached to the reducing-terminal core GlcNAc rather than on an outer arm. A series of fragment ions occurring at m/z 282, 486, 1578, and 1782 is due to two sets of complimentary ions representing cleavage of the nonreducing-terminal GlcNAc (m/z 282 and 1782) and Gal-GlcNAc (m/z 486 and 1578) residues on either arm of the monogalactosylated biantennary structure. The doubly charged species of the m/z 1578 and 1782 ions were observed at m/z 800 and 902. A related ion that occurred at low abundance was the double-cleavage product ion at m/z 1318, representing loss of both outer arms. In subsequent experiments, it was demonstrated that in structures D and E the additional hexose residues are on the nonreducing terminus. These spectra are straightforward and provide significant sequence information. In some cases, cross-ring fragmentation can also provide information about residue linkage positions. However, one piece of information these experiments cannot provide is the anomeric configuration of a particular linkage.

In this example it is not clear from the mass spectrometry data whether the terminal galactoses are alpha- or beta-linked to the penultimate galactose. This could not be determined using MS/MS alone, but this problem was easily addressed by incubating isolated CAMPATH glycopeptides with either α- or β-galactosidases, followed by MALDI-MS of the reaction mixture to determine which residues had been removed [18]. This approach has some limitations, particularly with respect to the characterization of linkages that have not been encountered previously, but in the case of a recombinant protein, the expression system is generally well characterized, obviating this concern. In this case, we were able to easily confirm that the terminal galactose residues were indeed α-linked, demonstrating that the difference in glycosylation between CHO and NS0-expressed CAMPATH was due to an α-galactosyltransferase present in the NS0 cells.

The preceding example points out the capabilities of mass spectrometry for the characterization of recombinant proteins and, in fact, for the process control and release specifications of therapeutic biomolecules. In the case of a molecule like a monoclonal antibody, which is glycosylated with a known structural motif and where the amino acid sequence has been determined, molecular weight profiling through LC/MS can contribute a great deal to characterizing the molecule. In this case, peptide mapping provided confirmation of the amino acid sequences of both protein subunits, as well as identifying and characterizing posttranslational modifications. The combination of site-specific glycosylation mapping and detailed oligosaccharide analysis provided a clear picture of carbohydrate structures, and a standard for comparison of subsequent profiles.

In addition to confirming structure, it is important to determine the stability of the molecules of interest, including the effects of heat, light, or pH. Many different modifications to recombinant proteins have been observed, such as oxidation, deamidation, disulfide scrambling, and internal cleavages. LC/MS peptide mapping quickly tracks and partially quantitates the oxidation of methionine residues in recombinant proteins. Similar techniques have been used to determine modifications after the fact as well as to monitor the effect of varying conditions on a protein's structure. For example, Keck [27] has used the reagent *t*-butyl hydroperoxide to determine the susceptibility of methionine residues to oxidation in recombinant interferon and tissue-type plasminogen activator. Oxidized peptides were separated by HPLC and identified by ES-MS. Chowdhury et al. [28] developed MS methods to identify sites of oxidation in performic acid-treated superoxide dismutase. During this study, MS was used to help confirm that tyrosine residues were modified to 3- and 3,5-dichlorotyrosine.

Deamidation of asparagine and glutamine residues, particularly Asn, can occur under relatively mild conditions and can cause microheterogeneity in recombinant proteins [29]. Since deamidation results in a mass shift of only 1 Da, this event is best detected after proteolysis of the protein, in an LC/MS map. Bischoff et al. [30] used HPLC purification followed by LC/MS to determine that recombinant hirudin was covalently modified. ES-MS of the Glu-C peptides of hirudin was used to determine that deamidation occurred at Asn-53, or Asn-33, followed by the formation of stable succinimide intermediates, which gave an ion 17 Da less than expected. Cacia et al. [31] identified the site of isoaspartic acid formation at an Asp-Gly sequence in a recombinant antibody by a combination of Edman sequencing (isoaspartate blocks Edman chemistry) and mass spectrometry.

In another case, Geoghegan et al. [32] used MS and Edman sequencing to show that a recombinant variant of glucagon (31 aa) produced in yeast was modified by the addition of dipeptides to the amino terminus of the peptide. Ross et al. [33] used MS to identify the sites of lysine acetylation in recombinant neurotrophin-3 expressed in *E. coli*. Field et al. [34] used MALDI-MS to identify neutral oligosaccharides from recombinant mutant tissue-type plasminogen activator (TNK-tPA). The glycans were studied because by high-performance anion-exchange chromatography (HPAEC) it appeared that some degradation was occuring during the purification of TNK-tPA from CHO cells. Neutral glycans were found to be mostly biantennary complex type structures with two, one, or zero galactose residues. It was discovered that a host-cell β-galactosidase was copurifying with the TNK-tPA at the first step, causing the variation in galactose content. All of the modifications just mentioned are amenable to detection by either LC/MS or nanoelectrospray in conjunction with other techniques (e.g., Edman sequencing or MS/MS for isoaspartate formation).

III. MASS SPECTROMETRIC SUPPORT OF SYNTHETIC CHEMISTRY

A. Introduction

Structural analytical chemistry in support of pharmaceutical drug discovery has undergone a radical change in the past 5 years. The areas of automation and combinatorial chemistry have changed the way that mass spectrometrists support pharmaceutical chemists and the tasks that were asked of them by the pharmaceutical chemists. In the "old" days, a chemist submitted a sample (usually a purified intermediate or final product) and requested mass spectral analysis. The mass spectrometrist decided the best analysis technique [electron impact (EI), fast atom bombardment (FAB), electrospray (ESI), etc.] and conducted the analysis. The data were interpreted and returned to the chemist in 2–3 days. In the changing environment of pharmaceutical drug discovery, this procedure has become untenable—first, because the number of samples has outstripped the available facilities and resources, and second, because the chemist is requesting faster turn-around time. This need for more and speedier analysis is the direct result of the changing nature of the pharmaceutical industry and drug discovery. To be competitive in the pharmaceutical industry requires the discovery of more drug candidates in a shorter period of time [35]. Not only has a faster turn-around time in analysis been required, but the introduction of combinatorial chemistry techniques has increased the number of samples. Some of these new combinatorial synthetic procedures place demands not only on the present technology in terms of numbers of samples for analyses, but also on the technology required for analysis. The newer solid-phase synthesis and multicomponent synthesis methods require new sampling and mass spectrometry techniques. This trend will only continue to grow in the coming years. To meet these challenges, several new procedures have been implemented to increase the throughput of samples in the laboratory without increasing the number of analysts needed. These new ideas have impact on both the traditional synthetic chemist and the combinatorial chemist by use of automation and what is termed open-access mass spectrometry. The following discussion describes what is being achieved at the present time in these two areas.

B. Open-Access Mass Spectrometry

Mass spectrometry from its inception was a tool used to support the synthetic organic chemist in his synthetic efforts [36,37]. Mass spectrometry was used to determine the structure of unknown compounds or verify the identity of a synthetic product. The basis for this was usually a molecular ion and several fragment ions. Other analytical data together with the reaction scheme were also used in this endeavor. A formal proof of structure requires that the analytical data from several techniques [infrared (IR), MS, nuclear magnetic resonance

(NMR), etc.] agree with a proposed structure. However, during the synthetic steps prior to the final product, chemists require a minimal amount of analytical data that will tell them whether a product is present and/or how pure the product is. Speed is essential, since the presence of the desired product will determine whether or not a reaction is taken to completion or a sample is purified. The sooner the analytical results are acquired, the more efficient and productive the chemist becomes. Any technique that can provide this information is used. Automated NMR has been successful [38], but suffers from complicated spectra that require interpretation and are prone to ambiguity, especially for mixtures. Also, depending on the technique employed and the amount of sample available, the acquisition time of NMR data can excessive. Certain mass spectrometry techniques can provide molecular ions for the species in a mixture in a matter of a few minutes, and mass spectrometry is readily amenable to automation. Because of this the chemist has gravitated toward mass spectrometry for quick results, and the mass spectrometrist has gravitated toward automation when the number of samples has outstripped manpower and equipment resources.

Early attempts at automation for the analysis of synthetic compounds borrowed from quantitative mass spectrometry and only automated the analysis of the sample. The use of autosamplers for GC and HPLC analysis is widespread, and their use by mass spectrometry manufacturers was easily accomplished for quantitative analysis in both the clinical and environmental fields. These coupled instruments were applied to the analysis of synthetic samples using a variety of ionization techniques [39]. The analysis usually required no chromatographic separation because the chemist required a simple answer to the question, "Is my product present?" Therefore, flow injection of the sample could be used. The samples were dissolved in a suitable solvent, deposited in a standard HPLC vial, and placed in the carousel or rack of an HPLC autosampler. Information concerning the submitter's name, notebook number, etc. was entered by hand or via a spreadsheet into the mass spectrometer's data system run list. The samples were then run in a batch mode overnight, with the chemist receiving the results of the analysis the following morning. Production of a mass spectrum from the data could be automated through various procedures. While samples were run faster, this procedure still required that the analyst perform a certain amount of sample handling, instrument setup, and analysis initiation. However, the time per analysis was reduced and the mass spectrometer was free during the day to perform other tasks such as LC/MS or nonroutine sample analysis. Several laboratories reported the introduction of such systems [40–43].

But the real savings in time for the chemist and analysts is a system where the samples are run on a continuous basis using a program that automatically logs and batches the sample, starts the mass spectrometer, analyzes the sample, and produces the final mass spectrum. These systems can be running at all times of the day to provide an answer for the chemist when needed without the intervention of an instrument operator. With the advent of relatively inexpensive mass spectrometers and HPLC automated instruments and the development of the computer programs required to run these systems unattended, most mass spectrometer manufacturers have provided systems to accomplish this procedure. The mass spectrometrist only has to provide maintenance for the instrument to ensure that quality mass spectra are produced. Time can now be spent on data interpretation and solving more complex structural problems rather than obtaining data for routine samples where the chemist only requires only a molecular ion.

The choice of an ionization method for the automated systems depends mainly on the type of sample that is analyzed. Analysts and/or instrument manufacturers have devised elaborate methods to automate the traditional solids probe methods of analysis for fast atom bombardment (FAB), chemical ionization (CI), and electron impact (EI), although most of the probe systems are still run in a batch mode process due to autosampler limitations. As discussed earlier, the use of flow injection with HPLC autosamplers is readily automated and presents the highest and fastest throughput of samples. In pharmaceutical drug discovery where most of the compounds are highly polar and/or water soluble, atmospheric-pressure ionization (API) methods [44–46] are usually used (both APCI and ESI), although the use of thermospray ionization (TSP) [47] was an early choice before the widespread availability of the API instruments. In a test of 60 typical compounds using ESI, APCI, and TSP prior to purchasing our first system, it was found that APCI afforded the highest success rate with the fewest interpretation problems [44]. Also, our experience with TSP was such that we felt more maintenance was necessary and therefore more downtime would be experienced. We therefore placed the first system in the APCI mode with the idea of occasionally switching the instrument to ESI when necessary. The success of this system was such that switching from one ionization technique to another was too disruptive of the chemists' routine analysis and we continued to analyze the ESI samples in a batch mode overnight. However, in a very short time we purchased a second system for open-access ESI analyses.

The introduction of these fully automated systems may reduce the tedium of routine sample analysis for the mass spectrometrist, but one must be careful to educate the chemist in the proper use of the techniques employed and the interpretation of the data produced. Sample preparation in the proper concentration is important to avoid weak or nonexistent ions for samples in too low a concentration and saturated data from samples too high in concentration. Saturated data is a common problem for the EI, CI, and TSP techniques, while the API methods do not suffer this problem due to the way ions are produced and sampled. But in all of the techniques, overloading of the sample will make source cleanup a more frequent chore. Another danger is that the chemist will take the absence of an ion as the absence of a particular component in the sample when in reality ion suppression phenomena may have caused the ion to be absent. Finally, the chemist must avoid the temptation to assign the relative concentration of components in a sample using the relative ion ratios in the mass spectrum. While some mass spectrometrists have found it necessary to implement a formal and lengthy education process on the use of the open-access systems [45], we have found that simple trial and error on the part of the chemist is the best teacher for their use. We tend to inform new chemists about the concentrations needed for the techniques and then ask them to seek help and advice from a mass spectrometrist if there are any problems in obtaining the data or with data interpretation. Problems with the open-access API instruments have been few and are mainly limited to problems with the autosamplers or with power outages and not with the chemists submitting inappropriate samples.

The use of both APCI and ESI has proven to be complimentary. Certain classes of compounds such as heavily halogenated nucleoside analogs and highly aromatic compounds will run readily on APCI while giving no data or a very weak response on ESI. Other classes of compounds such as peptides, tertiary-butyloxocarbonyl (t-BOC) substituted amines and amides, and aliphatic amines give the opposite results. Both techniques are widely used in the laboratory. However, there are compounds that do not ionize in either API technique because of their low molecular weight or the lack of a protonation site. Particle beam ionization (PB) has proven to be a technique useful for the flow injection analysis of these samples [45].

However useful flow-injection open-access API systems are, there are samples that are prone to ion suppression and do require some chromatographic separation. Therefore, open-access LC/MS has also been instituted in several laboratories [46, 47a] that provide data on complex mixtures in a timely and efficient manner in the same way that flow-injection samples are analyzed. Software has been developed that automatically processes the data using algorithms that detect peaks using either a total ion count threshold, the appearance of a particular ion, ultraviolet (UV) detection, or a combination of these methods. The turn-around time for the LC/MS systems is longer than for flow injection but is self-governing from the standpoint that chemists in a hurry for their data tend to use the flow-injection methods, leaving the LC/MS systems open to those who truly need the technique. Open-access GC/MS systems have also been developed that provide data in both the CI and EI modes. These chromatographic open-access systems provide useful data for the chemist but require more time for analysis and thereby increase the turn-around time for samples.

Fast turn-around time is essential to increase the productivity and efficiency of chemists. The need for a fast turn-around-time for most samples can be best illustrated by the need to monitor the course of reactions. In many cases, the starting material and product both appear as pseudo molecular ions in the flow-injection open-access analysis (either by APCI or ESI) and therefore can be monitored during the course of a reaction. In the best of all possible cases, the ion(s) representing the product(s) will increase and the ion(s) representing the starting material(s) will decrease, even to the point of disappearance. It is important to emphasize that the relative peak heights in the mass spectrum do not necessarily represent the true relative concentrations of the components. But if the reaction is monitored over time, the relative intensities of these ions should change, indicating that the reaction is still progressing. In most cases, the reaction is monitored until a steady state is reached (i.e., the relative intensity of the product and starting material molecular ions does not change) or until unwanted by-products begin to appear. Depending on the reaction, monitoring is done in half-hour to hour to multihour increments. Therefore increasing the turn-around time to 20–30 min would increase the wait time for sample analysis to a point where reaction monitoring would be infeasible in most instances.

To gain an appreciation for the impact that the open-access concept has had on mass spectrometry, we need only look at the number of samples submitted prior to its implementation. During the time when samples were submitted and analyzed in the traditional manner, approximately 70–80 samples per week were analyzed by two mass spectrometrists using a combination of FAB, EI, and CI techniques for approximately 70 chemists. Using the batch processing method and API techniques, this number doubled to about 150 per week. But with the institution of open-access methods, the number of samples has averaged about 300 per week—again for 70 chemists. Figure 11 shows a graph of the number of samples run per week on the open-access flow-injection systems. With traditional submission and analysis procedures this level of samples would not be maintained. While every sample submitted may not have been necessary to guarantee a reactions success or improve the yield of a product, the overall efficiency and productivity of the laboratory have increased since the inception of open-access mass spectrometry as measured by the yearly increase in the numbers of registered compounds.

C. Combinatorial Chemistry

Combinatorial chemistry is being driven by the need to produce large numbers of compounds to test against an expanding set of target enzymes. The goal is to find compounds that have activity against an enzyme target, or in other terms

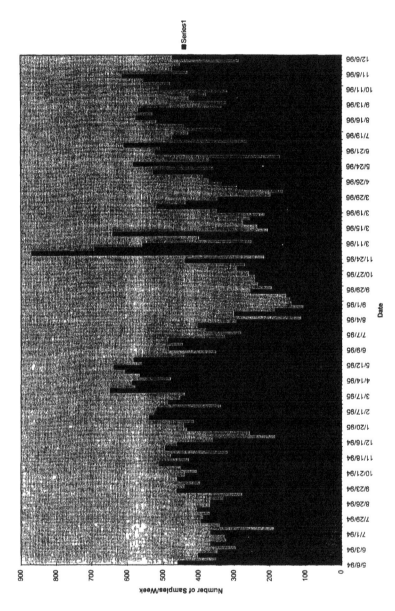

Figure 11 Number of samples analyzed by APCI from May 1994 to December 1996. The large dip in total number of samples in August 1995 occurred during a reorganization and move that disrupted work in the laboratories. Other weekly dips are due to holidays and vacation periods.

a "hit" against the target. The structure of these "hits" is further developed and refined to produce "lead" compounds. From this point the "leads" are developed and refined to produce the one or two compounds that have a pharmacologically significant action toward a disease state and that can be carried on to the testing and approval process required of new drugs. Tens of thousands of compounds are produced in libraries in microgram to milligram amounts by combinatorial methods for testing at the "hits" stage of the process, while several hundred to a thousand compounds are produced in milligram to gram amounts at the "leads" stage. The methods used to produce these compounds are are varied and wide-ranging as the compounds they produce. Several approaches are being used to produce the large numbers of compounds needed for testing at both the "hits" and the "leads" stages of development. Automated machines to run reactions have been and are being developed to produce these compounds in solution and on solid polymer supports. The solution-produced compounds can be analyzed in the same manner described earlier for the open-access systems but with sampling techniques adapted to the large array media used for their production and biological assay. The solid-phase-produced compounds can be cleaved from the solid-phase supports, dissolved in a suitable solvent, and tested and analyzed in the same manner as the solution-produced compounds. However, methods are being developed to both analyze and test these solid-phase compounds without the need for cleavage from the polymer resin.

To produce the chemical libraries, solution-phase and solid-phase synthetic procedures employing the same two- to five-step reaction for each member of the library but with varied reactants. The more reaction steps required to produce a library, the larger is the number of members in the library. In a one-step simple coupling reaction involving only two reactants X and Y with 10 different reactant X's and 10 different reactant Y's, the possible combinations lead to a library of 100 members (10×10). If a second step is implemented that couples each of the products of the first step with a set of 8 reactants Z, then a library with 800 members is produced ($10 \times 10 \times 8$). Libraries whose members contain only one compound of interest, as in the previous example, are called discrete libraries, while those with more than one compound are called pooled libraries. This is achieved by using more than one reactant in each step used to produce the library member. For instance, if three different reactants X and three different reactants Y are used for each member of the one-step library, then nine different possible final products are present for each member of the library. If three different reactants Z are used for each library member in a second step, then the number of possible compounds pooled in each member of the library is 27. The number of possible compounds produced by these methods grows geometrically as the number of steps and/or reactants is increased. The final product or products for each member of the library can be provided for testing, or they can be purified by, for example, recrystallization or chromatography. In most cases the reactions have very few by-products and a purification step is not necessary. In the case of solid-phase chemistry, the final products are cleaved from the support scaffold (beads, crowns, gears, etc.), dried, and supplied for testing. Most libraries, especially at the hit stage, are produced in 96-well microtiter plate formats. A large number of vendors supply automated equipment to handle this plate format. Once a library is manufactured, the presence of the desired product or products must be verified by analytical techniques.

The methods used to analyze the libraries borrow heavily from quantitative analysis and from the open-access methods described previously. The methods are highly automated and can sample from a variety of library formats, although most plates are produced in 96-well formats. The samples are then analyzed by flow-injection methods to produce one mass spectrum of each well or by chromatographic separation techniques (LC/MS, GC/MS) to produce a mass spectrum of each component in the well. The flow-injection methods require 1–2 min per well, while the LC/MS and GC/MS methods may require 7–30 min per well. The analysis of the data is automated to search for a molecular weight expected to be present in each mass spectrum. A report (either graphical or tabular) denotes the presence or absence of the desired compound in a particular well. The total amount of operator intervention during the analysis and data reduction is minimal, just as is the case for open-access mass spectrometry.

Time and resources are the limiting factor in the analytical procedure because large numbers of compounds are being produced. For a 50,000-member discrete library, flow-injection analysis of each well or member of the library would require over 800 hr of instrument time for a sample analysis time of 1 min per well. Two approaches are used to reduce the number of samples analyzed and at the same time assure the integrity of the library. First, extensive analysis of a test series of library members is made during the initial phase of library production. The chemist produces a small subset of the library under varying conditions to optimize the reaction conditions used to produce the library. Analysis of each member of the small subset will provide the chemist with the knowledge that the compounds of interest are produced with a variety of different reactants. Second, when the total library is produced a random number of library members (5–7%) are analyzed by mass spectrometry. If the desired compounds are present, then the chemist can assume that no errors occurred and that the rest of the library contains the desired components. Of course for smaller libraries, especially at the lead optimization stage, all members of the library are analyzed. Finally, whenever a hit is made against a target enzyme, an appropriate mass spectrometric technique is used to analyze the hit to verify the identity and structure of the active library member.

Solid-phase chemistry offers many advantages in the production of libraries, especially in the ability to wash away reactants and nonbound byproducts. Through the use of the mixed-reactant and portioning–mixing [48] methods, large numbers of compounds can be produced on polymeric beads, with each bead containing a single final product. These advantages can be offset because of handling and purification difficulties in cleaving the compounds from the solid support resin. However, strategies have been developed to test the compound while still bound to the solid support using the binding affinity of a compound to a target enzyme. Mass spectrometry can then be used to identify the active component after a particular bead has been identified as containing an active component. The active component is cleaved from the bead and identified using a variety of ionization technique. Direct analysis of the active component on the bead has been reported using MALDI-TOF/MS [49,50]. Other researchers have developed a strategy of encoding each bead with a tag compound that describes the synthetic history of the active component on the bead [51]. The tag is identified after cleavage from the resin or bead by a variety of means depending on its nature [51,52]. This is a particularly useful technique when the active compound binds covalently to the target enzyme or the active compound structure is destroyed or altered during cleavage.

The major failing with these methods lies in the handling of beads prior to analysis. No automated procedures now exist to select the active beads and then deliver them to the analytical instrument for analysis. To circumvent this problem, several researchers have reported the use of immunoaffinity extraction (IAE) columns [53,54] where the target enzyme is attached to a column support. The compounds from a library are passed through the IAE column and those that are active are bound to the enzyme through noncovalent (or affinity) interactions. The active compounds are then eluted from the IAE column and analyzed by LC/MS [54]. This is a highly automated system and can analyze and test large numbers of compounds in one step without the need to separate mixtures. Its main disadvantage is obtaining and maintaining IAE columns containing the target enzyme. Another researcher [55] has suggested using the capabilities of FT-MS to determine which compounds bind to a target enzyme in solution without the use of any solid support for either the enzyme or compounds of interest. Mixtures of the compounds and the enzyme are sampled and analyzed by electrospray FT-MS. Complexes of enzyme and active compound can then be identified using a combination of MS/MS and MS^n techniques.

IV. MASS SPECTROMETRY APPLIED TO DRUG METABOLISM STUDIES IN THE DRUG DISCOVERY PROCESS

A. Quantitative Mass Spectrometry for Support of Pharmacokinetic Studies: Introduction

After a molecule has been identified as a "hit" for a biological target, its affinity for the target protein is optimized through structure–activity relationship (SAR) studies. These studies often yield several different molecules, each of which has sufficiently high biological activity in vitro to be a drug candidate. Identification of the lead molecule is then made on the basis of its in vivo behavior, principally its bioavailability, half-life, and therapeutic index. Bioavailability is the extent to which the compound reaches general circulation from the administered dose form, usually expressed as a percentage of the administered dose. The half-life of a compound is the time required for 50% of the peak plasma concentration of the compound to be removed by excretion or biotransformation (metabolism). The therapeutic index expresses the selectivity of the compound between the desired therapeutic activity and the undesired toxic side effects. From a clinical perspective, the most promising drug is one with a high therapeutic index, and whose oral bioavailability and biological half-life permit an oral dose formulation with once-a-day dosing. Thus a lead compound should have a high oral bioavailability (>30%) and a relatively long biological half-life (>4 h).

Essential to all of these determinations is an accurate and fast analytical method for measuring the concentration of the compound in serum or plasma. Measurement of the concentration as a function of time after dosing will yield the pharmacokinetics of the compound. Determination of the pharmacokinetics after both intravenous (iv) and oral administration will permit the determination of its oral bioavailability and biological half-life. These data will allow the determination of the proper dose level, which is a level that leads to a sufficiently high plasma concentration to exceed the minimal effective response level but does not exceed the minimal toxic dose level. Determination of these three parameters will allow the medicinal chemists to further optimize the structure–activity relationships (SARs) of the drug.

In the not-distant past the standard method for measuring the circulating concentration of a drug used LC with UV or fluorescence detection to quantitate the levels of the drug in serum or plasma. Protein precipitation has been the cleanup method of choice for these methods due to its simplicity and (usually) high recovery. However, this high recovery is due in part to the lack of selectivity of the method, leading to numerous endogenous components being retained in the sample. These endogenous components must be chromatographically separated from the drug in order for successful quantitation using conventional detection. Moreover, the production of metabolites produces additional "interferences" that must be chromatographically separated from the drug compound for accurate quantitation. Complicating the chromatographic separation is the variety of animal species used for pharmacokinetic testing, each of which can have different endogenous compounds and different metabolites, and which requires the chromatographic method to be revalidated and, if necessary, redeveloped for each species. Given adequate time (2 weeks to several months), skilled chromatographers can produce analytical methods giving the desired separation for a given species. However, in the fast-paced, competitive world of drug discovery, this amount of time is no longer available.

Professor Fred McLafferty has defined an analytical method as being comprised of four "S" components: sensitivity, selectivity, speed, and $. This definition certainly holds true in the drug discovery process. As medicines become more potent, the dose becomes lower, and this pushes the required detection limit lower. In order to accurately describe the pharmacokinetics of a compound, one needs to be able to accurately quantitate the drug across 5 half-lives; thus, if the maximum circulating of the drug is 10 ng/ml, at the end of the fifth half-life the concentration would be 0.31 ng/ml. Since the majority of discovery pharmacokinetic work is performed using rodents, the maximum sample volume that can be obtained at each time point is 0.1 ml, and thus the analytical method must be capable of accurately quantitating the compound at a level of 31 pg on-column (assuming 100% of the sample is injected; typically only one-half of the sample is injected, reducing the limit of quantitation to 16 pg on-column). Because of the need to have an analytical method that can be used in multiple species, the method should have high selectivity, ideally only responding to the compound of interest and an internal standard. This high degree of selectivity would also permit the analysis time to be very short (1–3 min), and thus the selectivity of the method directly impacts the speed of the analysis. As pharmaceutical companies come under ever-increasing pressure to discover and develop drugs in an expedient manner, the speed of the analytical method becomes ever more important. Moreover, the selectivity directly impacts on not only the analysis time but also the time required to develop the method. As the selectivity increases, the time required to develop the method decreases: a very important aspect of methods with high selectivity. In the early stages of a project, many different compounds must be screened, and the time required to develop methods using UV detectors often is far longer than the time the method is actually used to quantitate samples.

The final "S" of analysis is $, or the cost of the method on a per-sample basis. The most cost-effective methods must be employed; however, the typical patent-protected drug on the market can be expected to earn at least $100,000,000 per year in order to justify the cost of drug discovery and development. The limited patent life of a drug means that getting the drug to the market quickly is more important than minimizing the analysis costs. The more quickly the pharmacokinetic information can be returned to the medicinal chemists, the more quickly the SAR can be optimized and the drug moved from discovery into development.

B. Application of "Fast" LC/API/MS/MS to Drug Quantitation

In 1986 Covey, Lee, and Henion [56] demonstrated an analytical method that has proven to meet the requirements of a quantitative method appropriate for drug discovery and development, and that has become the method of choice in the pharmaceutical industry. The method is fast LC/API/MS/MS (liquid chromatography coupled with tandem mass spectrometry using atmospheric-pressure ionization interfaces). The high selectivity of tandem mass spectrometry had previously been utilized for the analysis of crude mixtures of biological samples [57,58]. The innovation of Covey and Henion was their use of fast LC columns and a atmospheric-pressure chemical ionization (heated nebulizer) interface in conjunction with tandem mass spectrometry. The fast LC columns used were short (33 mm long) columns packed with 3-μm stationary phase particles. By using 3-μm particles rather than the standard 5-μm particles, the analysis time is decreased, because the optimal linear mobile phase velocity is inversely proportional to the particle diameter [59]. This also leads to a significant increase in the backpressure of the column (per unit length) because the optimum flow rate is inversely proportional to the cube of the particle diameter. The separation efficiency of a column increases as the particle diameter decreases, due to a reduction in the height equivalent of a theoretical plate of the separation caused by a decrease in the resistance to mass transfer of the analyte in the mobile phase. Finally, the use of 33-mm columns rather than the standard 250-mm columns minimizes the increase in mobile-phase backpressure caused by the smaller particles, but, more importantly, it decreases analysis time by a factor of 7.5, along with a corresponding decrease in resolution between analytes. This loss is resolution is only partially offset by the increase in separation efficiency due to the use of the smaller particles.

It is the extraordinarily high selectivity of tandem mass spectrometry that permits the use of these fast LC columns for the quantitation of drugs from biological tissues and fluids. Operation of a triple quadrupole in multiple reaction monitoring mode (for the analyte and internal standard) with unit mass resolution on Q1 and Q3 leads to, in most cases, extraordinarily low or no response from any compounds other than the analyte of interest. In essence, the mass spectrometer is being used for the bulk of the "separation" process, and the LC is serving in no small part as an automated sample introduction technique. Furthermore, the high selectivity resulting from the decrease in chemical noise leads to very high signal-to-noise ratios and thus very low detection limits for the analytes of interest. In addition, this high selectivity permits the use of crude sample preparation techniques, such as protein precipitation, as the sample cleanup method, and the LC column is providing enough separation to separate the analyte from the void volume of the column, where the salts will elute. Thus, sample cleanup time is minimized by use of the high-throughput protein precipitation technique. Development time for the sample cleanup method is reduced to zero, and overall method development time is reduced from weeks or months to less than 1 day. This dramatic decrease in method development time is a very important advantage to LC/MS/MS quantitation, because it allows the analytical chemists to develop methods in a time frame matched to the medicinal chemists needs.

Atmospheric-pressure ionization interfaces (APCI and electrospray) are ideally suited for interfacing with tandem mass spectrometry. Both produce pseudo-molecular ions, typically $[M + H]^+$, with very little, if any, fragmentation. This places all of the ion current from the analyte into only one ion—an ideal situation for tandem mass spectrometry experiments. The rationale for choosing one interface over the other is most often based on the solution-phase proton affinity of the analyte—electrospray requires preprotonated (or deprotonated) ions in solution and thus works best with fairly basic (or acidic) compounds, whereas APCI is also well suited for the analysis of more neutral compounds. In general, APCI has proven to be more rugged than electrospray, requiring less stringent sample cleanup, and APCI is more easily interfaced with the mobile phase flow rates of conventional chromatography columns (1 ml/min), although recent advances in electrospray interface design have made this flow rate compatibility issue less of a concern. One important difference between the two API interfaces is that electrospray is a concentration-dependent detector, whereas APCI is a mass-flux-sensitive detector. The concentration-dependent response of electrospray leads to its optimal use with small inner diameter LC columns, because the concentration of an analyte eluting from an LC column is inversely dependent on the square of the column inner diameter. Thus, an analyte eluting from a 1-mm (microbore) LC column will have a concentration 21-fold higher than the same injection of analyte on a 4.6-mm column. Covey [60] has shown that the response of an ion-spray interface (pneumatically assisted electrospray) on an API-III triple-quadrupole mass spectrometer does follow this predicted concentration dependence, giving a 21-fold higher response to an injection of 2 pmol of two peptides using a 1-mm-ID column compared with a 4.6-mm column. Note that the use of a 180-μm ID capillary LC column gave the best response in Covey's experiments, 300-fold higher response than the 4.6-mm column. For this reason capillary columns have received widespread use in qualitative mass spectrometry experiments. While capillary columns have been successfully applied in some quantitative works, in general they have not proven to be sufficiently rugged for routine use in the quantitative analysis of drugs from biological tissues and fluids. One-millimeter ID columns have proven to be sufficiently rugged for this application, and they have been widely used for such analysis.

C. Quantitation of a Modified 1,5-Benzodiazapine from Rat Serum Samples from an In Vivo Pharmacokinetic Study

The objective of this work [61] was to determine the pharmacokinetic behavior of a modified 1,5-benzodiazepine in rats. The rats were dosed at 4 mg/kg body weight via iv bolus dosing through a jugular-vein cannula, using propylene glycol as the drug vehicle. The circulating levels of the drug were sampled by taking a 0.1-ml aliquot of blood (10 time points over 24 hr). Since this work was part of a long-term study, a solid-phase extraction (SPE) method had been developed and automated on a sample preparation robot. This SPE method used C_{18} solid-phase extraction columns and a fluoridated internal standard that gave a precursor ion and a product ion 18 m/z units different from the corresponding ions of the target analyte.

A Sciex API-III triple-quadrupole mass spectrometer was used with a pneumatically assisted electrospray source. Multiple reaction monitoring was used to monitor the precursor ion/product ions of the analyte and the internal standard. A "fast" LC column was used (1 mm ID×50 mm long column packed with 3 μm Hypersil BDS-C_1^8 particles). The mobile phase was 60% acetonitrile in 2 mM ammonium acetate (pH 4.5) at a flow rate of 80 μl/min. Under these conditions the analyte eluted with a retention time of 2.5 min and a k' of 4. The overall sample throughput was 15 samples per hour.

Mobile phase standards over the range of 1.22–5000 ng/ml were analyzed in triplicate, and spiked serum standards over the range of 1–7500 ng/ml were analyzed in duplicate (before and after approximately 50 in vivo rat serum samples). Note that in order to extend the dynamic range of the analysis, the upper level standards were diluted to preclude saturation of the electrospray source (5-fold for the 625- and 1250-ng/ml standards, 10-fold for the 1000- and 5000-ng/ml standards). A comparison of normal resolution (unit mass) and open resolution (4 m/z peak width at half height) (Fig. 12) conditions on the triple quadrupole showed opening the resolution gave a very significant increase in absolute signal, and a 4-fold improvement in the limit of quantitation (LOQ). The LOQ is defined as the lowest level standard at which there is less than 20% difference between the experimental value and the value calculated from the response function (the calibration curve, Fig. 13). For these mobile-phase standards a first-order polynomial fit was used across the range of 2.44 ng/ml to 5000 ng/ml using $1/(X^2)$ weighting. This gave a correlation coefficient of 0.996, which exceeds the value required for Good Laboratory Practices (GLP)

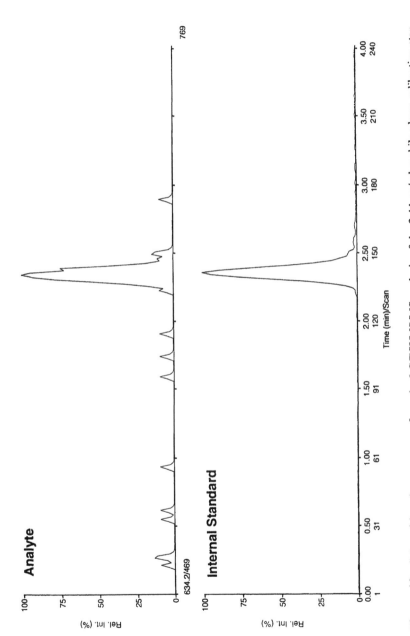

Figure 12 Selected ion chromatograms from the LC/ESI/MS/MS analysis of the 2.44-ng/ml mobile-phase calibration standard (12.2 pg injected on-column) corresponding to the limit of quantitation for the drug under "open-resolution" conditions.

Moseley et al.

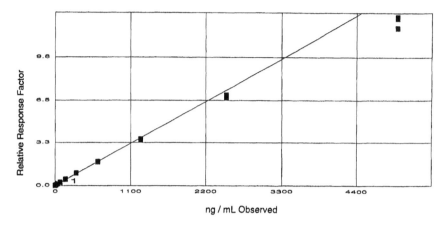

ng / mL Observed

Figure 13 Calibration curve from the quantitation of mobile phase standards across the concentration range of 2.44–5000 ng/ml using LC/ESI/MS/MS. Note that each standard was analyzed and plotted in triplicate. The curve was generated using a first-order polynomial fit with $1/(x^2)$ weighting, giving a correlation coefficient of 0.996.

analysis (0.990). Note that the LOQ of 2.44 ng/ml corresponded to an on-column quantitation limit of 12.2 pg.

When it is necessary to maximize the instrument response to obtain the LOQ required by the pharmacokineticists, open-resolution conditions for LC/MS/MS quantitation on a triple quadrupole may be used. Quantitation of the spiked serum standards and rat serum samples in this study was performed under open-resolution conditions. Note that the use of open-resolution conditions reduces the specificity of the analysis due to near-isobaric interferences, such as could occur from a doubly metabolized form of the drug (e.g., a de-methylated and oxidized metabolite would lose 14 m/z for demethylation, and add 16 m/z for oxidation, giving an overall change of $+2$ m/z). Thus, this "open-resolution" approach must be used with extreme caution. Previous metabolism studies with this particular analyte had shown that there were no interferences due to its metabolites or endogenous compounds under the conditions of this analysis.

The rat serum samples were quantitated in an analysis set where spiked serum samples were run in duplicate, once before and after the actual rat serum samples. These calibration standards were fitted to a first-order polynomial using $1/(X^2)$ weighting, yielding a correlation coefficient of 0.997 across the range of 5.00–7500 ng/ml. The LOQ with these spiked serum standards was found to be 5.0 ng/ml (25 pg on-column). Note that even though the triple quadrupole was operated under open-resolution conditions there were no interferences from endogenous compounds at the 5.00 ng/ml level (Fig. 14). The rat serum sam-

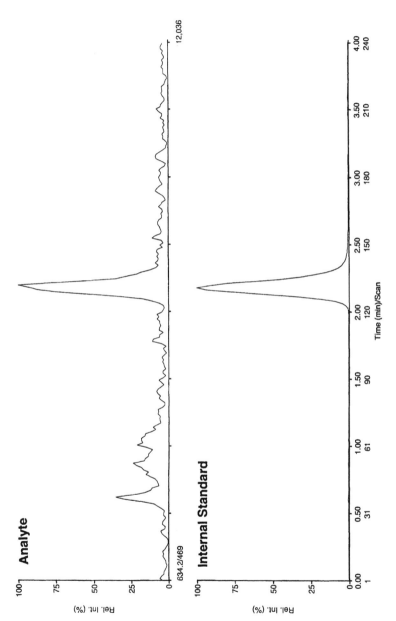

Figure 14 Selected ion chromatograms from the LC/ESI/MS/MS analysis of the 5.00-ng/ml spiked rat serum calibration standard (25 pg injected on-column) corresponding to the limit of quantitation for the drug under "open-resolution" conditions using "real" standards.

ples were quantitated using this curve, and the resulting concentrations of these samples were plotted as a function of time (Fig. 15) to give the pharmacokinetic curve. From this data set the pharmacokinetic parameters were calculated: a maximum circulating concentration of 4674 ng/ml, a half life of 12.2 hr, a clearance of 7.4 ml/min/kg, and a volume of distribution of 4356 ml/kg.

D. Qualitative Identification of Drug Metabolites Using Mass Spectrometry

In the drug discovery process it is not uncommon for a compound that has good in vitro activity to have poor in vivo activity because of a short half-life, due either to rapid excretion or to first-pass metabolism. If the short half-life is due to a high rate of metabolic clearance, identification of the sites of metabolic activity may allow the medicinal chemists to modify the chemical structure to retain the desired biological activity while minimizing the metabolic clearance. There are five principal classes of metabolic biotransformations: oxidation, reduction, conjugation, hydrolysis, and "miscellaneous" [62]. Oxidation is the most common biotransformation, noted by the addition of an oxygen to form a hydroxyl group, carboxylic acid, sulfoxide, or sulfone, or leading to N-, O-, or S- dealkylation, which involves oxidation of the alkyl group. Reduction is a much less common biotransformation. Conjugation (often referred to as Phase II metabolism) is a biotransformation involving the addition of a large polar

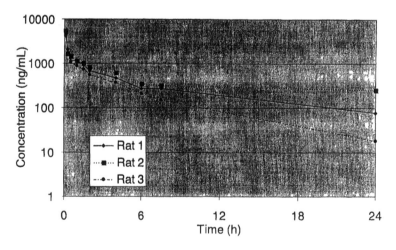

Figure 15 Concentration–time curve from the pharmacokinetic experiment. Note that biological pharmacokinetic variation from rat to rat is not uncommon.

group that is highly water soluble. The conjugated metabolites most commonly observed are glucuronides and sulfates. Hydrolysis is a common biotransformation of amides and esters, and occurs less frequently with phosphates and sulfates.

The structural information inherent in mass spectrometry and tandem mass spectrometry has been exploited for the rapid characterization of drug metabolites from a variety of biological fluids, principally plasma, urine, and bile. In order to expedite the identification of drug metabolites, in vitro incubation of the drug with liver microsomes or liver slices to generate metabolites has become a widespread practice in the pharmaceutical industry. Liver microsomes are the vesicular structures formed from fragments of the smooth endoplasmic reticulum produced by homogenization of liver tissue. While liver microsome incubations are more widely used for metabolism studies, liver slices offer several advantages over microsomes [63]:

1. They permit the study of whole-cell coupled Phase I and Phase II metabolism.
2. The architecture of the tissue and intercommunications between different cell types are maintained; thus, slices accurately mimic normal in vivo metabolism.
3. The system is directly applicable to human tissues: (a) It permits study of human drug metabolism prior to clinical trials, (b) comparisons of human and animal liver slice metabolism permits the identification of the animal species that most closely parallels human metabolism for a particular drug, and (c) small amounts of liver are required, permitting a single human liver to be used for a large number of different experiments.

Slices of transplant-quality livers (for which no matching recipient could be found) are sliced into 250-μm-thick slices, which are then mounted on a stainless steel screen. The screen is bent into a cylinder and placed into a scintillation vial partially filled with culture media spiked with drug. The scintillation vials are placed on a roller that rotates the vial, causing the liver slice to be rotated in and out of the culture media, keeping the slice moist while still supplying oxygen to the cells. Immediately following slicing, all biological functions are depressed, but are recovered within 2 hr, and they maintain physiological viability for at least 24 hr after preparation. Protein synthesis is maintained at levels comparable to the perfused organ, and cytochrome P-450 levels do not decrease over this 24-hr time period.

This in vitro liver slice approach offers the ability to rapidly and inexpensively generate metabolites from animal and human systems in a relatively clean matrix, and characterization of the resultant samples by LC/MS and LC/

MS/MS provides a similarly rapid means of identifying the metabolites. Omeprazole, an inhibitor of gastric acid secretion, was chosen as a test compound to demonstrate the utility of the combined techniques of in vitro liver slice metabolism and LC/ESI/MS/MS identification of the resultant metabolites [62]. Previous MS analysis of in vitro omeprazole metabolites from rat urine [64] had shown that this compound generates a number of both Phase I and Phase II metabolites that have proven to be very difficult to analyze by either direct probe insertion electron ionization (even with derivatization) or online thermospray LC/MS.

The microbore LC/ESI/MS and LC/ESI/MS/MS analyses were performed on a Sciex API-3 equipped with a pneumatically assisted electrospray interface. To assist in the MS/MS data interpretation, constant infusion of a $5 \times 10 - 5$ M solution of omeprazole was used to generate an MS/MS reference spectrum comprised of 12 daughter ions. Use of a high orifice bias voltage (100 V) generated the same fragments in the ion source, permitting the acquisition of MS/MS spectra of the eight most abundant ions observed in the MS/MS spectrum of omeprazole, in effect yielding MS/MS/MS spectra of omeprazole.

Human liver slices were incubated with 156 μM omeprazole for 4 hr. Solid-phase extraction was used to remove salts and concentrate the drug and its metabolites. Five-microliter aliquots of the $10\times$ concentrated samples were injected onto a Hypersil BDS C18 LC column (1 mm ID, 250 mm long), corresponding to an injection of 12 nmol or 4 μg omeprazole. The drug and its metabolites were eluted with a gradient elution program of 5% to 80% (over 75 min) acetonitrile in 2 mM ammonium acetate (pH 6.8), with a mobile-phase flow rate of 40 μl/min (Fig. 16). Unmetabolized omeprazole was the major component of the sample; however, numerous metabolites were also observed, including both Phase I (oxidation to alcohols, oxidation to carboxylic acids, oxidation to sulfone, O-dealkylation, reduction to sulfide) and Phase II (conjugation with glucuronide) metabolites. LC/ESI/MS/MS spectra permitted the identification of most metabolites, including the differentiation of isobaric metabolites. For example, a selected ion trace for the [M + H + O] + ions of omeprazole metabolites (Fig. 17) reveals three major mono-oxidized metabolites. The alcohol metabolites (Fig. 18) and the sulfone metabolite (Fig. 19) are clearly distinguished on the basis of their LC/ESI/MS/MS spectra. While the majority of the previously reported rat metabolites were found in this study to also be human metabolites, several rat metabolites were not found, and several of the human metabolites were not found to be rat metabolites. Moreover, this current work identified all of the human metabolites that have been found to be reported in the literature, as well as several additional human metabolites not yet reported (Table 1).

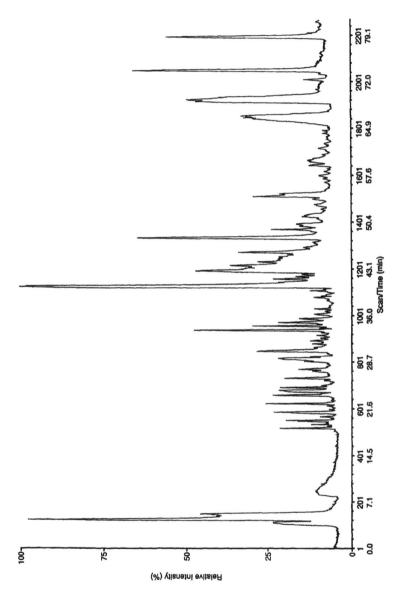

Figure 16 Total ion chromatogram from the LC/ESI/MS analysis of media from the incubation of 156 μM omeprazole with human liver slices for 4 hr. A 1×250 mm Hypersil BDS-C_{18} column was used with a gradient of 5–80% (75 min) acetonitrile in 2 mM ammonium acetate (pH 6.8).

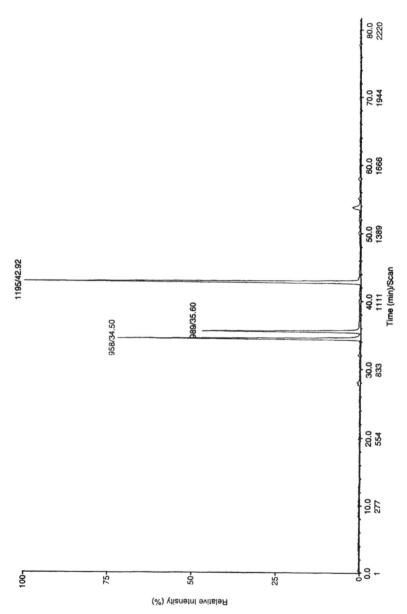

Figure 17 Selected ion chromatogram of the mono-oxidized metabolites from the LC/ESI/MS analysis of media from the incubation of 156 μM omeprazole with human liver slices for 4 hr.

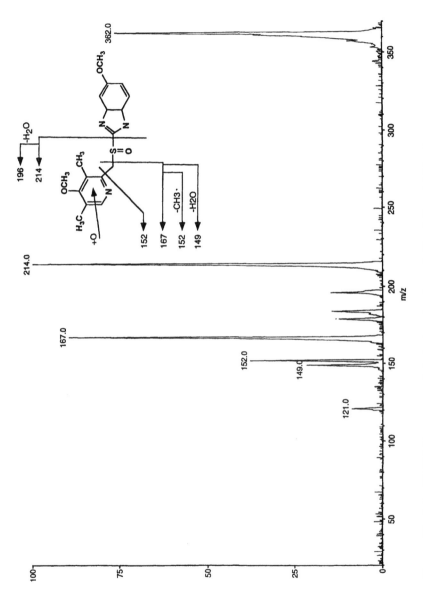

Figure 18 LC/ESI/MS/MS spectrum of the hydroxylated metabolite eluting at 34.5 min.

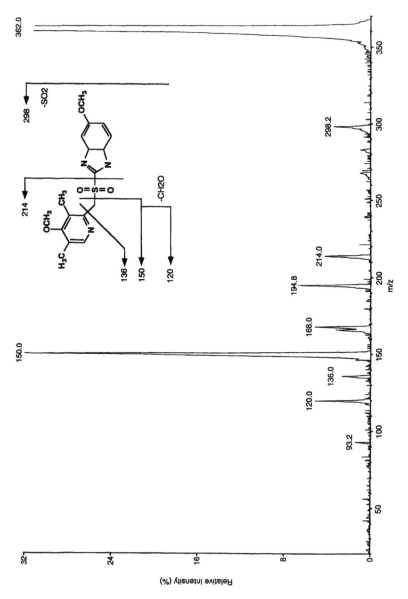

Figure 19 LC/ESI/MS/MS spectrum of the sulfone metabolite eluting at 42.9 min.

Table 1 Summary of Metabolites Found in the Human Liver Slice Study and Comparison with Known Omeprazole Metabolites

	Metabolites of Omeprazole		
	Rat urine in vivo LC/MS, no LC/MS/MS	Human in vivo	Human liver slices in vitro (this work)
Phase I metabolites			
Addition of one oxygen	Imidazole		Not found
	Pyridine	Pyridine	Pyridine-#1
	Sulfone	Sulfone	Sulfone
	Not found		Pyridine-#2
Addition of carboxylic acid	Pyridine	Pyridine	Pyridine
Loss of one oxygen	Sulfide	Sulfide	Sulfide
O-demethylation	Imidazole		Imidazole
	Pyridine		Pyridine (a)
Phase II metabolites			
Glucuronide	Imidazole (with demethylation)		Imidazole (with demethylation)
	Imidazole (with oxidation)		Not found
Sulfate	Pyridine-#1 (with demethylation)		Pyridine-#1 (a) (with demethylation)
	Pyridine-#2 (with oxidation)		Pyridine-#2 (a) (with oxidation)
Glutathione	Fragment from conjugate		Fragment from conjugate

V. CONCLUSIONS

Revolutionary changes in mass spectrometry during the past decade, particularly the introduction of the robust, broadly applicable ionization techniques of ESI, APCI, and MALDI, have made mass spectrometry an essential tool in every step of the drug discovery process. The first step in this process, characterization of the proteins and other biological molecules that define the phenotype of a disease state, has been accelerated and focused by the availability of rapid and exquisitely sensitive mass spectrometric techniques that have allowed the precise characterization of subtle changes in very large molecules. The impact of these advances in mass spectrometry instrumentation has been extended to the characterization of "hit" compounds generated by medicinal chemistry or combinatorial chemistry methods, through the implementation of simple-to-use, automated systems that provide immediate answers in real time for the optimization of chemistry and selection of leads, integrating mass spectrometry into the daily routine of synthetic chemistry. Finally, mass spectrometry has become the most versatile and powerful analytical technology for the optimization of the in vivo pharmacokinetic activity of lead molecules. The rapid expansion of the versatility of mass spectrometry has made it applicable to an exponentially larger range of problems in biology and chemistry. Not surprisingly, as a result, mass spectrometry has become the most important analytical tool in the drug discovery process.

REFERENCES

1. TC Burnette, JA Harrington, JE Reardon, BM Merrill, P de Miranda. Purification and characterization of a rat liver enzyme that hydrolyzes valacyclovir, the L-valyl ester prodrug of acyclovir. J Biol Chem 270:15827–15831, 1995.
2. E Watson, B Shah, R DePrince, RW Hendren, R Nelson. Biotechniques 16:278–281, 1994.
3. M Wilm, A Shevchenko, T Houthaeve, S Breit, L Schweigerer, T Fotsis, M Mann. Nature 379:466–469, 1996.
4. M Wilm, M Mann. Anal Chem 68:1–8, 1996.
5. M Mann, M Wilm. Anal Chem 66:4390–4399, 1994.
6. JS Andersen, B Svensson, P Roepstorff. Nature Biotechnol 14:449–457, 1996.
7. RN Wine, WW Ku, L Li, RE Chapin. Biol Reprod 56:439–446, 1997.
8. A Shevchenko, M Wilm, O Vorm, M Mann. Anal Chem 68:850–858, 1996.
9. W Wilm, G Neubauer, M Mann. Anal Chem 68:527–533, 1996.
10. K Biemann. Methods Enzymol 193:886–887, 1990.
11. SA Carr, MJ Huddleston, RS Annan. Anal Biochem 239:180–192, 1996.
12. JR Yates, JK Eng, AL McCormack. Anal Chem 67:3202–3210, 1995.
13. A Shevchenko, ON Jensen, AV Podtelejnikov, F Sagliocco, M Wilm, O Vorm, P Mortensen, A Shevchenko, H Boucherie, M Mann. Proc Natl Acad Sci USA 93:14440–14445.
14. CL Brummel, JC Vickerman, SA Carr, ME Hemling, GD Roberts, W Johnson, J Weinstock, D Gaitanopoulos, SJ Benkovic, N Winograd. Anal Chem 68:237–242, 1996.
15. DJT Porter, BM Merrill, SA Short. Identification of the active site nucleophile in nucleoside 2-deoxyribosyltransferase as glutamic acid 98. J Biol Chem 270:15551–15556, 1995.
16. E Delorme, T Lorenzini, J Giffin, F Martin, F Jacobsen, T Boone, S Elliott. Biochemistry 31:9871–9876, 1992.
17. NR Thotakura, RK Desai, LG Bates, ES Cole, BM Pratt, BD Weintraub. Endocrinology 128:341–348, 1991.
18. DM Sheeley, BM Merrill, LCE Taylor. Characterization of monoclonal antibody

glycosylation: Comparison of expression systems and identification of terminal alpha-linked galactose. Anal Biochem 247:102–110, 1997.

19. FM Rudd, RJ Leatherbarrow, TW Rademacher, RA Dwek. Diversification of the IgG molecule by oligosaccharides. Mol Immunol 28:1369–1378, 1991.

20. TW Rademacher, RB Parekh, RA Dwek. Ann Rev Biochem 57:785–838, 1988.

21. RB Parekh, TP Patel. Comparing the glycosylation patterns of recombinant glycoproteins. TIB Tech 10:276–280, 1992.

22. DA Cumming. Glycobiology 1:115–130, 1991.

23. CE Costello, JE Vath. In: J McCloskey, ed. vol 193. San Diego: Academic Press, 1990, pp 738–768.

24. RB Reinhold, VN Reinhold. 42nd ASMS Conference on Mass Spectrometry and Allied Topics, ASMS, Chicago, IL, 1994, pp 1156a–b.

25. U Galili. Interaction of the natural anti-Gal antibody with α-galactosyl epitopes: A major obstacle for xenotransplantation in humans. Immunol Today 14:480–482, 1993.

26. CAK Borrebaeck, A-C Malmborg, M Ohlin. Does endogenous glycosylation prevent the use of mouse monoclonal antibodies as cancer therapeutics? Immunol Today 14:477–479, 1993.

27. RG Keck. The use of t-butyl hydroperoxide as a probe for methionine oxidation in proteins. Anal Biochem 236:56–62, 1996.

28. SK Chowdhury, J Eshraghi, H Wolfe, D Forde, AG Hlavac, D Johnston. Mass spectrometric identification of amino acid transformations during oxidation of peptides and proteins: Modifications of methionine and tyrosine. Anal Chem 67:390–398, 1995.

29. DT Liu. Deamidation: A source of microheterogeneity in pharmaceutical proteins. TIB Tech 10:364–369, 1992.

30. R Bischoff, P Lepage, M Jaquinod, G Cauet, M Acker-Klein, D Clesse, M Laporte, A Bayol, A Van Dorsselaer, C Roitsch. Sequence-specific deamidation: Isolation and biochemical characterization of succinimide intermediates of recombinant hirudin. Biochemistry 32:725–734, 1993.

31. J Cacia, R Keck, LG Presta, J Frenz. Isomerization of an aspartic acid residue in the complementarity—Determining regions of a recombinant antibody to human IgE: Identification and effect on binding affinity. Biochemistry 35:1897–1903, 1996.

32. KF Geoghegan, CA Strick, S Guhan, ME Kelly, AJ Lanzetti, KE Cole, SB Jones, DA Cole, KJ Rosnack, RM Guinn, AR Goulet, T-P I, LW Blocker, DW Melvin, JA Funes. In: D Marshak, ed. Techniques in protein chemistry VII. San Diego: Academic Press, 1996, pp 193–200.

33. FE Ross, T Zamborelli, AC Herman, C-H Yeh, NI Tedeschi, ES Luedke. In: D Marshak, ed. Techniques in protein chemistry VII. San Diego: Academic Press, 1996, pp 201–208.

34. M Field, D Papac, A Jones. The use of high-performance anion-exchange chromatography and matrix-assisted laser desorption/ionization time-of-flight mass spectrometry to monitor and identify oligosaccharide degradation. Anal Biochem 239:92–98, 1996.

35. WH Moos. Inter. Symp. Lab. Auto. Robot. Proc. (ISLAR '96), Boston, 1996, pp 1–14.
36. KL Busch, GL Glish, SA McLuckey. Mass spectrometry/mass spectrometry: Techniques and applications of tandem mass spectrometry. New York: VCH, 1988.
37. A Benninghoven, FG Rudenauer, HW Werner. Secondary ion mass spectrometry: Basic concepts, instrumental aspects, applications and trends. New York: John Wiley and Sons, 1987.
38. M Levenberg. Digging out with a robot. J Autom Chem 18:149–152, 1996.
39. G Donato. Proceedings of the 41st Annual Conference on Mass Spectrometry and Allied Topics, San Francisco, CA, 1993, p 1104a.
40. S Mueller, P Fruehan, S Spanton. Proceedings of the 38th Annual Conference on Mass Spectrometry and Allied Topics, Tucson, AZ, 1990, p 451.
41. C Lindberg, J Paulson, A Blomqvist. J Chromatogr 554:215–226, 1991.
42. GS Kath, WJ McKeel, JL Smith, JM Liesch. Automatic sample loader for LKB model 9000 mass spectrometer. Rev Sci Instrum 57:3114, 1986.
43. MJ Hayward, JT Snodgrass, ML Thomson. Rapid Commun Mass Spectrom 7:85–91, 1993.
44. LCE Taylor, RL Johnson, R Raso. J Am Soc Mass Spectrom 6:387–393, 1995.
45. FS Pullen, GL Perkins, KI Burton, RS Ware, MS Teague, JP Kiplinger. Putting mass spectrometry in the hands of the end user. J Am Soc Mass Spectrom 6:394–399, 1995.
46. RC Spreen, LM Schaffter. Open access MS: A walk-up MS service. Anal Chem 68:414A–419A, 1996.
47. FS Pullen, DS Richards. Automated liquid chromatography/mass spectrometry for chromatographers. Rapid Commun Mass Spectrom 9:188–190, 1995.
47a. DV Brown, M Dalton, FS Pullen, GL Perkins, D Richards. An automated, open-access service to synthetic chemists: thermospray mass spectrometry. Rapid Commun Mass Spectrom 8:632–636, 1994.
48. A Furka. History of combinatorial chemistry. Drug Dev Res 36:1–12, 1995.
49. BJ Egner, GJ Langley, M Bradley. Solid phase chemistry: Direct monitoring by matrix-assisted laser desorption/ionization time of flight mass spectrometry. A tool for combinatorial chemistry. J Org Chem 60:2652–2653, 1995.
50. NJ Haskins, DJ Hunter, AJ Organ, SS Rahman, C Thom. Combinatorial chemistry: Direct analysis of bead surface associated materials. Rapid Commun Mass Spectrom 9:1437–1440, 1995.
51. S Brenner, RA Lerner. Encoded combinatorial chemistry. Proc Natl Acad Sci USA 89:5381–5383.
52. GM Geysen, RH Meloen, SJ Barteling. Use of peptide synthesis to probe viral antigens for epitopes to a resolution of a single amino acid. Proc Natl Acad Sci USA 81:3998–4002, 1986.
53. ML Nedved, S Habibi-Goudarzi, B Ganem, JD Henion. Characterization of benzodiazepine "combinatorial" chemical libraries by online immunoaffinity extraction, coupled column HPLC-ion spray mass spectrometry-tandem mass spectrometry. Anal Chem 68:4228–4236, 1996.
54. DB Kassel, TG Consler, M Shalaby, P Sekhri, N Gordon, T Nadler. Tech Protein Chem VI:39–46, 1995.

55. JE Bruce, GA Anderson, R Chen, X Cheng, DC Gale, SA Hofstadler, BL Schwartz, RD Smith. Bio-affinity characterization mass spectrometry. Rapid Commun Mass Spectrom 9:644–650, 1995.

56. TR Covey, ED Lee, JD Henion. High speed liquid chromatography/tandem mass spectrometry for the determination of drugs in biological samples. Anal Chem 58:2453, 1986.

57. RJ Perchalski, RA Yost, BJ Wilder. Anal Chem 54:1466, 1982.

58. HO Brotherton, RA Yost. Anal Chem 55:549, 1983.

59. LR Snyder, JJ Kirkland. Introduction to modern liquid chromatography. New York: John Wiley and Sons, 1979.

60. TR Covey. PESciex Application Note. 1993.

61. MA Moseley, T Tippin, R Hart, T Lloyd, R St. Claire III, K Halm, J Schwartz, A Land, I Jardine. Proceedings of the 43rd ASMS Conference on Mass Spectrometry and Allied Topics, Atlanta, GA, 1995.

62. MA Moseley, JS Walsh, JE Patanella, KA Halm, SE Unger. Proceedings of the 41st ASMS Conference on Mass Spectrometry and Allied Topics, San Francisco, CA, 1993.

63. PF Smith, G Krack, RL McKee, DG Johnson, AJ Gandolfi, VJ Hruby, CL Krumdieck, K Brendal. In Vitro Cell Dev Biol 22:706, 1986.

64. L Weidolf, TR Covey. Rapid Commun Mass Spectrom 6:192, 1992.

9

Analysis of Glycoproteins

Ron Orlando and Yi Yang*
University of Georgia, Athens, Georgia

We have attempted to write this chapter as an instructional text based on our general approach for analyzing glycoproteins. We have also tried to highlight some of the typical problems encountered, but rarely published, in the mass spectrometric analysis of these samples. Because complete books could be written on the structural characterization of glycoproteins, we have limited our fo-

Current affiliation: SmithKline Beecham Pharmaceuticals, King of Prussia, Pennsylvania.

cus to mammalian glycoproteins and their "typical" carbohydrate side chains. We therefore have not attempted to include every situation that could be encountered when analyzing glycoproteins, nor every mass spectrometric approach that has been used for their analysis.

I. INTRODUCTION

A. The Function of Carbohydrate Side Chains in Glycoproteins

The attachment of a carbohydrate to a protein (glycosylation) is one of the most common posttranslational modifications encountered in eukaryotic systems [1]. It has been estimated that 60–90% of all mammalian proteins are glycosylated at some point during their existence [1]. These carbohydrate chains play critical roles in numerous biological systems [1–3]. Often these carbohydrate side chains play a direct physiological role in a glycoprotein's biological activity, while in other cases they maintain physical properties [2,3]. The fact that carbohydrate side chains may also play a role in the immune response to glycoproteins increases the difficulty of obtaining glycoprotein therapeutics that do not elicit an adverse immunological response [1–4]. Consequently, structural information concerning the carbohydrate chains of glycoprotein therapeutics is often required for regulatory review of new therapeutics entering the market [1–4].

A major biological function of a glycoprotein's carbohydrate side chains is to serve as recognition markers for cell–cell and cell–molecule interactions. In this role, glycoproteins are fundamental to many important biological processes, including fertilization, immune response, viral replication, parasitic infection, cell growth, cell–cell adhesion, degradation of blood clots, and inflammation [1,5,6]. For example, sperm/egg recognition in mammals occurs when specific proteins in the sperm bind to certain carbohydrates present on the egg surface [7]. Carbohydrate side chains also act as receptors for viruses, bacteria, and parasites. Certain proteins found on the cell surface of the common cold virus (influenza) recognize specific types of sialic acid [8], a sugar commonly found on the glycoproteins that line the respiratory tracts of mammals. This interaction is thought to be responsible for an early step in the infection process, and thus, sialic acid analogs offer potential cures for the common cold [8]. The carbohydrate chains of glycoproteins are also involved in cell–cell interactions. The best documented example to date is that of the selectin family of receptor proteins. Selectins mediate the adhesion of leukocytes to endothelial cells (L-selectins), the recognition of leukocytes by stimulated or wounded endothelium (E-selectins), and the interactions of activated platelets or endothelium with leukocytes (P-selectins) [9,10].

Carbohydrate side chains may also have profound effects on the physico-chemical properties of glycoproteins. Previous studies by NMR spectroscopy, x-ray crystallography, and molecular dynamics simulations have clearly shown that the large size of oligosaccharides may allow them to cover functionally important areas of proteins, to modulate the interactions of glycoconjugates with other molecules, and to affect the rate of processes involving conformational change [5,11]. It has been observed that the distance across a single hexose residue (a sugar containing six carbon atoms) is approximately 5.4 Å [6], whereas a single amino acid adds only a relatively small distance of approximately 2.5 Å to the length of a polypeptide chain (R. Woods, personal communication, 1997). Keeping in mind the fact that carbohydrates can also carry multiple chains in dynamic motion, it becomes apparent that large areas of the surface of a protein can be shielded by relatively small oligosaccharides. With their relatively large size, the carbohydrate side chains may significantly alter the physicochemical properties of glycoproteins [1,6] by changing solubility, maintaining proper folding, and/or protecting against proteolysis and heat denaturation [7,8].

B. Unique Problems in Analyzing the Carbohydrate Chains of Glycoproteins

1. Structural Complexity

Carbohydrates can possess enormous structural complexity compared with other classes of biopolymers, which makes their structural elucidation a challenging task. For example, a tetrasaccharide composed of four different sugars theoretically could have any of more than 10,000 different structures due to the variety of possible sugar ring forms and linkages between component sugars. By comparison, a tetranucleotide or a tetrapeptide containing four different monomer subunits can each have only 24 different structures, as nucleotides and peptides are almost exclusively linear in form and have only a single type of linkage between each unit (amide bonds for proteins and $3'-5'$ phosphodiester bonds for nucleic acids). Since the carbohydrate chains attached to glycoproteins can contain 10 or more monosaccharide units, their possible structural multiplicity is enormous.

The structural diversity exhibited by carbohydrates results from numerous stereochemical centers, an anomeric center, multiple sites of linkages between monosaccharides, and the ability of an oligosaccharide to be branched. Each of these points is discussed in detail in the following sections.

Stereochemistry.   A hexose normally has four stereochemical centers located at carbons 2-5 (Figure 1). Consequently, there are 2^4 (or 16) possible hexose structures, which leads to the multiplicity of C_6 sugars, that is, galactose, glucose,

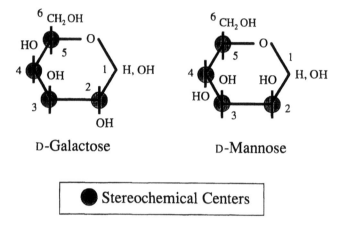

D-Galactose   D-Mannose

● Stereochemical Centers

Figure 1   The stereochemical centers of two common hexoses.

mannose, etc. The stereochemistry of the chiral atom farthest from the anomeric center (C1 for a hexose, see the following section on anomeric centers) determines whether a monosaccharide has a D- or L-configuration. Taking this fact into consideration, there are eight D- and eight L-hexoses. Most naturally occurring monosaccharides are members of the D-series, with fucose and iduronic acid being the most notable exceptions found in mammalian glycoproteins.

The Anomeric Center.   In solution, the acyclic form of a monosaccharide exists in equilibrium with four cyclyzed forms. Two of these are five-membered rings (furanose forms), while two are six-membered rings (pyranose forms). However, the two pyranose forms are generally the most abundant forms found in solution. For example, in dilute glucose solutions at 20°C, over 99% of the molecules are in the two pyranose forms [12]. This cyclization introduces a stereochemical center at C1 of the aldohexoses, which is defined as the anomeric center. Consequently, there are two possible stereoisomers for each ring form, called anomers, which are designated as α or β. This classification depends upon the relative orientation of the hydroxyl group of the anomeric carbon to the hydroxyl group of the chiral carbon farthest from the anomeric center. For the hexoses in the pyranose ring form that are typically found on mammalian glycoproteins, the β anomer has the C1 OH cis to the CH₂OH on C5, while these two groups are trans in the α anomer. Anomers can interconvert through the acyclic form by a process called mutarotation (Figure 2). However, mutarotation cannot occur when the anomeric carbon participates in a glycosidic bond, since the acyclic form of the sugar no longer exists in dynamic equilibrium with the cyclic forms.

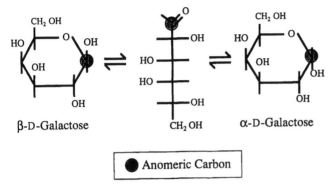

Anomeric Carbon

β-D-Galactose   CH₂OH   α-D-Galactose

Figure 2 The anomeric carbon of a hexose (pyranose form), showing the ability to interconvert between the α and β anomer.

Glycosidic Linkages. The anomeric carbon can form a glycosidic linkage via any of the hydroxyl groups on another monosaccharide (Figure 3). This characteristic allows for a range of different glycosidic linkages (i.e., 1,2; 1,3; 1,4; etc.) and branching of oligosaccharides. Historically, the terminus with a "free" anomeric center has been called the "reducing" terminus, as this center can act as a reducing agent, while the other end of the oligosaccharide is called the "nonreducing" terminus. An oligosaccharide can have only a single reducing terminus; however, multiple nonreducing termini are encountered in branched oligosaccharides. Technically, the reducing terminus is present only in free oligosaccharides. The anomeric carbon at the former reducing terminus

Galactose β 1-4 (Galactose β 1-2) Galactose α 1-6 Galactose

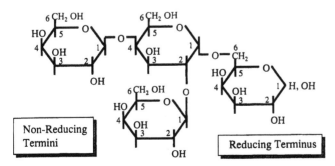

Figure 3 The variability in glycosidic linkages and their ability to have multiple nonreducing termini.

of a carbohydrate chain attached to a glycoprotein participates in the chemical bond to the protein. However, the site of carbohydrate attachment to the peptide backbone is often referred to as the reducing terminus of the oligosaccharide, as if the oligosaccharide had been released by hydrolysis.

2. Heterogeneity

One definition of a "pure" protein sample is one in which all of the polypeptide chains possess the same amino acid sequence. This definition is also typically applied to glycoproteins, without considering the carbohydrate portion of the molecule whose side chains are typically heterogeneous—that is, different carbohydrates can be found at any single glycosylation site [1–4]. For example, in numerous cases over 30 different carbohydrate side chains have been found at a single glycosylation site of a glycoprotein [1–4]. Consequently, a fairly complex mixture is often encountered when analyzing the carbohydrate chains of a "pure" glycoprotein.

3. Small Sample Quantities

The carbohydrate side chains of glycoproteins have tissue-, organ-, and species-specific structures. Further structural diversity may exist between glycoproteins isolated from normal versus transformed (malignant) cells derived from the same organ or tissue type [3,4]. Glycosylation is also highly sensitive to alterations in cellular function, and changes in glycosylation often result from various disease states [5,13,14]. Consequently, a glycoprotein isolated from different tissue sources in a single individual or even from a single tissue in that individual at different times can possess different carbohydrate side chains. These temporal changes in glycosylation exacerbate the analytical problems associated with heterogeneity and limit the benefit of isolating a glycoprotein from a population of individuals or from the same individual at different times to provide sufficient material for analysis.

Another approach used to increase the amount of glycoprotein available for analysis is to overexpress it in a cell line. However, glycoproteins that are overexpressed in a particular cell line may have carbohydrate chains and heterogeneity different from the native glycoprotein. Therefore, recombinant glycoproteins are of limited use in studying naturally occurring glycoproteins.

C. Typical Structures of Glycoprotein Glycans

Two main classes of carbohydrate chains are found attached to glycoproteins (Figure 4). This subdivision is based upon the type of glycosidic linkage to the protein: *N*-linked carbohydrate chains are attached through the nitrogen atom on the side chain of an asparagine (Asn) residue, while *O*-linked chains are

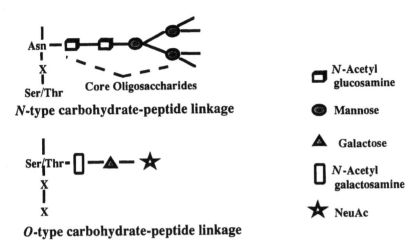

N-type carbohydrate-peptide linkage

O-type carbohydrate-peptide linkage

Figure 4 Examples of N- and O-linked carbohydrate side chains.

attached through the oxygen on the side chain of serine (Ser) and threonine (Thr) residues.

1. N-Linked Glycans

A glycosylated asparagine residue must form part of the tripeptide Asn-X-Ser/Thr (consensus sequence), where X is any amino acid except proline. However, the presence of this sequon by itself does not ensure glycosylation. All N-glycans share the same pentasaccharide core structure (Figure 5), but a wide variety and complexity of oligosaccharides and modifications can be attached to this core. N-Linked oligosaccharides can be divided into three subclasses based on the carbohydrate residues attached to the core pentasaccharide: high mannose (oligomannose), complex, and hybrid, as shown in Figure 6.

High-mannose N-glycans contain only α-mannose (Man) residues attached to the core. In mammalian glycoproteins, this class of carbohydrate

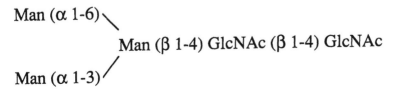

Figure 5 The core pentasaccharide of an N-linked glycoprotein.

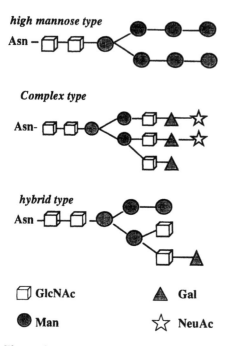

high mannose type

Complex type

hybrid type

☐ GlcNAc   ▲ Gal

● Man   ☆ NeuAc

Figure 6   Examples of high-mannose, complex, and hybrid type *N*-linked carbohydrate side chains.

chains possesses from five to nine Man residues, while glycoproteins overexpressed in yeast cells can have carbohydrate chains composed of 60 or more mannose residues.

Complex-type glycans contain no α-mannose residues other than those in the core pentasaccharide and have the largest number of structural variations. In this class, the core oligosaccharide is extended by the addition of one or more lactosamines (disaccharides composed of galactose and *N*-acetylated-glucosamine, Gal-GlcNAc) to form "antennae" or branches. Complex-type *N*-glycans can have up to five antennae, resulting in the formation of mono-, bi-, tri-, tetra-, and pentaantennary glycans. These antennae are commonly capped by a sialic acid, although a number of other possibilities exist. The presence or absence in the core region of an α-fucosyl residue to either of the *N*-acetylglucosamine residues (GlcNAc) in the core pentasaccharide further contributes to the structural variation of the complex-type glycans, as does fucosylation at other sites in the carbohydrate chains. Lastly, complex-type glycans can possess a range of other sugars and modifications, such as α-linked galactose and *O*-sulfation.

Hybrid *N*-glycans have the characteristic features of both the complex and high-mannose glycans. In this class of carbohydrate chains, several α-man-

nose residues are typically linked to the Man α-1,6 branch of the core pentasaccharide, and one or two complex antennae are linked to the Man α-1,3 branch.
The presence or absence of a bisecting GlcNAc residue also produces the structural variations found in the hybrid-type N-glycans.

2. O-Linked Glycans

O-Glycans are glycosidically attached through the oxygen atom on the side
chain of a serine or threonine residue. Unlike the potential sites for N-glycosyl-
ation, there is no known consensus sequence to predict possible sites of
O-glycosylation. Consequently, any Ser or Thr residue in a protein can be glycosylated. O-Linked glycans also do not share a common core structure. They
currently are categorized according to their different core structures into at least
six groups, which are:

Core 1: Galβ-1,3GalNAcα-Ser/Thr
Core 2: GlcNAcβ-1,6(Galβ-1,3)-GalNAcα-Ser/Thr
Core 3: GlcNAcβ-1,3GalNAcα-Ser/Thr
Core 4: GlcNAcβ-1,6(GlcNAcβ-1,3)α-Ser/Thr
Core 5: GalNAcα-1,3GalNAcα-Ser/Thr
Core 6: GlcNAcβ-1,6GalNAcα-Ser/Thr

These cores can be elongated by addition of Gal, fucose, sialic acids, and
GlcNAc residues and can contain a variety of modifications. Although the glycans are usually linked to serine or threonine residues through GalNAc, the
linkages may occur through other residues, such as fucose. O-Glycans may also
consist of a single monosaccharide, typically fucose or GlcNAc.

D. Traditional Methods to Structurally Characterize the Glycoprotein Glycans

In order to deduce the primary structure of the carbohydrate chain of a glycoprotein, a number of factors must be determined, which include:

1. The glycosylation sites and the type of glycosidic linkage at each site
    (N- or O-linked)
2. The sequence of the different sugar residues, including their
    branching points
3. The position of the glycosidic linkages [(1,2), (1,3), (1,4), (1,6), etc.]
4. The anomeric configuration of each sugar (α or β)
5. The identity of each sugar (Gal, Man, etc.)

Until recently, the carbohydrate components of glycoproteins were not
generally analyzed in the context of their sequence location due to the experimental difficulty in identifying and isolating glycopeptides containing specific

glycosylation sites. Instead, analysis of protein glycosylation was performed by analyzing the glycans after their release from their conjugate peptide. A common strategy for protein glycosylation analysis involves an elaborate combination of separation and analytical techniques [15]. The carbohydrate is first released in toto by endoglycosidase digestion, hydrazinolysis, or reductive elimination cleavage (the latter two methods being destructive of the protein backbone). The freed oligosaccharide side chains are then separated from the peptide fragments. Once isolated, the carbohydrate chains are separated by high-pH anion-exchange chromatography [16] or Bio-Gel P4 column chromatography [17]. If gel chromatography is used, the reducing terminus of the glycan is usually labeled first to enable its detection in subsequent procedures. Two methods commonly used to label glycans are reductive amination with a fluorescent compound, such as 2-aminobenzamide [18], and reduction with alkaline sodium borotritiide [19] to give the radiolabeled derivative. A number of techniques are then employed to deduce the structure of each purified oligosaccharide side chain, including chemical reaction/derivatization procedures followed by gas chromatography/mass spectrometry (GC/MS) to determine the linkage and composition of each oligosaccharide side chain [12]; mass spectrometry and tandem mass spectrometry (MS/MS) to identify the molecular weight, nonstereochemical sequence, and branching points [20–26]; and nuclear magnetic resonance (NMR) spectroscopy to deduce linkages, glycosyl sequence, and configuration at anomeric centers [29].

More recently, several techniques have been developed that allow the individual glycopeptides to be simultaneously identified and purified (as discussed later). This development allows the glycosylation site-specific characterization of the attached carbohydrate chains, which has become the norm.

The approach just outlined is extremely useful when ~100 nmol of pure glycoprotein is available [29]. However, when dealing with naturally occurring glycoproteins, this quantity of glycoprotein is not often obtainable, which prevents analysis of the carbohydrate side chains [29]. In other instances, isolating this amount of material can require exorbitant time and effort. With such large sample requirements, analyzing the carbohydrate side chains of all but the most abundant naturally occurring glycoproteins is a long and arduous task, when possible, which has severely limited the range of glycoproteins that have been studied.

II. ANALYSIS OF GLYCOPROTEINS

We typically employ a three-step process to characterize the carbohydrate chains of glycoproteins. First, the molecular weight (MW) of the intact glycoprotein is determined. Second, the peptide backbone of the glycoprotein is cleaved enzymatically and the resulting glycopeptides are identified and isolated. Third, a series of endo- and exoglycosidase digestions are performed to determine the primary structure of the carbohydrate side chains.

A. Determining the Molecular Weight of Intact Glycoproteins

The first analysis we perform is to determine the MW of the intact glycoprotein so that we can evaluate its purity. The MW tells us whether we have a mixture of glycoproteins; a mixture would cause problems for subsequent analyses.

Carbohydrate heterogeneity (see Section I.B.2) causes a problem for direct MW analysis of an intact glycoprotein, because the presence of heterogeneity often forces the analysis to be performed on a mixture of species with fairly high MW. The two MS approaches that can handle the MW range necessary to analyze intact glycoproteins are matrix-assisted laser desorption/ionization MS (MALDI-MS) and electrospray ionization MS (ESI-MS). Mixtures, however, are fairly difficult to analyze by ESI-MS [30]. For a single glycoform, ESI-MS produces a series of multiply charged molecular ions with different charge states, $[M + nH]^{n+}$, that typically appear in the 600 to 1700 mass-to-charge (m/z) region of the spectrum, regardless of the glycoform's MW [30]. Hence, ESI-MS gives a series of ions from each glycoform. The MW of the glycoform can then be determined from the spacing between these multiply charged molecular ions. When ESI-MS analysis is performed on a typically heterogeneous glycoprotein, a series of molecular ions is detected for each glycoform, and all of these ions appear in the same region of the spectrum. An example of this result is shown by the ESI-MS analysis of ribonuclease B (Figure 7A), which was analyzed in a 1:1 mixture of water:acetonitrile with 0.1% trifluoroacetic acid (TFA). Ribonuclease B has only a single glycosylation site with five different MW glycoforms; it is a relatively uncomplicated glycoprotein in structure. Although many ions are seen in its ESI-MS spectrum, it is difficult to identify the spacing between members of the same series because of the presence of a molecular ion series for the nonglycosylated protein and for each glycoform. As the number of glycosylation sites on a protein increases, so do the problems associated with heterogeneity, and eventually no information can be obtained from ESI-MS analysis of an intact glycoprotein. Furthermore, the problem caused by carbohydrate heterogeneity cannot usually be solved by high-performance liquid chromatography/mass spectrometry (HPLC/MS), since the different glycoforms typically do not separate.

MALDI-MS, on the other hand, produces primarily singly charged molecular ions that enable the MW of intact heterogeneous glycoproteins to be determined [see the MALDI-MS spectrum of ribonuclease B analyzed with sinapinic acid (3,5-dimethoxy-4-hydroxycinnamic acid) as the MALDI matrix in Figure

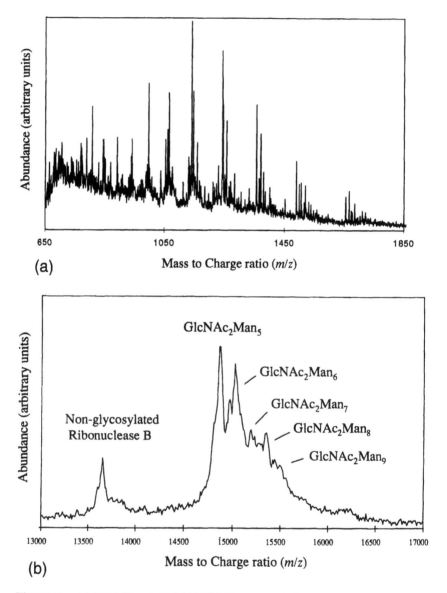

Figure 7 (a) ESI-MS and (b) MALDI-MS spectra of ribonuclease B.

7B]. Consequently, MALDI-MS is our technique of choice for the initial MW analysis of an intact glycoprotein. As we proceed with the characterization and see that the glycoprotein has only minimum carbohydrate heterogeneity, we would then perform ESI-MS, since ESI-MS currently provides more accurate MWs at higher resolution.

Carbohydrate heterogeneity also leads to problems with MALDI-MS, since heterogeneity distributes the molecular ion envelope over a wider m/z range than that of a similarly sized protein. This problem is clearly shown by the MALDI-MS analysis of avidin and myoglobin (Figures 8A and 8B). Avidin is a glycoprotein with a single N-linked glycosylation site, while myoglobin is a nonglycosylated protein. Despite a similar MW for both of these samples, the breadth of avidin's mass-to-charge range at half height is approximately eight times that of myoglobin, a direct result of carbohydrate heterogeneity. This phenomenon reduces the intensity of a glycoprotein's molecular ion by spreading its abundance over a broader m/z range. As peak-broadening increases with the increase in the number of glycosylation sites and carbohydrate heterogeneity, the molecular ions can spread over such a broad range that they become indistinguishable from the background. Consequently, highly heterogeneous glycoproteins are extremely difficult to analyze.

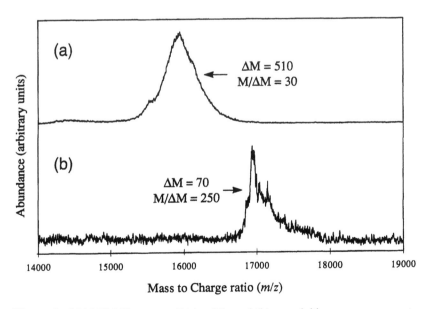

Figure 8 MALDI-MS spectra of (a) avidin and (b) myoglobin.

B. Identifying the Sites of Glycosylation

The second step in our approach is to identify which amino acid(s) in the glycoprotein are glycosylated, a process typically referred to as "glycosylation site mapping." This process is greatly simplified if we have in hand (or obtain) the amino acid sequence of the glycoprotein. For the purpose of this discussion, we assume that the amino acid sequence is known. We first cleave the peptide backbone so that we can identify the multiple glycosylation sites of the glyco-protein. Enzymatic cleavage is preferred, as chemical cleavage of the peptide backbone (CNBr and mild acid hydrolysis) can cleave glycosidic linkages as well as the peptide backbone, particularly those linkages involving sialic acids and fucose. After digestion, the resulting mixture is separated and the glycopep-tide fractions identified.

A number of chromatographic techniques have been developed to purify and identify N-linked glycopeptides from complex digests of glycoproteins. Several of these use "comparative mapping," which is based on comparing HPLC ultraviolet (UV) traces or total ion current (TIC) traces in LC or LC/MS analysis of a proteolytic digest of a glycoprotein before and after treatment with N-glycanase to identify putative N-linked glycopeptides [24]. This technique appears to be successful for glycoproteins containing relatively few glycosyla-tion sites but becomes increasingly less reliable as the size of glycoprotein and the number of glycosylation sites increase.

A second approach for isolating glycopeptides from a glycoprotein digest uses a column composed of resin-bound lectins (carbohydrate-binding proteins) [32]. Lectins with known binding specificities are employed in these experi-ments to ensure that those carbohydrates will be detected that have the struc-tural types of interest. Unfortunately, since there are no lectins that bind all types of glycoprotein glycans, multiple lectin columns are often used. However, several glycopeptides and/or glycoforms of the same glycopeptide may still not bind to any of the columns.

Glycopeptides may also be identified by LC/MS based upon the presence of a heterogeneous population of oligosaccharides at a given glycosylation site [33]. On reversed-phase HPLC, these glycoforms partially separate, with the more heavily glycosylated glycoforms eluting first, as they are less hydropho-bic. Glycopeptide ions thus appear in a diagonal ladder of peaks in plots of m/z versus time, while peptides appear as a vertical line. This method will not work if the glycoprotein under study does not have significant heterogeneity at each glycosylation site.

A recently developed LC/MS technique has become very useful in map-ping both N- and O-linked proteolytic fragments of glycoproteins [34–36]. In this method, the electrospray mass spectrometer selectively detects glycopep-tides by monitoring diagnostic sugar oxonium ions produced by collision-induced dissociation (CID) of glycopeptide ions. A variety of carbohydrate-specific marker ions is observed including those at m/z 366 [Hex-HexNAc$^+$], m/z 292 [NeuAc$^+$], m/z 204 [HexNAc$^+$], m/z 163 [Hex$^+$], and m/z 147 [deoxy-Hex$^+$]. The m/z 204 ion is the most general carbohydrate marker ion, because it is produced by most structural types of both N- and O-linked carbohydrate chains.

Dissociation of the molecular ions can be achieved in two ways: either in the electrospray source by increasing the collision excitation potential between the sampling orifice and the skimmer (typically referred to as "stepped orifice voltage" scanning), or in the collision cell (Q2 of a triple-quadrupole mass spectrometer) by scanning the first mass analyzer while monitoring the abundance of a diagnostic sugar oxonium ion in the second (referred to as "precursor ion scanning"). Stepped orifice voltage scanning requires the use of only a single-quadrupole spectrometer and is more sensitive, whereas precursor ion scanning requires a triple-quadrupole spectrometer and is more selective because only those precursor ions are detected that decompose to yield the fragments of interest. Also, when the LC effluent is split so that only a portion of it enters the mass spectrometer, both of these procedures allow the glycopeptide-containing fractions to be simultaneously purified for further study.

Once the glycopeptide-containing fractions have been identified, we must next identify the glycosylated amino acids. For N-linked glycopeptides, we must identify which of the Asn residues of the consensus tripeptides are glycosylated. In cases in which the peptide backbone can be cleaved between each of the consensus tripeptides, we can identify the peptide by the MW of the purified glycopeptide after the carbohydrate chain has been released with an N-glycanase, which in turn tells us which Asn residue is glycosylated. Remember that when using PNGase F (one of the common N-glycanases) for this procedure, the Asn is converted to an Asp, producing a peptide that is one mass unit higher than the predicted mass of the unmodified peptide. Situations in which the peptide backbone cannot be cleaved between each of the consensus tripeptides are a bit more problematic, and analysis in this case is similar to that for O-glycosylation, which is discussed next.

O-Linked glycosylation is one of the more difficult posttranslational modifications to identify. Two of the problems associated with this analysis are the lack of a known consensus sequence and the presence of contiguous sites of O-glycosylation between proteolytic cleavage sites. Although O-glycosylation can be inferred from a blank cycle during Edman degradation, additional experiments are usually needed to confirm such results.

A number of mass spectrometric techniques have been used to directly identify glycosylation sites in glycoproteins. MS/MS can occasionally identify glycosylation sites by analyzing the fragment ions produced from CID of the glycopeptide [37–39]. However, this technique does not appear to be a generally applicable approach, particularly when the linkage between the modifying group and the protein is fragile, such as in a glycosidic linkage [37]. Consequently, the fragment ions in the MS/MS spectrum primarily result from losses from the carbohydrate chain. The few fragment ions arising from peptide backbone cleavage have typically also lost their complete carbohydrate chain, thus preventing the modified residue from being directly identified. The MS/MS spectrum of a glycopeptide does contain, however, a wealth of information on carbohydrate structure, including information on nonstereochemical sequence and branching [20–28,37–39], and hence MS/MS should be performed even though it may not identify the glycosylation site.

Another approach that can be used to identify modified amino acids combines carboxypeptidase or aminopeptidase digestion with MALDI-MS [40]. Here, the N- and/or C-terminus of the peptide is degraded enzymatically, while MALDI-MS spectra are acquired at various digestion times. This process produces a ladder of peptide fragment ions that differ by the time-based degradation of the terminal amino acids. However, it is often difficult to identify glycosylation sites by this approach, since the catalytic functions of the proteolytic enzymes are often impaired at or near the modified amino acid for steric and/or electronic reasons so that analysis cannot proceed beyond the modified site. For this reason this approach is limited to peptides with a single modification site.

On-probe pronase digestion/MALDI-MS has also been used to identify glycosylation sites [41]. In this scheme, the glycosylated peptide is digested with pronase, which cleaves the peptide backbone in a random fashion and thus produces a ladder of peptide fragments. MALDI-MS analysis of these digestion mixtures gives us the masses of the peptide fragments, from which we can determine the modification sites.

Many instances still remain in which the glycosylation sites cannot be determined, despite the effort that has been directed to developing this analysis. These limitations are particularly true for mucins, since mucins are highly glycosylated and have very large regions that contain contiguous sites of O-glycosylation.

C. Deducing the Primary Structure of Carbohydrate Side Chains

The primary structures of the carbohydrate side chains of glycoproteins can be extremely complicated. Fortunately, the majority of the carbohydrate side chains attached to glycoproteins have known structures, which have been compiled into a central, public domain database, the Complex Carbohydrate Structure Database

(CCSD) with its search program, CarbBank.* For example, only 132 of the ~7000 articles published in 1994 on glycoproteins contained an N-linked carbohydrate structure that had not been previously included in the CCSD (S. Doubet, CCSD/CarbBank Director, personal communication, 1995). Hence, new N-linked structures can be expected to be encountered in less than 2% of the current studies on glycoproteins. A similar case can be made for the O-linked carbohydrate side chains of glycoproteins. The analytical problem, therefore, is reduced in most instances to differentiating among known structures.

To determine the primary structures of the carbohydrate side chains of a glycoprotein, we first perform a series of experiments on the glycopeptide-containing fractions identified and purified by LC/MS to determine the MW of each oligosaccharide side chain present. Second, we search the CCSD for these experimental MWs to produce a "short" list of possible carbohydrate side chain structures. Third, we conduct a series of exoglycosidase digestions followed by MALDI-MS analysis to differentiate among the potential structures obtained from the CCSD search. Each of these procedures is described in detail in the following sections.

1. Determining the Molecular Weight of the Carbohydrate Side Chains

We can determine the MWs of the N-linked carbohydrate chains present in the sample by analyzing the glycopeptide before and after release of the side chain with an endoglycosidase. An example of this process is shown for a purified tryptic glycopeptide with two glycoforms from bovine desialylated (asialo) fetuin (Figure 9A and 9B, respectively). After deglycosylation, the molecular ions of the glycopeptides have decreased by 1623 and 1988. The difference in mass between the glycosylated and deglycosylated species is converted to the MW of the carbohydrate chains by the addition of 19 Da to the experimental mass shift. This addition is made to account for the loss of H_2O during the formation of the N-glycosidic bond and for the change of Asn (residue mass 114 Da) to Asp (residue mass 115 Da) due to the modification from the N-glycanase deglycosylation. In this example, these MWs correspond to the losses of carbohydrate chains consisting of $Hex_5HexNAc_4$ and $Hex_6HexNAc_5$, indicating that they are complex bi- and triantennary carbohydrate side chains.

Determining the MWs of O-linked carbohydrate chains is a bit more problematic, as there are no endoglycosidases that will release all O-linked side

---

*Ultimately, one will be able to perform a CCSD search over the World Wide Web through the CCRC's home page (www.ccrc.uga.edu). Until this service becomes available, both CCSD and CarbBank can be downloaded over the Internet from ftp://ncbi.nlm.nih.gov/repository/carbbank/.

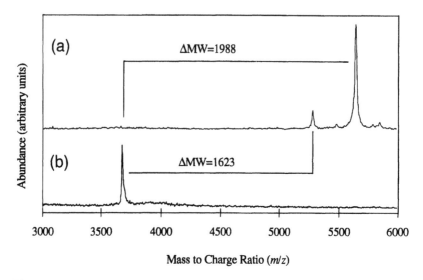

Figure 9 MALDI-MS spectra of the purified tryptic glycopeptide T7 (a) before and (b) after release of its N-linked carbohydrate side chain with PNGase F.

chains. The most general O-glycanase will only release a disaccharide composed of Gal(β1,3)GalNAc [42]. Fortunately, MS/MS can provide the MWs of O-linked carbohydrates along with their nonstereochemical sequence and branching sites (discussed earlier).

2. Searching the CCSD for Experimental MWs

The experimental MW of a carbohydrate side chain can be used as a search parameter for CarbBank to provide a reasonable amount of structural information from the CCSD, as demonstrated here for the carbohydrate side chains attached to the tryptic glycopeptide T7 of bovine asialofetuin. As mentioned earlier, this glycopeptide has at least two different carbohydrate side chains attached to it: a biantennary chain and a triantennary chain. When we use the experimental MW of the triantennary chain to search the >1000 N-linked carbohydrate structures contained in the CCSD, we find that only 10 structures (summarized in Figure 10) have an asialo MW within $\pm 0.1\%$ of the experimental value of interest to accommodate the mass accuracy of MALDI-MS operated in the linear mode. We performed similar searches on the biantennary chain, and 11 structures were found (for brevity, structures are not shown). It is important to reiterate that in 1994 new N-linked structures were encountered in less than 2% of the studies on glycoproteins. Thus, the probability is fairly small of encountering a structure that is not included in the CCSD.

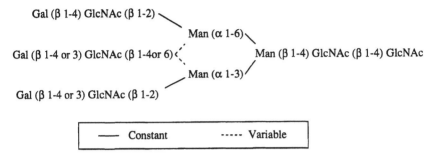

Figure 10 Summary of the 10 N-linked carbohydrate side chains found in the CCSD with a molecular weight equal to that of the triantennary chain attached to the purified tryptic glycopeptide T7.

The list of only 10 possible triantennary carbohydrate side chain structures attached to this glycopeptide demonstrates that, despite the enormous possible complexity introduced by stereochemical differences among sugars, multiple linkages, branching, and anomeric centers, very few naturally occurring structures have a particular MW. For example, all of the triantennary structures found (Figure 10) are of complex structure and have identical monosaccharide compositions. Furthermore, variability in these structures is limited to the nonreducing terminal Gal-GlcNAc disaccharide units and their points of attachment to the core pentasaccharide. Therefore, the MWs of the carbohydrate side chains of a glycoprotein provide a significant amount of information concerning their structures.

3. Identifying the Primary Structure of Each Carbohydrate Side Chain

The stereochemistry, linkage, and anomeric configurations of each individual monosaccharide in a carbohydrate side chain must be ascertained once the MW has been obtained. For N-linked oligosaccharide side chains, this information is limited to those monosaccharides located outside the invariant core region. These particular structural features are currently elucidated by NMR spectroscopy and GC/MS analyses, the analytical procedures that require the greatest amount of sample [12,21]. Consequently, significant reductions in sample requirement depend upon the development of new methods to determine the stereochemistry, linkage, and anomeric configuration of each monosaccharide.

Sequential exoglycosidase digestion coupled with MS detection has been demonstrated as an alternative method to provide the stereochemistry, linkage, and anomeric configuration of each monosaccharide [43–50]. A range of commercially available exoglycosidases (Table 1) releases the nonreducing terminal monosaccharide residues in a manner that is highly selective for the particular stereochemistry, anomeric configuration, and linkage of the released monosaccharide (i.e., Gal β1,4; Gal β1,3; etc.). After each exogylcosidase treatment, a portion of the sample is analyzed by MS to determine the shift in MW, which tells us the number of monosaccharides released. By using a series of these enzymes, one can determine the stereochemical sequence, linkages, and anomeric centers and thereby establish the structure of the oligosaccharide side chain.

The use of exoglycosidases limits the kind of carbohydrate side chains that can be analyzed by this procedure to mammalian glycoproteins, since the majority of commercially available exoglycosidases are specific for those monosaccharides typically found on these glycoproteins (see Table 1). The current lack of exoglycosidases for the less common monosaccharides and for the modifications that are occasionally found on mammalian glycoproteins prevents full elucidation of the primary structures of oligosaccharide side chains with these moieties. However, modified N-linked side chains have been encountered in just under 10% of the mammalian glycoproteins whose carbohydrate side chains have been analyzed to date (S. Doubet, CCSD/CarbBank Director, personal communication, 1995). Even when only the common nonmodified sugars

Table 1 Commercially Available Exoglycosidases and Their Specificities

Enzyme/source	Specificity	Enzyme/source	Specificity
Sialidase		β-GlcNACase	
Arthrobacter ureafaciens	α 2–(6>3, 8)	Streptococcus pneumoniae	β 1–2
Clostidium perfringens	α 2–3, 6, 8	Jack bean	β 1–2, 3, 4, 6
Vibrio chloerae	α 2–3, 6, 8	Chicken liver	β 1–3, 4
Salmonella typhimurium	α 2–(3 >6>8, 9)		
Newcastle virus	α 2–3, 8	α-Mannosidase	
		Jack bean	α 1–(2, 6)>3
α-Galactosidase		Aspergillus saitoi	α 1–2
Coffee bean	α 1–3, 4, 6		
		β-Mannosidase	
β-Galactosidase		Helix pomatia	β 1–4
Jack bean	β 1–(6>4>3)		
Streptococcus pneumoniae	β 1–4	α-Fucosidase	
Bovine testes	β 1–(3, 4)	Chicken liver	α 1–2, 3, 4, 6
Chicken liver	β 1–3, 4	Bovine epididymis	α 1–(6>2, 3, 4)
		Almond meal	α 1–3, 4, or 2
α-GalNACase			
Chicken liver	α 1–3 (Ser, Thr)		

are present, sequential exoglycosidases cannot always provide full characterization, particularly when analyzing high-mannose oligosaccharides whose structures are similar. Despite these limitations, the exoglycosidase/MS approach with its low sample requirement offers perhaps the best method currently available to obtain this level of structural information (linkage, stereochemistry, and anomeric configuration) from the wide range of naturally occurring mammalian glycoproteins that cannot be isolated in sufficient quantities for analysis by other methodologies.

To demonstrate how sequential exoglycosidase digestions can elucidate the primary structure of a carbohydrate chain, we analyzed a purified tryptic glycopeptide (T7) from asialofetuin with this process. Although this glycopeptide has both bi- and triantennary chains, we focus on the triantennary chains to simplify the discussion. A series of eight exoglycosidase digestions (Figure 11) ultimately identified the triantennary carbohydrate chains as the two structures shown in Figure 12. It is important to reiterate that all variability in the triantennary structures is limited to the three Gal-GlcNAc disaccharides and their points of attachment to the core pentasaccharide. Hence, one need only determine the linkages within these three disaccharides and how they are attached to the core to obtain the complete triantennary structure.

β1,4-Linked Gal is the most common moiety present at the non-reducing termini of all possible triantennary structures (Figure 10). Thus, in the first digestion we used β-galactosidase from *Streptococcus pneumoniae*, which selectively releases β1,4-linked Gal (Table 1) [51,52]. A 20-pmol sample of this fraction was digested, and the MALDI spectrum from ~2 pmol of this sample after digestion (Figure 13) reveals that two different triantennary carbohydrate side chains are attached to this peptide, one with three nonreducing terminal Gal residues with β1,4 linkages (Tri-1) and the other with only two β1-4-linked Gal residues (Tri-2). The carbohydrate side chain of Tri-2 presumably contains one β1,3-linked Gal, as this is the only other nonreducing terminal monosaccharide observed to date in naturally occurring mammalian structures (Figure 10). This experiment also identified that both of the Gal residues on the biantennary chain are attached by β1,4-linkages.

At this point, digestion with a different β-galactosidase could have been used to identify the linkage of this terminal residue. However, if this terminal Gal is removed, we would also lose the ability to differentiate between these two triantennary chains (Tri-1 and Tri-2), since both species would have the same structure and MW after this second digestion (Figure 14). Hence, information concerning which antenna(e) was capped with the non-β1,4-linked Gal residue would be lost. To avoid this problem, we retained the Gal residue that was not β1,4-linked by not digesting with another β-galactosidase at this point, and thus protected this antenna from future exoglycosidase digestion. As seen in later discussion, after the Gal β1,4-capped antennae were fully characterized,

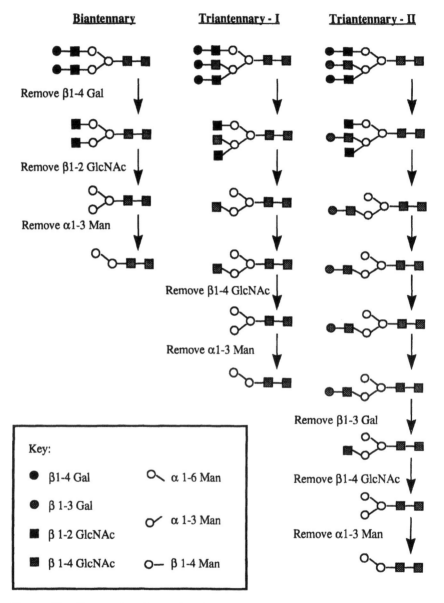

Figure 11   The exoglycosidase digestion scheme used to elucidate the primary structures of the carbohydrate chains attached to the purified tryptic glycopeptide T7.

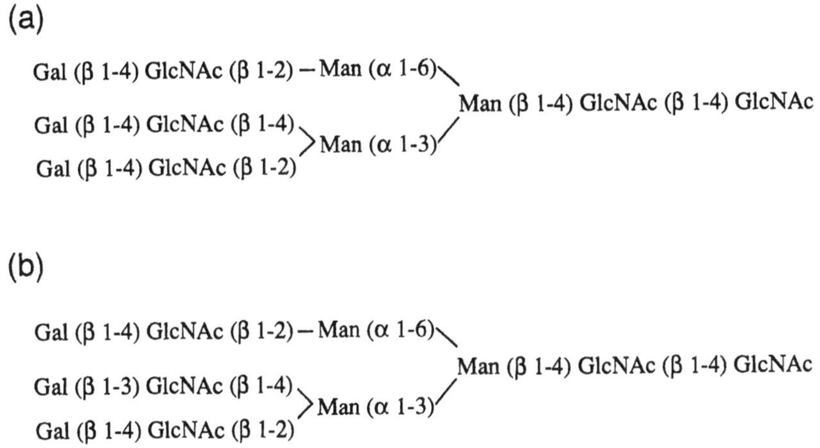

(a)

Gal (β 1-4) GlcNAc (β 1-2) — Man (α 1-6)

Gal (β 1-4) GlcNAc (β 1-4)   Man (β 1-4) GlcNAc (β 1-4) GlcNAc
                                   Man (α 1-3)
Gal (β 1-4) GlcNAc (β 1-2)

(b)

Gal (β 1-4) GlcNAc (β 1-2) — Man (α 1-6)

Gal (β 1-3) GlcNAc (β 1-4)   Man (β 1-4) GlcNAc (β 1-4) GlcNAc
                                   Man (α 1-3)
Gal (β 1-4) GlcNAc (β 1-2)

Figure 12   The two triantennary structures (a and b) attached to the purified tryptic glycopeptide T7.

the focus shifted to characterization of the protected antenna. With this procedure we can deduce the oligosaccharide structures of both Tri-1 and Tri-2, despite their obvious similarity and simultaneous presence in the same sample.

Because of our desire to leave the non-β1,4 Gal-capped antennae intact, the focus shifted to the GlcNAc resides exposed by release of the β1,4 Gal residues. Since β1,2-linked GlcNAc is the most common moiety present at this position, the remaining 18 pmol of sample was treated with an exoglycosidase to release this residue. MALDI analysis after this digestion reveals that all of the glycopeptides have lost two GlcNAc residues. These results combined with those from release of the β1,4-linked Gal residues demonstrate that all of the carbohydrate chains in this sample contain two antennae consisting of Gal β1,4 GlcNAc β1,2. Consequently, these two digestions fully elucidated the complete primary structure of the biantennary carbohydrate chain as well as two of the three antennae of the triantennary chains.

At this stage we encountered a dilemma similar to that discussed earlier concerning the release of the non-β1,4 Gal residue: that is, if a second GlcNAcase is employed to identify the linkage of the non-β1,2 GlcNAc, then the site of attachment of this residue to the core pentasaccharide is lost. Because this piece of information is required to differentiate between the possible triantennary structures, our attention turned to the site of attachment of the non-β1,2-linked antenna to the core pentasaccharide. The remaining 16 pmol of sample was treated with an exoglycosidase that releases α1,3 Man. MALDI analysis of this sample after digestion reveals that this treatment did not affect the two trianten-

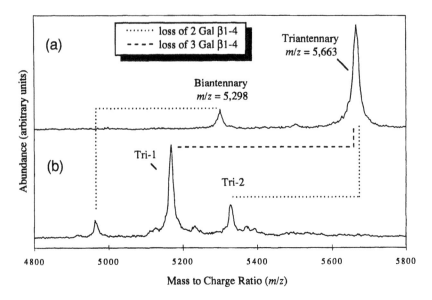

Figure 13 MALDI-MS spectra of the purified tryptic glycopeptide T7 (a) before and (b) after release with β-galactosidase of β1,4-linked galactose residues.

nary structures. This result indicates that the third antenna of both triantennary structures is attached to the α1,3 Man residue of the core pentasaccharide, since in this instance the α1,3 Man would not be released by exoglycosidase treatment as it would not be located at a nonreducing terminus.

We next probed the linkage between the non-β1,2 GlcNAc and the α1,3 Man residue by digestion with an exoglycosidase specific for β1,4 GlcNAc residues followed by MALDI-MS. The results from this experiment demonstrated that Tri-1 had lost a single GlcNAc, and thus the non-β1,2 GlcNAc of this carbohydrate chain is attached by a β1,4-linkage to the α1,3 Man branch of the core pentasaccharide. The results from these four digestions fully elucidated the primary structure of Tri-1.

The only carbohydrate chain that has not been fully characterized at this point is Tri-2. In fact, only the Gal-GlcNAc and GlcNAc-Man linkages in the third antenna of this chain are still unknown. We designed a series of exoglycosidase digestions targeting this antenna to provide this missing information. The only known non-β1,4 Gal-GlcNAc linkage is β1,3, as discussed earlier. MALDI analysis of the sample after release of β1,3 Gal demonstrated that the Tri-2 glycopeptide had lost a single Gal, which identified the Gal-GlcNAc linkage as being β1,3. The GlcNAc-Man linkage was then probed by release of

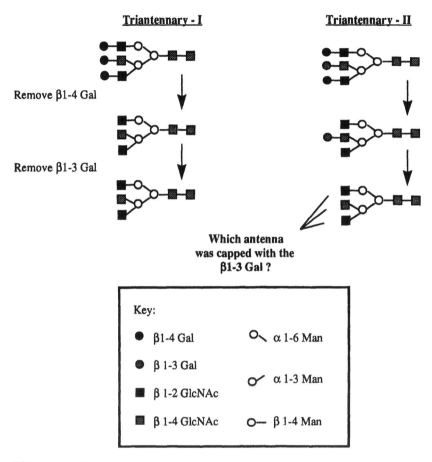

Figure 14   Results expected from performing sequential exoglycosidase digestions to release the β1,4 Gal residues followed by release of the β1,3 Gal residue on the triantennary carbohydrate chains attached to the purified tryptic glycopeptide T7.

β1,4 GlcNAc followed by MALDI-MS. This treatment identified that the GlcNAc residue was attached to the core pentasaccharide by a β1,4-linkage. Thus, the complete structure of Tri-2 had been fully elucidated.

The combined results of all these experiments demonstrate that Tri-1 has structure A and Tri-2 has structure B (Figure 11). These digestions simultaneously permitted the full characterization of the biantennary carbohydrate side chain attached to this glycopeptide. These structural assignments are identical to those from previous studies on the N-linked chains of bovine fetuin [53,54]. These eight digestions followed by MALDI-MS, therefore, elucidated the complete primary structure of each of the carbohydrate side chains present in this sample from a mere 20 pmol of glycopeptide. This sample quantity is approximately 5000 times less material than the amount needed to obtain these carbohydrate structures using other methodologies [12,21].

4. Experimental Conditions for Glycosidase Digestion

The initial work on exoglycosidase carbohydrate sequencing was performed with size-exclusion chromatographic or gel electrophoretic analysis of the digestion products [52]. These two techniques are immune to the buffers used in these enzymatic digestions; these buffers contain 100 mM of either sodium acetate or sodium citrate/phosphate. Since the activity and specificity of exoglycosidases have only been established in these buffers, they are currently used for exoglycosidase sequencing with MS detection [43–49], even though these buffers are detrimental to MS analysis. The problems associated with these buffers are easily seen in the MALDI-MS spectrum (Figure 15) of the glycopeptide T7 of bovine asialofetuin before and after digestion with β-galactosidase in the buffer suggested for this enzyme (100 mM sodium acetate at pH 6). The peak broadening caused by the sodium-containing buffer decreases the

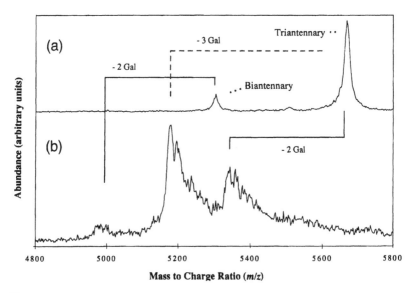

Figure 15 MALDI-MS spectra of purified tryptic fragment 7 from bovine asialofetuin (a) before and (b) after release of the nonreducing terminal β1,4-linked Gal residues by digestion with β-galactosidase from *Streptococcus pneumoniae* in the suggested buffer (100 mM sodium acetate at pH 6).

MW accuracy of these measurements and causes problems when analyzing gly-coforms with similar MWs. By broadening the molecular ion envelope, the presence of excess sodium reduces the signal-to-noise ratio of the molecular ions. In fact, the new biantennary structure is observed just above the matrix background, and after subsequent exoglycosidase digestions with the suggested buffers, this glycopeptide could not be detected above the background. There-fore, desalting is required before performing MALDI-MS analysis [43–49], but desalting creates additional experimental procedures for the researcher and can result in sample losses.

A more efficient approach involves the use of non-sodium-containing buffers that eliminates the need to desalt the sample prior to MS analysis. We have demonstrated that we can replace the usual exoglycosidase digestion buffers with a 25 mM ammonium acetate solution adjusted to the proper pH [50]. The success of this replacement is shown by direct MALDI-MS analysis of the T7 glycopeptide from bovine asialofetuin when β-galactosidase digestion is per-formed in a 25 mM ammonium acetate solution adjusted to pH 6 (Figure 13). Here, the molecular ion peak widths and signal-to-noise ratios are comparable to those observed before the treatment. The use of ammonium acetate solution at this concentration also eliminates the peak-broadening problems discussed earlier, because this buffer is volatile and therefore most of it evaporates when the sample is placed into the MALDI-MS. However, at buffer concentrations over 25 mM, sufficient ammonium acetate is present in the sample, even under vacuum, to produce ammonium adducts in the MALDI-MS spectrum. Both of these digestions yield identical products, demonstrating that this exoglycosidase has the same specificity in both solutions. However, the use of ammonium acetate eliminates the problems associated with the sodium buffer. Furthermore, ammonium acetate solutions permit direct MALDI-MS analysis of the exo-glycosidase digestion mixture without prior desalting, thus saving time and minimizing sample loss.

The conditions for endoglycosidase digestion can be modified similarly to permit direct MALDI-MS analysis of the digestion products. In these in-stances, 25 mM ammonium carbonate is a suitable buffer. It is also important to obtain and use a glycerol-free glycosidase. Enzyme manufacturers typically add glycerol as a stabilizing agent, but this additive is detrimental to MALDI-MS analysis.

III. CONCLUSIONS

Glycoproteins are one of the more difficult types of biological polymers to analyze, and it is not currently possible to obtain every structural detail from every glycoprotein. Consequently, there is need for the continued development of new and more refined procedures for their analysis. This ongoing development has made available a range of analytical strategies from a variety of laboratories. We have focused in this chapter on describing the general analytical strategy of our laboratory, which may be different from that of others. Therefore, we have not been able to discuss every approach currently used in the field, but the reader may like to explore some of these through several review articles covering these subjects [26,55].

ACKNOWLEDGMENTS

This work was supported by the U.S. National Institutes of Health (grant 2-P41-RR05351-07), the National Science Foundation (grant CHE-9626835), and the University of Georgia Research Foundation. We would also like to thank Carl Bergmann, Rosemary Nuri, Michele Ritter, Jennifer White, and Scot Weinberger for their helpful comments.

REFERENCES

1. A Varki. Biological roles of oligosaccharides: all of the theories are correct. Glycobiology 3:97–130, 1993.
2. CF Goochee, MJ Gramer, DC Andersen, JB Bahr, JR Rasmussen. The oligosaccharides of glycoproteins—Bioprocess factors affecting oligosaccharide structure and their effect on glycoprotein properties. Bio/Technology 9:1347–1355, 1991.
3. M Fukuda, B Bothner, P Ramsamooj, A Dell, PR Tiller, A Varki, JC Klock. Structures of sialylated fucosyl polylactosaminoglycans isolated from chronic myelogenous leukemia cells. J Biol Chem 260:12957–12967, 1985.
4. K Yamashita, T Ohkura, Y Tachibana, S Takasaki, A Kobata. Comparative study of the oligosaccharides released from baby hamster kidney cells and their polyoma transformant by hydrazinolysis. J Biol Chem 259:10834–10840, 1984.
5. TW Rademacher, RB Parekh, RA Dwek. Glycobiology. Annu Rev Biochem 57:785–838, 1988.
6. RA Dwek. Glycobiology—Toward understanding the function of sugars. Chem Rev 96:683–720, 1996.
7. N Sharon, H Lis. Carbohydrates in cell recognition. Sci Am 268:82–89, 1993.
8. GD Glick, PL Toogood, DC Wiley, JJ Skehel, J Knowles. Ligand recognition by influenza virus. The binding of bivalent sialosides. J Biol Chem 266:23660–23669, 1991.
9. GR Larsen, D Sako, TJ Ahern, M Shaffer, J Erban, SA Sajer, RM Gibson, DD Wagner, BC Furie, B Furie. P-selection and E-selection—Distinct but overlapping leukocyte ligand specificities. J Biol Chem 267:11104–11110, 1992.
10. Q Zhou, RD Cummings. Cell Surface Carbohydrates and Cell Development. Boca Raton, FL: CRC Press, 1992, pp 99–125.

11. RB Parekh, RA Dwek, BJ Sutton, DL Fernandes, A Leung, D Stanworth, TW Rademacher, T Mizuochi, K Taniguchi, K Matsuta, T Nagano, T Miyamoto, A Kobata. Association of rheumatoid arthritis and primary osteoarthritis with changes in the glycosylation pattern of total serum IgG. Nature 316:452–457, 1985.

12. K Yamashita, H Ideo, T Ohkura, K Fukushima, I Yusa, K Ohno, K Takeshita. Sugar chains of serum transferrin from patients with carbohydrate deficient glycoprotein syndrome evidence of asparagine-N-linked oligosaccharide transfer deficiency. J Biol Chem 268:5783–5789, 1993.

13. K Furukawa, K Matsuta, F Takeuchi, E Kosuge, T Miyamoto, A Kobata. Kinetic study of a galactosyltransferase in the B cells of patients with rheumatoid arthritis. Int Immunol 2:105–112, 1990.

14. RD Cummings, RK Merkle, N Stults. Laboratory Methods in Vesicular and Vectorial Transport. New York: Academic Press, 1989, pp 329–371.

15. YC Lee. High-performance anion-exchange chromatography for carbohydrate analysis. Anal Chem 189:151–162, 1990.

16. A Kobata, K Yamashita, S Takasaki. BioGel P-4 column chromatography of oligosaccharides: effective size of oligosaccharides expressed in glucose units. Methods Enzymol 138:84–94, 1987.

17. JC Bigge, TP Patel, JA Bruce, PN Goulding, SM Charles, RB Parekh. Nonselective and efficient fluorescent labelling of glycans using 2-amino benzamide and anthranilic acid. Anal Biochem 230:229–238, 1995.

18. H Yoshima, T Mizuochi, M Ishii, A Kobata. Structure of the asparagine-linked sugar chains of α-fetoprotein purified from human ascites fluid. Cancer Res 40:4276–4281, 1980.

19. CJ Biermann, GD McGinnis. Analysis of Carbohydrates by GLC and MS. Boca Raton, FL: CRC Press, 1989.

20. A Dell, JE Thomas-Oates. Fast atom bombardment-mass spectrometry (FAB-MS): sample preparation and analytical strategies. In: CJ Biermann, GD McGinnis, eds. Analysis of Carbohydrates by GLC and MS. Boca Raton, FL: CRC Press, 1989, pp 217–235.

21. R Orlando, CA Bush, C Fenselau. Structural analysis of oligosaccharides by tandem mass spectrometry: fragmentation of the sodium adduct ions. Biomed Environ Mass Spectrom 19:747–754, 1990.

22. L Poulter, AL Burlingame. Desorption mass spectrometry of oligosaccharides coupled with hydrophobic chromophores. Methods Enzymol 193:661–689, 1990.

23. SA Carr, GD Roberts. Carbohydrate mapping by mass spectrometry: a novel method for identifying attachment sites of Asn-linked sugars in glycoproteins. Anal Biochem 157:396–406, 1986.

24. L Poulter, R Karrer, AL Burlingame. n-Alkyl p-aminobenzoates as derivatizing agents in the isolation, separation, and characterization of submicrogram quantities of oligosaccharides by liquid secondary ion mass spectrometry. Anal Biochem 195:1–13, 1991.

25. A Dell. Preparation and desorption mass spectrometry of permethyl and peracetyl derivatives of oligosaccharides. Methods Enzymol 193:647–660, 1990.

26. VN Reinhold, BB Reinhold, CE Costello. Carbohydrate molecular weight profiling,

sequence, linkage, and branching data: ES-MS and CID. Anal Chem 67:1772–1784, 1995.

27. MC Huberty, JE Vath, W Yu, SA Martin. Site-specific carbohydrate identification in recombinant proteins using MALD-TOF MS. Anal Chem 65:2791–2800, 1993.

28. BL Gillece-Castro, AL Burlingame. Oligosaccharide characterization with high-energy collision-induced dissociation mass spectrometry. Methods Enzymol 193:689–712, 1990.

29. H van Halbeek. Methods in Molecular Biology. Vol. 17. Totowa, NJ: Humana Press, 1993, pp 115–148.

30. RD Smith, JA Loo, CG Edmonds, CJ Barinaga, HR Udseth. New developments in biochemical mass spectrometry: electrospray ionization. Anal Chem 62:882–899, 1990.

31. K-L His, L Chen, DH Hawke, LR Zieske, P-M Yuan. A general approach for characterizing glycosylation sites of glycoproteins. Anal Biochem 198:238–245, 1991.

32. ME Hemling, GD Roberts, W Johnson, SA Carr, TR Covey. Analysis of proteins and glycoproteins at the picomole-level by online coupling of micro-bore high-performance liquid chromatography with flow fast atom bombardment and electrospray mass spectrometry: a comparative study. Biomed Environ Mass Spectrom 19:677–691, 1990.

33. KL Duffin, JK Welply, E Huang, JD Henion. Characterization of N-linked oligosaccharides by electrospray and tandem mass spectrometry. Anal Chem 64:1440–1448, 1992.

34. SA Carr, MJ Huddleston, MF Bean. Selective identification and differentiation of N- and O-linked oligosaccharides in glycoproteins by liquid chromatography-mass spectrometry. Protein Sci 2:183–196, 1993.

35. MJ Huddleston, MF Bean, SA Carr. Collisional fragmentation of glycopeptides by electrospray ionization LC-MS and LC-MS-MS methods for selective detection of glycopeptides in protein digests. Anal Chem 65:877–884, 1993.

36. KF Medzihradszky, BL Cillece-Castro, RR Townsend, AL Burlingame, MR Hardy. Structural elucidation of O-linked glycopeptides by high-energy collision-induced dissociation. J Am Soc Mass Spectrom 7:319–328, 1996.

37. TN Krogh, E Mirgorodskaya, E Mortz, P Hojrup, P Roepstorff. Characterization of protein glycosylation by post source decay and exoglycosidase digestions. Proceedings of 44th ASMS Conference on Mass Spectrometry, Portland, OR, 1996, p 1336.

38. MJ Kieliszewski, M O'Neill, J Leykam, R Orlando. Structure determination by tandem mass spectrometry of glycopeptides in a pronase digest of the proline-hydroxyproline-rich glycoprotein from Douglas fir. J Biol Chem 270:2541–2549, 1995.

39. DH Patterson, GE Tarr, FE Regnier, SA Martin. C-terminal ladder sequencing via matrix-assisted laser-desorption mass-spectrometry coupled with carboxypeptidase-Y time-dependent and concentration-dependent digestions. Anal Chem 67:3971–3978, 1995.

40. Y Yang, R Orlando. Unpublished data.

41. J Umemoto, VP Bhavanandan, EA Davidson. Purification and properties of an endo-α-N-acetyl-D-galactosaminidase from *Diplococcus pneumoniae*. J Biol Chem 252:8609–8614, 1977.

42. CW Sutton, JA O'Neill, JS Cottrell. Site-specific characterization of glycoprotein carbohydrates by exoglycosidase digestion and laser desorption mass spectrometry. Anal Biochem 218:34–46, 1994.

43. PA Schindler, CA Settineri, X Collet, CJ Fielding, AL Burlingame. Site-specific detection and structural characterization of the glycosylation of human plasma proteins lecithin:cholesterol acyltransferase and apolipoprotein D using HPLC/electrospray mass spectrometry and sequential glycosidase digestion. Protein Sci 4:791–803, 1995.

44. CA Settineri, AL Burlingame. Strategies for the characterization of carbohydrates from glycoproteins by mass spectrometry. In: JW Crabb, ed. Techniques in Protein Chemistry V. San Diego: Academic Press, 1994, pp 97–104.

45. KF Medzihradszky, DA Maltby, SC Hall, CA Settineri, AL Burlingame. Characterization of protein N-glycosylation by reversed-phase microbore liquid chromatography/electrospray mass spectrometry, complementary mobile phases and sequential *exo*-glycosidase digestion. J Am Soc Mass Spectrom 5:350–358, 1994.

46. CW Sutton, AC Poole, JS Cottrell. Carbohydrate characterization of a glycoprotein by matrix assisted laser desorption mass spectrometry. In: RH Angeletti, ed. Techniques in Protein Chemistry IV. San Diego: Academic Press, 1993, pp 109–116.

47. DJ Harvey. Matrix-assisted laser desorption/ionization of oligosaccharides. American Laboratory 26:22–28, 1994.

48. GD Roberts, WP Johnson, S Burman, KR Anumula, SA Carr. An integrated strategy for structural characterization of the protein and carbohydrate component of monoclonal antibodies: application to anti-respiratory syncytial virus Mab. Anal Chem 67:3613–3625, 1995.

49. Y Yang, R Orlando. Simplifying the exoglycosidase digestion/MALDI-MS proceduresfor sequencing N-linked carbohydrate side chains. Anal Chem 68:570–572, 1996.

50. JJ Distler, GW Jourdian. The purification and properties of β-galactosidase from bovine testes. J Biol Chem 248:6772–6780, 1973.

51. A Kobata. Use of endo- and exoglycosidases for structural studies of glycoconjugates. Anal Biochem 100:1–14, 1979.

52. ED Green, A Gabriela, J Baenziger, S Wilson, H van Halbeek. The asparagine-linked oligosaccharides on bovine fetuin. Structural analysis of N-glycanase-released oligosaccharides by 500-megahertz ^{1}H NMR spectroscopy. J Biol Chem 263:18253–18268, 1988.

53. DA Cumming, CG Hellerqvist, M Harris-Brandts, SW Michnick, JP Carver, B Bendiak. Structures of asparagine-linked oligosaccharides of the glycoprotein fetuin having sialic acid linked to N-acetylglucosamine. Biochemistry 28:6500–6512, 1989.

54. AL Burlingame. Characterization of protein glycosylation by mass spectrometry. Curr Opin Biotech 7:4–10, 1996.

55. UN Reinhold, BB Reinhold, CE Costello. Carbohydrate molecular weight profiling, sequence, linkage, and branching data: ES-MS and CID. Anal Chem 67:1772–1784, 1995.

10

Analysis of Phosphorylated Proteins by Mass Spectrometry

Chhabil Dass
The University of Memphis, Memphis, Tennessee

I. INTRODUCTION

A. Traditional Approach for Analysis of Phosphoproteins

For expression of their biological activity, bioactive peptides are synthesized in the cell body via a long chain of events starting with transcription, from the peptide gene, of mRNA that encodes the precursor protein. The nascent precursor protein then undergoes proteolytic cleavage and posttranslational modifications during its transport through the endoplasmic reticulum, Golgi complex, and secretary vesicles to produce a bioactive peptide. The important posttranslational modifications include proteolytic cleavage, acylation (acetyl, formyl, and myristyl), carboxylation, glycosylation, lipidation, amidation, phosphorylation, and sulfation. The knowledge of this so-called "regulated pathway" for a specific precursor family is of practical significance because the inability of a regulatory mechanism to function normally leads to metabolic aberrations underlying various pathophysiological and clinical manifestations.

Next to proteolytic cleavage, protein phosphorylation is the most important and ubiquitous posttranslational event known to occur in proteins and peptides [1,2]. The amino acids involved are serine, threonine, and tyrosine. The significance of phosphorylation stems from its ability to control cellular signaling events. Phosphorylation is also critical to the function of many proteins, hormones, neurotransmitters, and enzymes. It is now clear that phosphorylation is mediated by certain substrate-specific phosphotransferase enzymes called protein kinases, which catalyze the transfer of the terminal-phosphate moiety of a nucleoside triphosphate (ATP or GTP) to the nucleophilic hydroxyl group of serine, threonine, and tyrosine residues of proteins and peptides. A large number of protein kinases have been identified, and their number is increasing [1]. Each kinase contains the consensus sequence for recognition and protein phosphorylation. One class of kinases has recognition motifs that have specificity for phosphorylation of the hydroxyl group of serine and threonine [1–3]. In recent years, tyrosine-specific kinases have attracted much attention [1,3,4]. They are known to participate in transmitting signals that lead to cell growth, proliferation, and differentiation. Phosphorylation is a reversible process. A group of enzymes collectively known as protein phosphatases is involved in dephosphorylation of phosphoproteins and peptides [3].

Because of the unusual importance of phosphorylation and the current high level of interest in this ubiquitous covalent modification, a highly sensitive and specific experimental approach is needed to determine the state of phosphorylation of peptides and proteins. The traditional approach for detection of phosphorylation sites typically starts with radiolabeling cells and tissues to a steady state by incubation with radioactive ^{32}P-phosphate [5–7]. This procedure is followed by isolation of the labeled protein by immunoprecipitation or acid extraction/precipitation, and separation using sodium dodecyl sulfate (SDS) polyacrylamide gel (PAGE) or chromatographic techniques. The isolated radio-labeled protein is then chemically or enzymatically cleaved into smaller fragments, which are resolved by reversed-phase (RP) high-performance liquid chromatography (HPLC) and subjected to Edman degradation for identification of individual phosphorylated amino acids [8]. Although the Edman procedure does not require radiolabeling for sequencing peptides, radiolabeling speeds up identification of phosphopeptide fragments in a protein digest. Detection of individual labeled amino acids is accomplished using thin-layer chromatography (TLC), HPLC, or electrophoresis. Immunoblotting and autoradiography are used to locate those labeled amino acids on the TLC and gel electrophoresis plates. Isoelectric focusing is also used to distinguish polar phosphorylated peptides from the corresponding unmodified derivatives.

This protocol, however, has several weak points. Beside being a potential radioactive hazard, this entire procedure is labor-intensive, prone to sample losses, and often fails to recognize the exact site of phosphorylation. In addition, sometimes complete incorporation of ^{32}P does not occur, and because of multiplicity of phosphorylated species within the cell, the steady state is rarely reached. Identification of phosphorylated amino acids by traditional Edman procedure is also fraught with problems. Under the harsh conditions employed for chemical degradation, the phosphate ester bonds to serine and threonine are less stable. Both undergo β-elimination to form phenylthiohydantoin (PTH) dithioerythritol by-products. In addition, low recoveries of PTH derivatives of phosphotyrosine have been encountered. Some refinements have been made to this procedure. For example, the phosphoserine residue is converted to S-ethylcysteine before the peptide is subjected to the Edman cycle [9]. This modified approach, however, is not applicable to phosphotyrosine and threonine residues.

B. Role of Mass Spectrometry in Characterization of Phosphorylated Proteins

Alternatively, mass spectrometric methods can offer a viable strategy to analyze proteins and peptides containing phosphorylated amino acid residues. Mass spectrometry is probably the most versatile and comprehensive analytical tool and an indispensable component in the arsenal of biomedical research. This distinction is the result of its capability to provide unsurpassed molecular specificity, high detection sensitivity, and unparalleled versatility in determining structures of unknown compounds. High molecular specificity is achieved because of incontrovertible genealogical relationship between the precursor and its product ions. A mass spectrum is usually the signature of an analyte because it contains both molecular mass and compound-specific fragment ions. The emergence of tandem mass spectrometry (MS/MS) provides an opportunity for mixture analysis and increases molecular specificity even further due to exclusive mass selection of the precursor ions of an analyte [10].

Traditionally, ionization in mass spectrometry is accomplished by bombarding the sample with an electron beam. Although this method is very simple and easy to use, the requirement that the analyte must be present in the ion source in the form of gaseous molecules restricts its use to low mass compounds only. A large number of biopolymers such as peptides and proteins are excluded from the protocol of electron ionization. The advent of fast atom bombardment (FAB) [11] and its variation, liquid secondary ionization mass spectrometry (liquid SIMS) [12], ^{252}Cf-plasma desorption ionization [13], electrospray ionization (ESI) [14,15], and matrix-assisted laser desorption ionization (MALDI) [16] has revolutionized the protein and peptide analysis approach. These novel concepts in ionization have made mass spectrometric analysis of macromolecules a reality. Currently, biopolymers with molecular mass over 100,000 Da are analyzed routinely. Furthermore, mass spectrometry has a unique advantage of being interfaced with a variety of chromatographic systems such as gas chromatography (GC), HPLC, capillary electrophoresis (CE), supercritical fluid chromatography (SFC), and TLC, allowing analysis of a variety of sample types from gaseous to highly involatile and thermally labile compounds.

Protein and peptide chemistry has greatly benefited from these impressive developments. Mass spectrometry is playing an ever-increasing role in solving research problems in biomedical sciences. Complete sequencing of peptides, determination of molecular mass of proteins and peptides, verification of the primary structure of proteins predicted from the cDNA sequence, characterization of natural mutants, and identification of posttranslational modifications are now routinely accomplished by mass spectrometry techniques.

Several strategies for characterization of proteins and peptides containing phosphorylated amino acid residues have been devised. Earlier work in this field used FAB (or liquid SIMS) with a varying degree of success [17–20]. FAB and liquid SIMS are conceptually similar except for the fact that FAB uses a beam of fast atoms and liquid SIMS a beam of high-energy ions (usually Cs^+ ions). Of the two, liquid SIMS provides better sensitivity and high mass capability due to the fact that the ion beam used in liquid SIMS is much more focused and of higher energy than the atom beam used in FAB. From here on in this chapter, the term liquid SIMS is used for both atom-beam and ion-beam techniques. The combination of conventional mass spectrometry and liquid SIMS has been successfully applied for sequence analysis of phosphopeptides [21]. However, a better approach for sequence determination of phosphopeptides is to use a combination of liquid SIMS and tandem mass spectrometry

[20]. Liquid SIMS, either alone or coupled with MS/MS, is now a well-established technique for sequencing peptides [22–25].

The last few years have witnessed the advent of methods based upon ESI-MS [26–30]. This technique is more sensitive than liquid SIMS and is uniquely qualified to act as an interface between HPLC (and CE) and mass spectrometry. Using this combination, a more efficient approach to screen the presence of phosphopeptides from the HPLC eluents has been developed [29,30]. The HPLC peaks are identified via selected-ion monitoring (SIM) of the phosphate group marker ions, such as PO_3^- and PO_2^- (and possibly $H_2 PO_4^-$). MALDI-time-of-flight (TOFMS) is another ideal mass spectrometry technique for the analysis of phosphopeptides. The molecular mass of each peptide fragment in a proteolytic digests can be easily determined by this technique. Applications of TOFMS have also been developed to establish the sequence of phosphopeptides [31]. These peptides are first ionized by MALDI and the products of their decomposition are then mass analyzed and detected in a reflectron TOF mass spectrometer by the postsource decay approach. MALDI-TOFMS, in combination with protein ladder sequencing technique, is also applied to identify phosphorylated residues in polypeptides [32]. The peptide ladders are generated by controlled stepwise chemical degradation of the peptide at the N-terminus. Mass measurement of the resulting peptide ladders helps in identifying phosphorylated residues.

Mass spectrometric methods for characterization of phosphorylated proteins and peptides have several potential advantages. These methods are inherently faster and require less sample. In some methods, sequence information can be obtained with subpicomolar amounts of proteins. When the sequence of a protein is known, the presence of phosphopeptides in the digest may be established quickly from the measured molecular masses. The mass of a phosphorylated peptide is higher by 80 Da for each phospho unit from the corresponding nonphosphorylated analog. The location of the phosphate group in the sequence of a peptide can be unambiguously pinpointed by tandem mass spectrometry. In addition, mass spectrometry is the only convenient method for blocked N-terminal peptides. Furthermore, hazardous radioactive isotope labeling is not required.

II. MASS SPECTROMETRY PROTOCOL

This chapter describes a mass spectrometry-based protocol for identification of the precise state of phosphorylation in proteins and peptides. Two mass spectrometric techniques, liquid SIMS and ESI-MS, are used in this protocol. The analysis begins with a completely pure protein. If the protein is not pure, standard procedures, involving homogenization of the target tissue or cell culture and separation of the target protein by using SDS-PAGE or gel permeation chromatography, may be followed for purification of proteins from biological specimens [33]. The homogenization step, however, should be performed under conditions that prevent phosphorylation and dephosphorylation. When the primary structure of the protein is known, which normally is the case, a simple mass measurement of the intact protein can determine whether the protein exists in the phosphorylated form. This analysis can be performed by ESI-MS. Some situations require that the primary structure of the protein be known to solve the problem. A standard procedure for determining the primary structure of a protein is to follow the peptide-mapping approach. This protocol is illustrated in Figure 1. The first step of this protocol is site-specific cleavage of the purified protein into smaller manageable peptide fragments by chemical degradation or proteolysis. The next step is molecular mass determination of the peptide fragments thus derived. Both liquid SIMS and ESI-MS can be used. This information can rapidly verify phosphorylation of the protein. For location of the exact site of phosphorylation, the amino acid sequence of the peptide fragments is determined. Peptide fragments are fractionated into individual components by using RP-HPLC and their sequence is determined by conventional mass spectrometry or tandem mass spectrometry following their ionization by liquid SIMS. Alternatively, the peptide fragments are analyzed by an

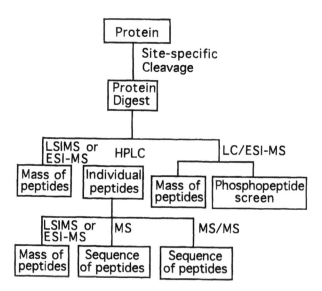

Figure 1   Mass spectrometry–based protocol for the analysis of phosphorylated proteins.

online combination of HPLC and ESI-MS. This procedure provides an elegant and rapid means to accurately determine the molecular mass of each fragment directly and for selective detection of phosphopeptides in the digest. The presence of phosphorylated residues in peptides is ascertained by determining the sequence of only those selected peptides.

A. Site-Specific Protein Cleavage

For structural analysis of a protein, the purified protein is broken into smaller segments by digestion with one or two selected endoproteases and/or by chemical digestion [5,33] (see Table 1). This approach is equally applicable to proteins containing phosphorylated residues. Cyanogen bromide (CNBr) is the most used chemical reagent. It cleaves a protein at the C-terminal side of methionine. In acidic conditions, most of the resulting C-terminal homoserine residues are in the lactone form. Because methionine is less frequently present in proteins, the peptide fragments generated by the CNBr digestion are very few but large. ESI and MALDI are usually the mass spectrometry techniques of choice for the analysis of those larger peptide segments.

Various amino acid-specific enzymes are available that will yield reproducible peptide maps of a protein [5,33,34] (see Table 1). The use of any particular enzyme chemistry depends upon the primary structure of the protein and the information desired. Trypsin is more or less a universal choice for most applications. This enzyme cleaves a protein at the C-terminal side of lysine and arginine residues. These residues are frequently present in proteins. However, the Lys-Pro bond is not affected by the trypsin action. The chemical purity of trypsin may be an important factor. Care should be taken that it does not contain chymotrypsin. Otherwise, additional cleavages at the peptide bonds C-ter-

Table 1 Protein Cleaving Agents

Cleaving agent	Specificity	Digestion conditions
Cyanogen bromide	Met-X	70% TFA
N-Chlorosuccinimide	Trp-X	50% Acetic acid
Trypsin	Arg-X, Lys-X	50 mM NH$_4$HCO$_3$, pH 8.5, 37°C
Chymotrypsin	Phe-X; Tyr-X; Trp-X; Leu-X	50 mM NH$_4$HCO$_3$, pH 8.5, 37°C
Thermolysin	X-Leu; X-Ile; X-Val; X-Met; X-Phe; X-Ala	50 mM NH$_4$HCO$_3$, pH 8.5, 40°C
Endoproteinase Asp-N	X-Asp	50 mM NH$_4$HCO$_3$, pH 7.6, 37°C
Endoproteinase Glu-C	Glu-X	50 mM NH$_4$HCO$_3$, pH 7.6, 37°C
Endoproteinase Arg-C	Arg-X	50 mM NH$_4$HCO$_3$, pH 8.0, 37°C
Endoproteinase Lys-C	Lys-X	50 mM NH$_4$HCO$_3$, pH 8.5, 37°C

minal to hydrophobic residues will occur. Larger and overlapping fragments can be generated by treatment with other endoproteases (Table 1). As an example, endoproteinase Lys-C can be used to generate peptide fragments by cleaving the protein at positions C-terminal to lysine. Similarly, endoproteinase Arg-C and endoproteinase Glu-C (also known as *Staphylococcus aureus* V8 protease) are used to cleave bonds C-terminal to arginine and glutamic acid residues, respectively. Endoproteinase Asp-N is specific to the bond N-terminal to aspartic acid.

A standard procedure can be used for digestion of a protein (discussed later). However, care should be taken to use volatile buffers; otherwise, ionic components of the buffer will interfere in both liquid SIMS and ESI-MS analyses. Ammonium carbonate and ammonium bicarbonate both can be used to adjust pH of the digestion mixture between 7.0 and 8.5. The volatile salts are removed by a lyophilization step. If a nonvolatile buffer is used or any other nonvolatile components or salts are present, a solid-phase extraction step may be incorporated.

1. Experimental Conditions for Digestion of Proteins

The digestion conditions and buffers used with various reagents are listed in Table 1. The typical experimental conditions for digestion of proteins with cynogen bromide and endoproteases are as follows.

Digestion with Cyanogen Bromide. Dissolve the protein (1 mg/100 μl) and CNBr (2 mg/100 μl) in 70% trifluoroacetic acid (TFA). To approximately 100–200 pmol of the protein, add 100-fold molar excess per methionine residue of CNBr solution, and digest the contents for 2 hr at ambient temperatures. Remove the excess CNBr by rotary evaporation or by using an SPE cartridge.

Digestion with Endoprotease. The following procedure is used for digestion of proteins with trypsin: Dissolve the protein (100 μg/100 μl) and *N*-tosyl-L-phenylalanine chloromethylketone (TPCK)-trypsin (20 μg/100 μl) in the digestion buffer (0.05 *M* ammonium bicarbonate). Add the trypsin solution to the protein solution in the 1:50 (w:w) enzyme:substrate ratio and digest the mixture at 37°C, preferably overnight. Vortex the solution. If needed, add a fresh portion of trypsin and digest at 37°C for another 3–5 hr. Lyophilize the content of the digest. The procedure employing other endoproteases is essentially the same as used here for trypsin digestion (see Table 1 for the pH and buffers used).

B. Fractionation of Protein Digest by RP-HPLC

Because of its speed, simplicity, and extremely high resolving power, RP-HPLC is the most suitable technique for fractionation of complex mixtures of peptides in a protein digest. The success of the HPLC separation largely depends upon the appropriate choice of column dimensions, column packing, mobile phase, and gradient. A 250×2.1-mm C-18 column packed with particles 3–5 μm having a pore diameter of 300 Å provides an excellent separation of tryptic peptides. This column is suitable for separation of 50-pmol samples. A C-4 column may be used for separation of larger peptides, such as those generated by the CNBr treatment. The water–acetonitrile mobile-phase system is invariably used for peptide separation with ultraviolet (UV) detection at 200–210 nm. A flow rate of 0.2–0.3 ml/min is used. For large sample amounts, a 250×4.6-mm column with a mobile-phase flow rate of 0.5–1.0 ml/min may be used.

Only volatile buffers and ion-pairing reagents should be used for purification and separation of peptides. TFA, acetic acid, formic acid, triethanol amine:formic acid, ammonium acetate, and ammonium bicarbonate are most suitable additives to the HPLC mobile phase. All these compounds can be used for separation of phosphopeptides, except the triethanol amine:formic acid buffer, because triethanol amine forms adducts with the phosphate group (unpublished work from our lab).

1. Experimental Procedure for Separation and Purification of Peptides

In our laboratory, separation and purification of synthetic peptides is achieved by RP-HPLC using a Hewlett Packard model 1050 liquid chromatograph. This LC system consists of a quaternary gradient pumping system, a diode array UV-visible detector, and a Chemstation data system. The samples are injected via a 20-μl sample loop onto an analytical column (250×4.6 mm) packed with C-18 stationary phase. For elution, a linear gradient of acetonitrile in deionized water is optimized (e.g., from 12 to 80% in 50 min). TFA (0.1%) is added to the mobile phase as the ion-pairing reagent. The flow rate is adjusted to 1.0 ml/min. The fractions are monitored at 200 nm. To minimize adsorption losses, the peptide fractions are collected in polypropylene tubes. The solvent is removed by lyophilization. For the separation of individual fragments of the protein digest, the use of a narrow-bore column and a flow rate of 0.3 ml/min is preferred. The fractions are collected in Eppendorf tubes.

2. Solid-Phase Extraction Procedure for Desalting the Sample

The use of a disposable SPE cartridge is a simple and convenient way to concentrate the sample and remove salts from it. A simple procedure outlined here is followed to purify a peptide sample: Wet the disposable C-18 SPE cartridge with 2–3 volumes of methanol, and load the protein digest onto the cartridge. Wash the cartridge with 5–6 ml of deionized water to remove highly polar materials. Elute the peptide mixture with 5–6 ml of methanol. Lyophilize the contents and save for further analysis.

C. Mass Spectral Analysis

The last several years have witnessed dramatic developments in instrumental design and refinements in mass spectral analysis techniques. Concurrent with these developments, protein chemistry has also undergone profound changes. The older methods of analysis are being replaced with newer, more sensitive and convenient methods. In this chapter, the use of liquid SIMS and ESI-MS, the two most modern mass spectrometry techniques, for the analysis of phosphorylation of proteins is described. Although liquid SIMS is relatively less sensitive and somewhat troubled by matrix and other interferences, the use of this technique in the mass spectrometry protocol is included because it is simple to use and readily available in most modern mass spectrometry laboratories. The exact procedure used in mass spectrometry analysis of protein digests depends upon the type of instrumentation available and the information desired. For structural characterization of a peptide, two types of information are obtained, the molecular mass and the sequence-determining fragment ions. Liquid SIMS and ESI-MS both are suitable for these measurements. The use of a combined RP-HPLC/ESI-MS approach is also described for characterization of peptides and selective identification of phosphorylated peptides in a mixture.

III. METHODS BASED UPON LIQUID SIMS TECHNIQUE

Liquid SIMS is used for the analysis of condensed-phase polar compounds. Ionization of a compound using this technique is accomplished by mixing the compound with a nonvolatile and usually polar liquid matrix that is bombarded by a beam of high-energy Cs^+ ions. The original work of Barber used a beam of keV-energy fast atoms [11]. The selection of a proper matrix is crucial to the success of liquid SIMS analysis. The intensity of the peptide molecular ion signal is influenced by the surface composition of the sample/matrix mixture. In a mixture of hydrophilic and hydrophobic peptides, the latter have a tendency to occupy the surface of the matrix and thus overwhelm the mass spectrum, whereas hydrophilic peptides exhibit a poor response. Glycerol, α-thioglycerol, 3-nitrobenzyl alcohol, and a 5:1 (w:w) mixture of dithiothreitol and dithioerythritol (DTT/DTE) are the most suitable matrices for peptide analysis. However, the information desired will dictate the choice of a particular matrix. A recent publication from our laboratory has demonstrated that the extent of fragmentation observed in peptides during liquid SIMS is dependent upon the matrix used [35]; glycerol promotes increased fragmentation, whereas the molecular ions formed using α-thioglycerol and the DTT/DTE mixture are relatively stable. Therefore, it is prudent to use α-thioglycerol or the DTT/DTE mixture when the object is to determine the molecular mass of a peptide or when its sequence is to be determined by MS/MS. However, when conventional mass spectrometry is used for sequence determination, it is best to dissolve the peptide in glycerol. The matrix surface composition may be altered by adjusting the pH or by addition of surfactants. For best results the sample should be pure. The presence of salts or any other metal cations is a cause of concern in liquid SIMS analysis. Care should be taken to remove these impurities; otherwise, the detection sensitivity will be lower due to adduct formation between these metal cations and peptides. The SPE procedure described earlier can be used to remove salts in a sample.

A. Molecular Mass Determination by Liquid SIMS

Although ionization of samples up to 15,000 Da mass has been observed with liquid SIMS, the use of this technique for molecular mass measurements of the peptide fragments in a protein digest should be restricted to fragments of 5–30 amino acids. Most tryptic fragments are within this mass range. These fragments can be analyzed directly without prior separation. The comparison between the peptide map of a normal protein and the corresponding phosphorylated protein in many cases will enable determination of the phosphorylation state. The phosphorylated protein digest will show the presence of new signals and the absence of expected signals. The new peaks will show an increase in mass of 80 Da for each phospho group. An amount of 10–100 pmol of a peptide is required for molecular mass determination by liquid SIMS. This amount of sample should be dissolved in less than 1 μl volume of α-thioglycerol or the DTT/DTE mixture. Addition of small amounts of trifluoroacetic acid (TFA; 1 μl of 0.1%) or acetic acid (1 μl of 5%) to the sample/matrix mixture will promote the formation of the $[M+H]^+$ ions. A well-focused beam of high-energy Cs^+ ions should be used for ionization of the peptide fragments. A double-focusing magnetic sector instrument equipped with a high-field magnet is a better choice for mapping the protein digest, although in some cases a high-mass-range quadrupole mass filter can also be used.

Because the phosphate group imparts hydrophilic character to a peptide, phosphopeptide mixtures can be a problem in liquid SIMS analysis. Although most of the peptides may be detected in a mixture, phosphorylation of small and hydrophilic peptides may escape detection by liquid SIMS due to suppression of their signal by other surface-active hydrophobic peptides. In addition, the signal intensity may vary from one peptide to another. Therefore, to negate this suppression effect, the spectrum should be acquired over an extended period. A representative spectrum is then obtained by summing and averaging all scans. Another alternative to avoid this situation is to generate larger peptide fragments by using a different degradation scheme (Table 1).

1. Experimental Procedure for Molecular Mass Determination
   of Peptide Fragments by Liquid SIMS

The molecular mass determination of individual members of the protein digest
with liquid-SIMS can be accomplished by using a double-focusing magnetic
sector or a quadrupole instrument. We use the front end (which consists of a
magnetic sector sandwiched between two electric sectors, i.e., EBE geometry)
of a hybrid tandem mass spectrometer of EBE-qQ geometry (Micromass AutoSpec Q). The instrument is equipped with Digital Vax Station 3100-based
data system and Opus software. Ionization is accomplished by bombarding the
peptide/matrix mixture with a beam of Cs^+ ions produced by typically operating the Cs^+ ion gun at voltages between 30 and 40 keV and an emission of
approximately 1.1 μA. The ions are accelerated out of the source at a potential
of 8 kV. α-Thioglycerol is used as a liquid matrix. The following steps should
be followed: Optimize the energy and emission current of the Cs^+ ions beam
to provide maximum ion signal. Calibrate the entire mass range of analysis
with CsI cluster ions. Dissolve the lyophilized digest in a 50:50 (v:v) wa-
ter:methanol solvent and mix thoroughly an aliquot of this solution (5–10 μl
containing about 5 μg of the protein digest) with 0.5 μl of matrix on the stainless steel target. Because of the possibility of sample loss due to sudden sput-
tering in the vacuum lock of the ion source, do not add the entire amount of
the sample solution at one time. Add 2 μl of the solution initially, followed by
heating the probe gently under an infrared (IR) lamp. Repeat the procedure
until all of the aliquot has been added to the target. Add 1–2 μl of 0.1% TFA
to the sample/matrix mixture. Acquire the mass spectra by scanning the magnet
in the appropriate mass range at a scan speed of 7 sec/decade and at a mass
resolution of 1500. Several scans should be acquired to obtain a representative
spectrum.

B. Sequence Determination by Liquid SIMS and Conventional Mass Spectrometry

Identification of the site of phosphorylation requires the knowledge of the
amino acid sequence of a peptide. A completely pure sample should be used
when the sequence is determined by conventional mass spectrometry. As discussed earlier, this requirement is met by separating the individual members of
the protein digest by using RP-HPLC.

To sequence a peptide by mass spectrometry it is essential to know the
molecular mass and the m/z values of sequence-specific fragment ions. The
molecular mass is determined from the m/z value of the $[M + H]^+$ or $[M - H]^-$
ions. Although the molecular ions of peptides produced by liquid SIMS usually
possess only a little excess energy over their ground state, often this energy is

sufficient to produce a meaningful fragmentation profile in a number of peptides. A systematic study from our laboratory of seven phosphorylated peptides has shown that extensive fragmentation is observed during liquid SIMS if a peptide is dissolved in glycerol [21]. Although we have not investigated the upper range of peptides that can be sequenced with this approach, we believe that useful sequence information may be obtained for peptides having 12–15 amino acids.

A discussion of the types of ions formed during fragmentation of peptides is pertinent here. The excess energy deposited during the desorption/ionization step causes cleavage of bonds all along the peptide backbone and in various side chains [22,23]. Three different types of bonds are broken in a peptide chain, namely, the alkyl carbonyl bond (CHR—CO), the peptide amide bond (CO—NH), and the amino alkyl bond (NH—CHR) (Scheme 1). The charge is retained either by the N-terminus or the C-terminus fragment. Thus, six different sequence-specific ion series may be formed. These ions are represented by symbols a, b, c, x, y, and z [36,37]. Further cleavage in the nth side chain of these newly formed ions produces secondary ions denoted by symbols w_n, v_n, and d_n. Although some studies have claimed that the latter type of ions is formed only during the collision-induced dissociation (CID) step [24,25], re-

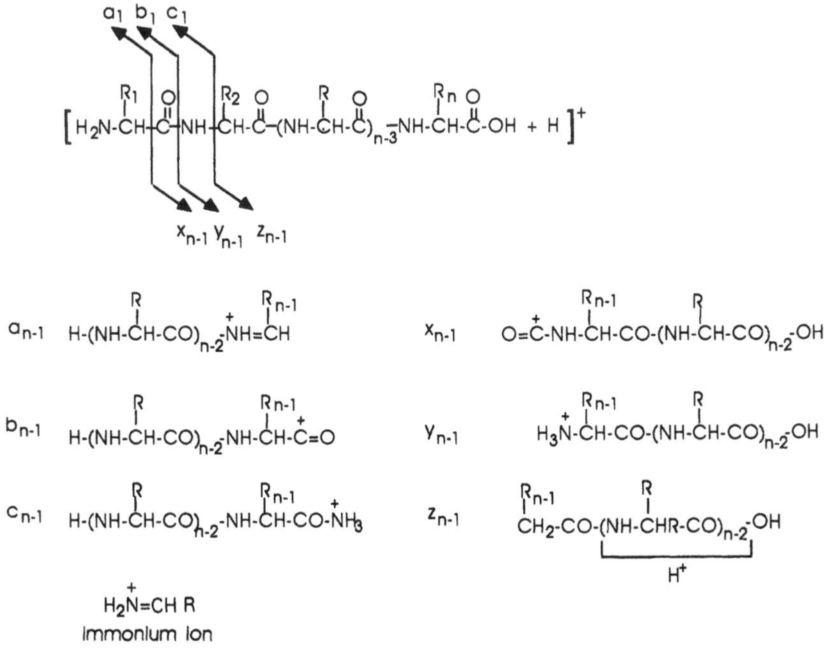

Scheme 1   The mechanism of formation of sequence-specific ions.

sults from our laboratory have shown that these ions are also present in a conventional mass spectrum [21] (see Figure 2). Cleavage of the peptide backbone also produces immonium ions and internal fragments (Scheme 1). Immonium ions may undergo further fragmentation (Scheme 2). Immonium ions and their fragments confirm the presence of certain amino acids, whereas the sequence of the peptide is derived from the backbone fragments.

The first important task in deriving the sequence a peptide from the mass spectral data is to recognize the molecular ion of the peptide. This task is easily accomplished due to the overwhelming abundance of the $[M+H]^+$ ions in the spectrum. From the m/z value of the $[M+H]^+$ ion, the molecular mass of the peptide can be determined. The next step is to identify a specific sequence ion series. Determination of the entire sequence of a peptide requires that all members of at least one of the six sequence ion series are present in the spectrum. Often, two partial overlapping sequence ion series, one each from both ends, will also suffice. Next, terminal amino acids in the sequence of a peptide are recognized. Identity of the remaining amino acid residues from either N- or C-terminus is established by the difference in masses between the successive ions of a specific sequence ion series. Phosphoserine, -threonine, and -tyrosine resi-

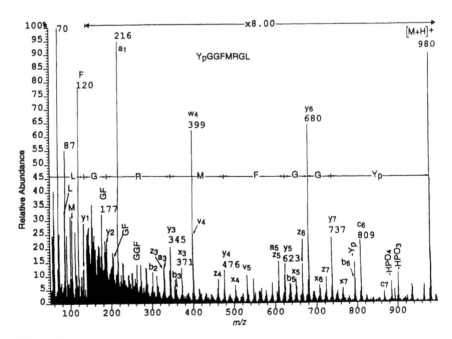

Figure 2 Positive-ion conventional mass spectrum of $Y_pGGFMRGL$. Glycerol was used as the matrix and liquid SIMS for ionization of the peptide. (From ref. 21. Copyright John Wiley & Sons Limited. Reproduced with permission.)

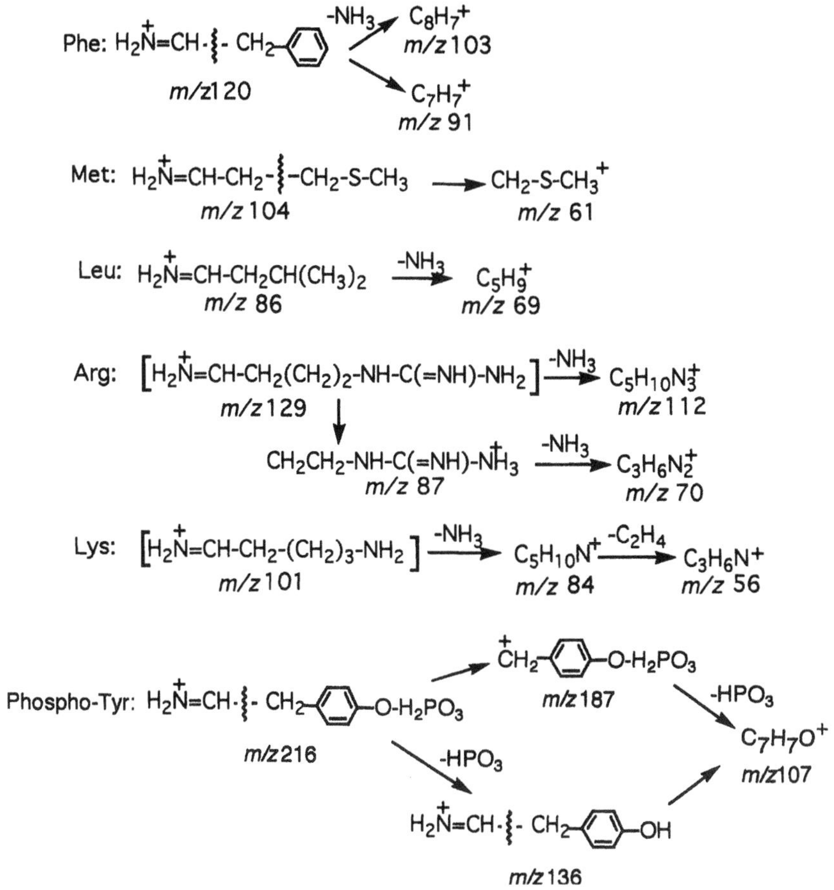

Phe: $H_2\overset{+}{N}=CH\cdot\{-CH_2-\langle\bigcirc\rangle$

$\xrightarrow{-NH_3} C_8H_7^+$ $m/z\,103$

$m/z\,120$

$\searrow C_7H_7^+$ $m/z\,91$

Met: $H_2\overset{+}{N}=CH-CH_2-\{-CH_2-S-CH_3 \longrightarrow CH_2-S-CH_3^+$

$m/z\,104$ $m/z\,61$

Leu: $H_2\overset{+}{N}=CH-CH_2CH(CH_3)_2 \xrightarrow{-NH_3} C_5H_9^+$

$m/z\,86$ $m/z\,69$

Arg: $\left[H_2\overset{+}{N}=CH-CH_2(CH_2)_2-NH-C(=NH)-NH_2\right] \xrightarrow{-NH_3} C_5H_{10}N_3^+$

$m/z\,129$ $\downarrow$ $m/z\,112$

$CH_2CH_2-NH-C(=NH)-N\overset{+}{H}_3 \xrightarrow{-NH_3} C_3H_6N_2^+$

$m/z\,87$ $m/z\,70$

Lys: $\left[H_2\overset{+}{N}=CH-CH_2-(CH_2)_3-NH_2\right] \xrightarrow{-NH_3} C_5H_{10}N^+ \xrightarrow{-C_2H_4} C_3H_6N^+$

$m/z\,101$ $m/z\,84$ $m/z\,56$

Phospho-Tyr: $H_2\overset{+}{N}=CH\cdot\{-CH_2-\langle\bigcirc\rangle-O-H_2PO_3$

$\overset{+}{C}H_2-\langle\bigcirc\rangle-O-H_2PO_3$ $m/z\,187$

$m/z\,216$

$\xrightarrow{-HPO_3} C_7H_7O^+$ $m/z\,107$

$\searrow -HPO_3$

$H_2\overset{+}{N}=CH\cdot\{-CH_2-\langle\bigcirc\rangle-OH$

$m/z\,136$

Scheme 2 The mechanism of formation of certain amino acid–specific low-mass ions.

dues are recognized when the mass of two ions in a series differ by 167, 181, or 243 Da, respectively.

Because of the presence of a strongly acidic group in phosphopeptides, negative-ion liquid SIMS analysis may also be useful for obtaining sequence information. If sufficient peptide is available, it is a good practice to obtain spectra in both ionization modes. Often, the negative-ion analysis provides complementary sequence ions data [22]. The structural features that stabilize a negative ion are usually different than those that stabilize a positive ion.

If this procedure does not provide the complete sequence of a phospho-peptide or the identity of the phosphorylated residues cannot be established from the partial sequence, one can use an alternative approach, in which the terminal amino acid is removed sequentially and the molecular mass of the truncated peptides is determined after each degradation cycle [25,38]. For removal of the C-terminal amino acids, the peptide is treated with carboxypeptidase. The N-terminal residues are removed by subjecting the peptide to controlled stepwise Edman degradation or treatment with aminopeptidase.

1. Experimental Procedure for Amino Acid Sequence Determination by Liquid SIMS and Conventional Magnet Scan

The instrument used in the previous section (the front end of the AutoSpec Q) is also used for this type of measurement. An aliquot (5–10 μl containing 3–5 μg of the peptide) of the reconstituted HPLC fraction of the protein digest is dispersed on the stainless steel target that already contains 0.5 μl of glycerol. The use of glycerol is preferred because it promotes fragmentation of the peptide. There is no need to acidify the peptide/glycerol mixture, because addition of an acid suppresses fragmentation [35]. A beam of Cs^+ ions (30–40 keV and 1.1 μA emission) is used to ionize the peptide. Acquire the spectrum by scanning the magnet from 20–30 Da above the m/z value of the $[M+H]^+$ ion of the peptide down to 100 Da at a scan rate of 7 sec/decade and at a resolution of 1500 Several scans (10–12) are acquired, summed, and averaged. The data is smoothed to provide a good signal-to-noise (S/N) ratio. If the peptide is large (12–15 amino acids long), acquire the complete mass spectrum in two overlapping mass ranges. This procedure improves detection sensitivity. If needed, the spectrum can also be acquired in the negative-ion mode using the same procedure.

2. A Specific Example of Sequence Determination by Liquid SIMS and Conventional Magnet Scan

The potential of liquid SIMS for sequence determination of phosphopeptides is illustrated in Figure 2 using the example of an octapeptide Y_pGGFMRGL of molecular mass 979 Da, where Y_p is the phosphorylated tyrosine residue. This spectrum was obtained in the positive-ion mode for less than 5 nmol of the peptide dissolved in glycerol. Several sequence-specific ions are recognized in this spectrum. The foremost are y-type C-terminal ions. A complete coverage of this ion series is observed. Most members of the x-type and z-type ion series are also present. Because the peptide analyzed in this example is of known sequence, the sequence ions are readily discerned. However, in the case of an unknown sequence, the presence of satellite ions may help in unambiguous identification of a specific sequence ion series. For example, identification of y-type ions will be less demanding if x- and z-type ions are also present. These ions are found in the positive-ion spectrum at 26 Da above and 15 Da below, respectively, the m/z value of y-type ions. The same strategy can be applied to recognize the N-terminal ions (e.g., b-type ions will be associated with a- and c-type ions). However, the N-terminal ions are fewer and less abundant in the spectrum of $Y_pGGFMRGL$. This spectrum also contains immonium ions and their low-mass fragments due to phenylalanine (m/z 120, 103, and 91), methionine (m/z 104 and 61), leucine (m/z 86 and 69), phosphotyrosine (m/z 216, 187, 136, and 107), arginine (m/z 112, 87, and 70), and lysine (m/z 84 and 56) (see Scheme 2). Another unique mass spectral feature of phosphotyrosine-containing peptides is the loss of HPO_3 and HPO_4 neutrals from the $[M+H]^+$ ions.

In order to compare the positive- and negative-ion modes of analysis, the negative-ion spectrum of the same octapeptide is shown in Figure 3. This spectrum is endowed with a number of sequence-specific fragment ions. The N- and C-termini ion series are both well defined. Thus, this spectrum may be used to provide sequence from both ends of the peptide. The sequence of a peptide is derived with a greater confidence when two overlapping N- and C-termini ion series are used.

C. Sequence Determination by Liquid SIMS and Tandem Mass Spectrometry

A conventional mass spectrum obtained by liquid-SIMS is often beset with chemical background noise caused by the matrix ions and sample impurities still present even after following the best possible purification steps. Therefore, unambiguous assignment of the sequence ions may be a risk. Furthermore, as the size of the peptide increases the extent of fragmentation is reduced because less energy per bond is available to induce fragmentation. One approach to overcome these drawbacks is to use tandem mass spectrometry. The unique feature of this technique is its capability to analyze a mixture of peptides, because two stages of mass analysis are used in this procedure. The first stage (MS1) exclusively mass selects the molecular ion of the desired peptide, which undergoes fragmentation in the intermediate region, and the second stage (MS2) mass analyzes those fragment ions. Tandem mass spectrometers are equipped with some provision to induce fragmentation of energetically cool (i.e., nondecomposing ions) mass-selected peptide ions. The most common method is collision activation, in which the rapidly moving sample ions collide with a neutral gas, usually helium, causing extensive fragmentation in a peptide. However, even with this approach, sequencing of peptides above mass 3500 Da is problematic, primarily due to decreased molecular ion abundance and reduced efficiency of CID of high-mass peptides.

In order to increase sensitivity of sequence analysis by tandem mass spectrometry, it is important to reduce the ion source fragmentation and increase

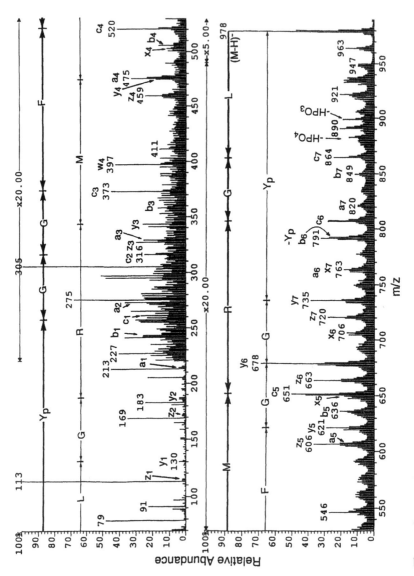

Figure 3 Negative-ion conventional mass spectrum of $Y_pGGFMRGL$. Glycerol was used as the matrix and liquid SIMS for ionization of the peptide. (From ref. 21. Copyright John Wiley & Sons Limited. Reproduced with permission.)

the molecular ion signal. As discussed earlier, this can be achieved to some extent by optimization of the matrix composition. The lack of fragmentation in α-thioglycerol and the DTT/DTE mixture is a deciding factor in favor of their use as a matrix. Further improvement in the $[M + H]^+$ ion signal can be achieved by acidifying the matrix/sample mixture with either acetic acid or TFA. The MS/MS spectrum can be acquired with <5 nmol of a phosphopeptide.

1. Equipment for Tandem Mass Spectrometry

Several options are available with respect to the instrumental configurations for MS/MS operation [10,39; also see this volume]. One configuration is a four-sector mass spectrometer, in which two double-focusing magnetic sector mass spectrometers are combined. In this arrangement, the collisions are usually performed in the field-free region located between MS1 and MS2. Both mass selection and mass analysis are at high resolution. Another benefit of the magnetic sector-based MS/MS approach is that fragmentation occurs at keV energies via single collisions, resulting in reproducible and well-understood mass spectra. However, these instruments are very expensive and out of the reach of most laboratories.

A cheaper alternative to the four-sector design is a double-focusing machine of either EB or BE configuration (where B and E are the magnetic and electric fields). In the former, CID of the mass-selected peptide molecular ion is performed in the first field-free region and the products are mass analyzed by using a linked-field scan, in which both electric and magnetic fields are scanned together while keeping their ratio constant. With this arrangement, mass analysis of the products is at high resolution, but the precursor ion selection is very poor. We have used this approach successfully for sequence analysis of several peptides [40]. In the BE design, the precursor ion is mass selected by adjusting the magnetic field and the products are analyzed by scanning the electric field. This mode of analysis is not an ideal means for sequencing peptides due to very poor mass resolution of the product ions.

A triple-sector quadrupole, consisting of three sequentially arranged quadrupole analyzers, is also adequate for many applications. The first and third quadrupoles are used for mass selection and mass analysis, respectively, whereas the rf-only middle quadrupole is used as the collision region. Dissociation of the mass-selected ions occurs via low-energy multiple collisions. The resulting spectrum, however, depends upon the pressure and nature of the collision gas, as well as upon the collision energy used. Thus, the low-energy CID spectrum may differ from instrument to instrument. Mass selection and mass analysis both are at unit resolution. The main advantages of this device are low cost, simplicity of instrumentation, and ease of coupling with high-performance separation devices.

Another viable option is a hybrid configuration, typically consisting of a double-focusing magnetic sector instrument as MS1 and two quadrupoles as MS2. The first quadrupole is operated in the rf-only mode and serves as the collision region. The best performance features of the two types of mass spectrometers are utilized in this design. One of the unique advantages of this configuration is that both low-energy and high-energy CID data can be acquired using the same instrument. The high-energy CID spectrum is acquired in a manner similar to that used in the EB instrument. For low-energy CID, the precursor ions are mass selected by MS1 into the rf-only quadrupole. In this experiment, mass selection is attained at high resolution, permitting selection of monoisotopic ions containing ^{12}C atoms only, and mass analysis of the fragments is at unit mass resolution. The potential of TOF machines with postsource decay option is also being explored for MS/MS analysis of phosphorylated peptides [31]. TOF mass spectrometers have also been incorporated as MS2 in newer generation of tandem mass spectrometers (e.g., in EBE-TOF and Q-TOF designs). In the future, the MS/MS applications of the quadrupole-based and ion cyclotron resonance-based (FT-MS) ion traps for the analysis of phoshopeptides may also become common.

2. Experimental Procedure for Amino Acid Sequence Determination by Liquid SIMS and Tandem Mass Spectrometry

The AutoSpec Q hybrid tandem mass spectrometer described earlier is used by us for MS/MS experiments. Both low-energy and high-energy CID data can be acquired with this instrument. The high-energy CID takes place in the first field-free region. Because the CID spectrum is acquired in the B/E linked-field scan mode, the instrument is calibrated by scanning down to m/z 5 with a mixture of LiI and NaI. This low-mass calibration allows to monitor the product ions from the molecular ion of the octapeptide (m/z 980) mentioned before down to a mass of 70 Da. In order to increase the abundance of the $[M + H]^+$ ions and to reduce the ion-source fragmentations, α-thioglycerol is used as a liquid matrix. For acquiring the spectrum, load an aliquot (containing 3–5 μg of the peptide) of the HPLC fraction onto the probe as described earlier. Also, deposit 1–2 μl of 0.1% TFA onto the probe. Mass select the $[M + H]^+$ ion of the peptide in the first field-free region and introduce helium in the collision cell until the signal is reduced to half its original value. Acquire the MS/MS spectrum in the continuum mode by using linked-field scan at constant B/E. Several scans should be acquired, summed, averaged, and smoothed to provide a good spectrum.

To acquire the low-energy CID spectrum, load the peptide sample on the probe as described earlier. Mass select the $[M + H]^+$ ion of the peptide by adjusting the MS1 (EBE) portion of the AutoSpec Q. Introduce argon in the collision region (rf-only quadrupole) to reduce the $[M+H]^+$ ion intensity by 60%. Acquire the MS/MS spectrum by scanning the mass analyzer quadrupole. The data can be manipulated as already described.

3. An Example of the Use of Tandem Mass Spectrometry for Sequencing Phosphopeptides

Tandem mass spectrometry can tolerate heterogeneous mixtures. Therefore, sequencing of peptide fragments can be accomplished without prior separation of the protein digest. However, if the mixture is too complex or the $[M+H]^+$ ion signal is poor due to reduced surface activity of phosphopeptides in the liquid matrix, a partial separation of the digest may be performed by RP-HPLC. This step assures that each HPLC fraction contains a few peptides of similar size and hydrophobicity.

An example of the use of tandem mass spectrometry for sequence determination of phosphopeptides is presented in Figure 4. Again, to compare this spectrum with the conventional mass spectrometry data, MS/MS spectrum of the same octapeptide is included. This spectrum is a high-energy CID spectrum. Less than 5 nmol of the peptide was dissolved in α-thioglycerol and ionized by positive-ion liquid SIMS. The spectrum was acquired by using the front end (EBE section) of the hybrid tandem mass spectrometer described earlier and the constant B/E linked-field scan.

In contrast to the positive-ion conventional mass spectrum (Figure 2), the MS/MS spectrum (Figure 4) of $Y_pGGFMRGL$ contains a complete series of b-type ions. The y-type ion series is also complete with the exception that y_1 and y_2 ions are relatively weak. In addition, nearly all z-type ions and a few a-type ions are also formed. Thus, the B/E linked-field scan MS/MS technique can be used to sequence this peptide from both ends, an advantage over the conventional positive-ion spectrum.

IV. ANALYSIS OF PHOSPHOPEPTIDES BY ELECTROSPRAY IONIZATION MASS SPECTROMETRY

Electrospray ionization is rapidly becoming an indispensable technique for structural analysis of proteins and peptides. In principle, ESI is a field-assisted process, in which ionization occurs by electrospraying a solution of the sample from the end of a stainless steel capillary electrode held at a high potential relative to a surrounding counter electrode [14,15]. Ejection of the sample ions from a charged droplet into gas phase takes place when the size of the droplet is reduced to a level where the electrical field on its surface becomes excessively high, with the result that it can no longer hold the ion. Unique features

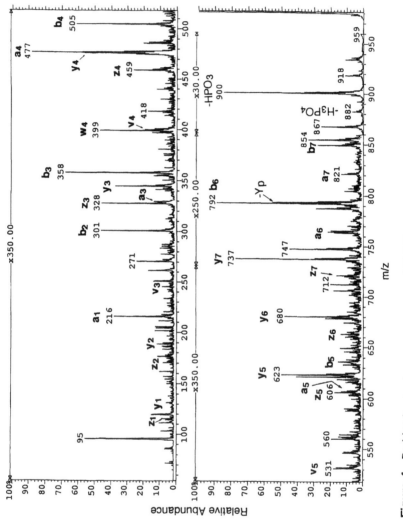

Figure 4 Positive-ion tandem mass spectrum of Y_pGGFMRGL. α-Thioglycerol was used as a matrix and liquid SIMS for ionization of the peptide.

of this technique are lack of fragmentation and formation of multiply charged ions. A wide mass range of compounds from small to large can be analyzed by ESI-MS. The basis of high-mass analysis is the formation of a series of multiply charged ions of a molecule. One of the benefits of multiple charging is to reduce significantly the m/z of the intact analyte, bringing very high-mass analytes within the usable mass range of an ordinary mass spectrometer. The technique is useful in determination of molecular mass of macromolecules to an accuracy of better than 0.01%.

A. Molecular Mass Determination by ESI-MS

Electrospray ionization-MS is the ideal choice for molecular mass determination of larger peptide fragments. This technique is at least an order of magnitude more sensitive than liquid SIMS and is better suited to handle mixtures of peptides due to its tolerance to sample impurities (except for excessive salts) and absence of matrix affects usually encountered in liquid SIMS. Therefore, no prior separation of the protein digest is required.

For determination of molecular mass of peptides in a mixture (e.g., a peptide digest) by ESI-MS, peptides are injected into a flowing solution, usually a 50:50 (v:v) mixture of water:acetonitrile containing 0.1% TFA. The simplicity of the spectrum depends upon the number of basic residues in each fragment. The tryptic fragments always contain a basic residue at the C-terminus, and therefore each fragment produces a single-charged and a double-charged ion. An example of this type of analysis is presented in Figure 5, which contains the spectrum of a mixture of 10 phosphorylated and nonphosphorylated peptides. Each peptide is clearly identified by the presence of singly and doubly charged ions.

1. Experimental Procedure for Obtaining the Molecular Ion Information of the Protein Digest

A Micromass Platform II mass spectrometer is used in our laboratory for molecular mass determination of biomolecules by ESI. The instrument consists of a single-quadrupole mass filter with a mass range for a singly charged ions of ~3000 Da. It is fitted with an atmospheric pressure ESI source, a megaflow ESI probe, and PC-based MassLynx data system. The megaflow probe uses an additional flow of nitrogen to assist evaporation of the charged droplets. To perform the analysis, deliver continuously a 50:50 (v:v) mixture of water:acetonitrile containing 0.1% TFA into the ESI source at a flow rate of 10 μl/min via a syringe pump (KD Scientific, model 200). Adjust the source temperature to 120°C and sample cone voltage to 30 V. Set the flow of the nebulizing and drying gases (both high purity nitrogen) to 10 L/hr and 250 L/hr, respectively.

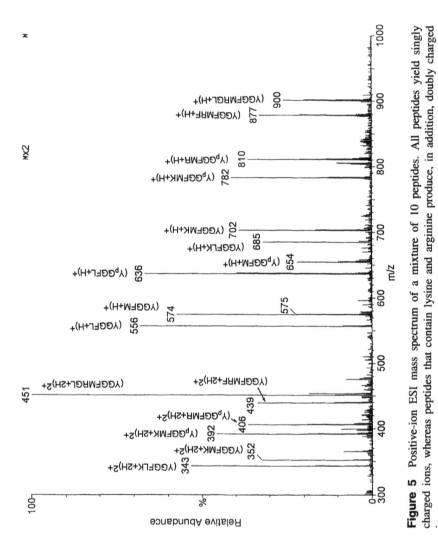

Figure 5 Positive-ion ESI mass spectrum of a mixture of 10 peptides. All peptides yield singly charged ions, whereas peptides that contain lysine and arginine produce, in addition, doubly charged ions.

Calibrate the entire mass range of analysis with CsI cluster ions. Dissolve the lyophilized peptide mixture in the same ESI solvent system (10–100 pmol in 100 μl). Inject a 10-μl sample into the flowing solution via a 10-μl sample loop injector (Rheodyne 7125). Acquire the spectrum by scanning the quadrupole in the mass range of 400–2500 Da until all of the sample solution has been electrosprayed. All scans are averaged and subtracted from the background to generate a representative spectrum.

B. Sequence Determination by ESI-MS

Apart from molecular mass determination, ESI-MS can also be used to obtain sequence ion information. It has been shown that large multiple-charged peptide ions can be more efficiently fragmented compared to single-charged ions [41]. Fragmentation of the molecular ion of a peptide can be induced either in the ion source or in the intermediate region of a tandem mass spectrometer. The ion source fragmentations are induced by adjusting the sample cone voltage (see Chapter 1). This step imparts additional kinetic energy to the ions, causing them to undergo CID in the ion source. This type of analysis can be readily performed in a single-quadrupole machine. However, unambiguous assignment of sequence by the ion source CID procedure requires that the sample is free from impurities. For ESI-MS/MS sequence analysis, the triple-sector quadrupole instrument discussed earlier is commonly used. Most tryptic fragments produce doubly charged ions, and therefore are good candidates for CID studies using tandem mass spectrometry.

C. Analysis of Protein Digests by Liquid Chromatography/ ESI-MS

The technique of LC/ESI-MS is one of the most exciting developments of recent times in analytical methodology. This combination provides benefits of the two most powerful stand-alone analytical techniques. Although HPLC is capable of resolving complex mixtures, on its own it does not provide unambiguous identity of the analyte. A drawback of mass spectrometry, on the other hand, is its limitation in terms of handling a large mixture of compounds, although, as discussed earlier, tandem mass spectrometry may be of limited help for mixture analysis. By combining HPLC with mass spectrometry, its potential is increased manyfold. The problems that are difficult to handle by mass spectrometry or HPLC alone are tackled with relative ease with LC/MS. The potential advantages of this combination are: (1) as the performance of both HPLC and MS are synergistically enhanced, both instruments can be operated at lower than their optimal performance, (2) measurement of retention time introduces an additional compound-specific parameter in the analysis, (3) some extra laborious purification steps and sample losses can be avoided, and (4) signal suppression due to impurities is minimized. All these factors increase detection sensitivity and molecular specificity and decrease analysis time.

Several characteristics of ESI make it an ideal interface for coupling HPLC with mass spectrometry. First, ESI is a liquid-phase ionization technique. Second, ionization occurs at atmospheric pressure, and third, the water–acetonitrile (containing 0.1% TFA) mobile-phase gradient system commonly used in HPLC elution is also the most appropriate solvent for stable electrospray ionization. In addition, current ESI source designs can accept a wide range of solvent flows from nanoliters to milliliters, enabling the use of a variety of HPLC column dimensions.

1. Chromatography Columns and Mobile Phases

Columns used in HPLC are available in different lengths and diameters [42]. Sensitivity, efficiency, sample loading capacity, and sample size are important criteria in the selection of column dimensions. Because of extra molecular specificity afforded by the LC/MS setup, shorter columns may be used to speed analysis. The longer columns offer some advantage in the separation of large peptides and proteins but have greater sample losses. A 20- to 50-mm-long column provides excellent separation of protein digests. For smaller peptides, a 100- to 150-mm-long column may be used. Columns are available ranging in internal diameter (ID) from 0.1 to 4.6 mm. Small diameter columns provide increased sensitivity and are used when the amount of sample is limited; sensitivity can be increased fourfold by decreasing the column diameter by a factor of 2. With regard to the particle size of the packing material, the smaller size particles offer better resolution; columns packed with 3–5 μm particles are appropriate for analytical separations. The average pore diameter of the packing is also critical: It should be roughly 10 times larger than the molecular diameter of analytes. A pore size of 100 Å or less should be used for very small hydrophilic peptides and 300 Å for separation of proteins and large peptides.

The C-18 bonded phase is suitable for separation of tryptic peptides. A C-4 packing may be used for separation of larger peptide fragments. The water:acetonitrile mobile-phase system commonly used in UV-visible detection is also ideal for HPLC/ESI-MS analysis. As mentioned earlier, this solvent system is compatible with the ESI operation. TFA can be used as the ion-pairing reagent. Nonvolatile buffers should not be used. The flow rate of the mobile phase is determined by the diameter of the column used (discussed later).

2. The LC/ESI-MS Setup

The actual experimental setup used in LC/ESI-MS analysis is dictated by the sample size, column dimensions, and the ion source available [42]. A packed capillary column (~0.3 mm ID) may be directly connected to an ESI source.

The rate of liquid flow (1–5 μl/min) of this column is compatible with the flow requirements of the electrospray process. Capillary columns have the potential of reduced sample load and high sensitivity. Major limitations of these columns are lack of reproducibility in flow rate, establishing a reproducible gradient, uniformity of packing, and the fact that they are applicable to nanoliter injection volumes only. Therefore, to obtain reliable gradients, precolumn flow splitting becomes necessary. In actual setup, the solvents are delivered by micro-LC pumps at higher flow rates (usually 100–500 μl/min) and the required flow is diverted to the capillary column via a flow splitter. Low-volume injector loops (60 or 500 nl; Valco, Houston, TX) are available for use with this system.

Microbore columns (0.8–1 mm, ID) are suitable for low-picomole sample amounts, and can be used with a pneumatically assisted ESI source without any precolumn or postcolumn flow splitting at the mobile-phase flow rates of 10–100 μl/min. A 5-μl sample loop injector is used with this column.

For analysis requiring high efficiency, reliability, and large sample loads, standard analytical columns (2.1–4.6 mm, ID) are employed. The 2.1-mm ID columns (popularly known as narrow-bore columns) operate at 200–500 μl flow rates. They can be connected without a flow splitter to a pneumatically assisted ESI source. The use of wide-bore columns (4.6 mm ID) is not recommended for the LC/MS analysis of proteins and peptides. Under circumstances where a wide-bore column needs to be used, it can be directly connected to a currently available mega-flow ESI source without flow splitting. A mobile-phase flow of up to 1 ml/min can be tolerated by this source without disturbing the ESI operation. However, we suggest the use of a post-column flow splitter to reduce the flow of the solvent entering the ESI source. At higher flow rates, increased solvent background is observed, resulting in lower detection sensitivity (discussed later). Post-column flow splitting has additional benefits; peptide fractions can be collected if needed for further applications. To reduce the solvent background, the ESI source may be operated at higher temperatures (e.g., at 150–180°C) with the precaution that the peptide does not fragment at that temperature.

The setup shown in Figure 6 is used in our laboratory for the analysis of phosphopeptides. Peptides are separated on a narrow-bore C-18 column with gradient elution at a flow rate of 200 μl/min. The UV-visible diode array detector is connected in line with the ESI source. Because the ESI source can accept large liquid flow rates, it is not necessary to split the flow. The eluents are monitored both by UV absorption (at 200 nm) and by the total ion current (TIC) chromatogram generated in the positive-ion mode. The advantage of this system is that the output from the two detectors can be compared directly.

3. Mass Spectrometry Monitoring of LC Eluents

Two types of scans can be acquired for LC/MS analysis of a protein digest. A full scan allows determination of the molecular mass of each eluting fragment.

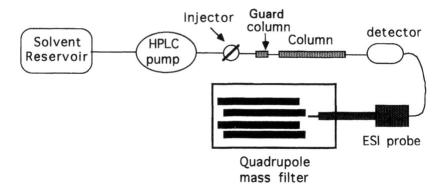

Figure 6   Schematic diagram of the HPLC/ESI-MS system.

For analyzing a protein digest, the mass spectrum is acquired in the mass range of 400–2500 Da in either a positive- or negative-ion mode. We have tested the utility of this approach by analyzing the same mixture of 10 enkephalin peptides. The setup shown in Figure 6 is used for this experiment. HPLC/ESI-MS separation of the peptide mixture is shown in Figure 7a, which contains the positive-ion TIC chromatogram of the peptide mixture. All peptides are well separated in this chromatogram. A comparison of the UV (not shown) and TIC profile shows that the chromatographic integrity is preserved in this LC/ESI-MS system.

The detection sensitivity of this system is in the femtomole range. However, it is our experience that the background ion current has a tremendous influence on detection sensitivity. The main contribution to the background is from water clusters. Two types of cluster ion series are formed from the electrosprayed solution; one ion series contains ions of the type $[H_3O^+ + nH_2O]$ and the other of the type $[CH_3CN + H^+ + nH_2O]$. The composition and flow rate of the mobile phase both determine the extent and nature of the cluster ions. The cluster formation can be reduced by increasing the source temperature, drying gas flow, and sampling cone voltage. These parameters must be optimized before sample analysis is attempted. In our set-up, a source temperature of 150°C, drying gas flow of 350 L/hr, and sampling cone voltage of 30 V are found optimum for mobile-phase flow of 200 μl/min. Higher source temperatures and sampling cone voltages are detrimental as those settings induce fragmentation in peptides, resulting in reduced molecular ion signal.

In the second type of scans, phosphopeptides are detected selectively by monitoring the phosphate group marker ions [29,30]. This method incorporates CID of the peptide in the ion source to produce these compound-specific ions. CID of phosphotyrosine-containing peptides yields two phosphate marker ions PO_3^- and PO_2^- at m/z 79 and 63, respectively. The peptides containing phos-

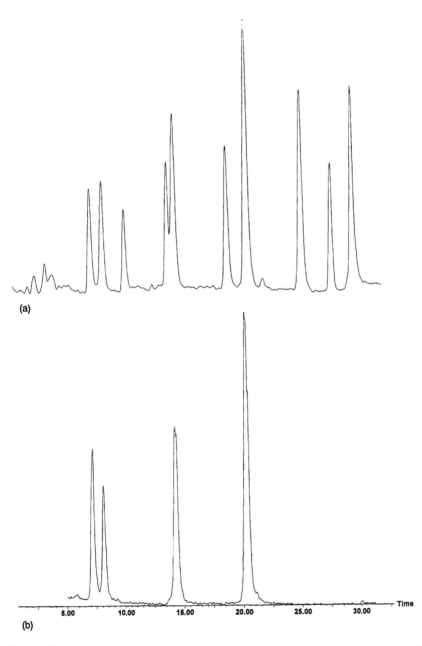

Figure 7 (a) Positive-ion TIC chromatogram of a mixture of 10 peptides. The peptides were ionized by ESI. (b) Negative-ion selected ion chromatogram of the same peptide mixture obtained by monitoring m/z 79. Only Y_pGGFM, Y_pGGFL, Y_pGGFMK, and Y_pGGFMR show the signal at their retention time of 7.0, 8.0, 14.2, and 20.1 min, respectively. For the gradient used, see experimental section.

phorylated serine and threonine also produce these ions and an additional phosphate group-specific ion at m/z 97 ($H_2PO_4^-$) [29]. Thus, monitoring of ion current at m/z 79 and 63 provides a method for simultaneous detection of peptides containing all three types of phosphorylated residues, namely, phosphoserine, -threonine, and -tyrosine. Another advantage of this technique is that the chromatogram is acquired in the SIM mode, which is at least 1000 times more sensitive than the full-scan technique. Also, once the phosphorylated peptides are identified, only those HPLC peaks need to be collected for further sequence analysis, resulting in saving of time and labor.

The potential of this approach for selective detection of phosphorylated peptides in a mixture is illustrated by analyzing the same mixture of 10 peptides. This mixture contains four phosphotyrosine-containing peptides. The HPLC/ESI-MS setup and the experimental conditions of separation are the same as those listed earlier for the full-scan technique. The sample cone voltage, however, is set at 200 V. Figure 7b contains the selected-ion chromatogram obtained by monitoring m/z 79. The peaks are observed only at the retention times of four phosphorylated peptides. Thus, this approach offers the possibility of selective detection of phosphorylated peptides in a protein digest. The voltage setting used in this separation for CID of those phosphopeptides is not optimum. Of the four phosphorylated peptides, only Y_pGGFM and Y_pGGFL yield signal at m/z 63, and the ion current due to m/z 79 from all four peptides is still increasing.

In the preceding analyses, the full scan and the SIM scan were acquired in separate chromatographic runs. However, with current advances in computer software, it is possible to obtain both types of data in a single run [43]. For this, the mass spectrometer is scanned in the complete mass range for 2–4 sec in the positive-ion mode at low sample cone voltage setting. The two marker ions are then sequentially monitored for 200 msec each in the negative-ion mode at higher sample cone voltages. This cycle is repeated again.

4. Experimental Procedure for Liquid Chromatography/ ESI-MS Analysis

In our studies, a Hewlett Packard model 1050 liquid chromatograph is interfaced to the Platform II quadrupole mass spectrometer via a megaflow ESI probe. A narrow-bore (250 × 2.1 mm) Vydac 201HS54 C-18 column packed with 5 μm and 90 Å pore size particles is used for separation of peptide mixtures. A solvent mixture of HPLC-grade acetonitrile in deionized water containing 0.1% TFA is used as the mobile phase at a flow rate of 200 μl/min. For separation of the peptide mixture in Figure 7, the following nonlinear gradient is used: 15–19% acetonitrile in 0–10 min, 19–21% in 10–20 min, 21–40% in 20–30 min, and 40% from 30–40 min. The sample is injected via a 20-μl volume sample loop. The ESI source temperature is adjusted to 150°C and sampling cone voltage to 30 V. The flow of nebulizing and drying gasses is adjusted to 15 L/hr and 350 L/hr, respectively. To monitor the HPLC eluents, scan the quadrupole in the full-scan mode in the mass range of 400–2500 Da. The chromatograms shown in Figure 7 were acquired in the 550–990 Da mass range.

For selective detection of phosphorylated peptides, switch the instrument to the negative-ion mode, and set the data system to record the ion current due to m/z 63 and 79 in the SIM mode. The dwell time at each ion can usually be set to 200 msec with the sample cone voltage at 250 V. When sample amount is limited and faster analysis is required, both full-scan and SIM scan can be acquired in the same HPLC run.

ACKNOWLEDGMENTS

The author highly appreciates the financial support of the National Institutes of Health (grant NS 28025) and The University of Memphis, Memphis. The assistance of the author's colleagues Dr. P. Mahalakshami and Xirong Zhu is also highly appreciated.

REFERENCES

1. T Hunter. Protein kinase classification. In: T Hunter, BM Sefton, eds. Methods in Enzymology. San Diego: Academic Press, 1991, Vol 200, pp 3–37.
2. AM Edelman, DK Blumenthal, EG Krebs. Protein serine/threonine kinases. Annu Rev Biochem 56:567–613, 1987.
3. BE Kemp, ed. Peptides and Protein Phosphorylation. Boca Raton, FL: CRC Press, 1990.
4. T Hunter, JA Cooper. Protein-tyrosine kinases. Annu Rev Biochem 54:897–930, 1985.
5. IM Rosenberg. Protein Analysis and Purification, Benchtop Techniques. Boston: Birkhauser, 1996.
6. L Engstorm, P Ekman, EM Humble, U Ragnarsson, O Zetterquist. Detection and identification of substrates for protein kinases: use of proteins and synthetic peptides. In: F Wold, K Moldave, eds. Methods in Enzymology. San Diego: Academic Press, 1984, Vol 107, pp 130–154.
7. T Hunter, BM Seften, eds. Methods in Enzymology. San Diego: Academic Press, 1991, Vol 201.
8. HE Meyer, E Hoffmann-Posorske, H Korte, LMG Heilmeyer, Jr. Sequence analysis of phosphoserine-containing peptides. Modification for picomolar sensitivity. FEBS Lett 204:61–66, 1986.
9. CFB Holmes. A new method for selective isolation of phosphoserine-containing peptides. FEBS Lett 215:21–24, 1987.

10. KL Busch, GL Glish, SA McLuckey. Mass Spectrometry/Mass Spectrometry: Techniques and Applications of Tandem Mass Spectrometry. New York: VCH Publishers, 1988.

11. M Barber, RS Bordoli, RD Sedgwick, AN Tyler. Fast atom bombardment of solids (FAB): a new ion source for mass spectrometry. J Chem Soc Chem Commun 325–327, 1981.

12. W Aberth, K Straub, AL Burlingame. Secondary ion mass spectrometry with cesium ion primary ion beam and liquid target matrix for analysis of bio-organic compounds. Anal Chem 54:2029–2034, 1982.

13. B Sundqvist, RD Macfarlane. ^{252}Cf-Plasma desorption mass spectrometry. Mass Spectrom Rev 4:421–460, 1985.

14. JB Fenn, M Mann, CK Meng, SF Wong, CM Whitehouse. Electrospray ionization for mass spectrometry of large biomolecules. Science 246:64–71, 1989.

15. RD Smith, JA Loo, RR Ogorzalek Loo, M Busman, HR Udseth. Principles and practice of electrospray ionization-mass spectrometry for large polypeptides and proteins. Mass Spectrom Rev 10:359–451, 1991.

16. M Karas, F Hillenkamp. Laser desorption ionization of proteins with molecular masses exceeding 10,000 daltons. Anal Chem 60:2299–2301, 1988.

17. C Fenselau, DN Heller, MS Miller, HB White III. Phosphorylation in riboflavin-binding protein characterized by fast atom bombardment mass spectrometry. Anal Biochem 150:309–314, 1985.

18. BW Gibson, AM Falick, AL Burlingame, L Nadasdi, AC Nguyen, GL Kenyon. Liquid secondary ionization mass spectrometry characterization of two synthetic phosphotyrosine-containing peptides. J Am Chem Soc 109:5343–5348, 1987.

19. H Michel, DF Hunt, J Shabanowitz, J Bennett. Tandem mass spectrometry reveals that three photosystem II proteins of spinach chloroplasts contain N-acetyl-O-phosphothreonine at the NH$_2$ termini. J Biol Chem 263:1123–1130, 1988.

20. BW Gibson, P Cohen. Liquid secondary ionization mass spectrometry of phosphorylated and sulfated proteins and peptides. In: JA McCloskey, ed. Methods in Enzymology. San Diego: Academic Press, 1990, Vol 193, pp 480–501.

21. C Dass, P Mahalakshmi. Amino acid sequence determination of phosphoenkephalins using liquid secondary ionization mass spectrometry. Rapid Commun Mass Spectrom 9:1148–1154, 1995.

22. C Dass, DM Desiderio. Fast atom bombardment mass spectrometry analysis of opioid peptides. Anal Biochem 163:52–66, 1987.

23. K Biemann, S Martin. Mass spectrometry determination of amino acid sequence of peptides and proteins. Mass Spectrom Rev 6:1–75, 1987.

24. K Biemann. Sequencing of peptides by tandem mass spectrometry and high-energy collision-induced dissociation. In: JA McCloskey, ed. Methods in Enzymology. San Diego: Academic Press, 1990, Vol 193, pp 455–479.

25. K Biemann. Primary structure of peptides and proteins. In: T Matsuo, RM Caprioli, ML Gross, Y Seyama, eds. Biological Mass Spectrometry, Present and Future. New York: John Wiley & Sons, 1994, pp 275–297.

26. AJ Rossomando, J Wu, H Michel, J Shabanowitz, DF Hunt, MJ Weber, TW Sturgill. Identification of Tyr-185 as the site of tyrosine autophosphorylation of recom-

binant mitogen-activated protein kinase p42^{mapk}. Proc Natl Acad Sci USA 89:5779–5783, 1992.

27. T Covey, B Shushan, R Bonner, W Schröder, F Hucho. LC/MS and LC/MS/MS screening for the sites of post-translational modification in proteins. In: H Jörnvall, JO Höög, AM Gustavsson, eds. Methods in Protein Sequence Analysis. Basel: Birkhäuser Press, 1991, pp 249–256.

28. LM Nuwaysir, JT Stults. Electrospray ionization mass spectrometry of phosphopeptides isolated by on-line immobilized metal-ion affinity chromatography. J Am Soc Mass Spectrom 4:662–229, 1993.

29. MJ Huddleston, RS Annan, MF Bean, SA Carr. Selective detection of phosphopeptides in complex mixtures by electrospray liquid chromatography/mass spectrometry. J Am Soc Mass Spectrom 4:710–717, 1993.

30. J Ding, W Burkhart, DB Kassel. Identification of phosphorylated peptides from complex mixtures using negative-ion orifice-potential stepping and capillary liquid chromatography/electrospray ionization mass spectrometry. Rapid Commun Mass Spectrom 8:94–98, 1994.

31. RS Annan, SA Carr. Phosphopeptide analysis by matrix-assisted laser desorption time-of-flight mass spectrometry. Anal Chem 68:3413–3421, 1996.

32. BT Chait, R Wong, RC Beavis, SBH Kent. Protein ladder sequencing. Science 262:89–92, 1993.

33. WJ Boyale, P van Der Geer, T Hunter. Phosphopeptide mapping and phosphoamino acid analysis by two dimmensional separation on thin-layer cellulose plates. In: T Hunter, BM Seften, eds. Methods in Enzymology. San Diego: Academic Press, 1991, Vol 201, pp 110–149.

34. TD Lee, JE Shively. Enzymatic and chemical digestion of proteins for mass spectrometry. In: JA McCloskey, ed. Methods in Enzymology. San Diego: Academic Press, 1990, Vol 193, pp 361–374.

35. C Dass. The role of a liquid matrix in controlling FAB-induced fragmentation. J Mass Spectrom 31:77–82, 1996.

36. P Roepstorff, J Fohlman. Proposal for common nomenclature for sequence ions in mass spectrometry of peptides. Biomed Mass Spectrom 11:601, 1984.

37. K Biemann. Contributions of mass spectrometry to peptide and protein structure. Biomed Environ Mass Spectrom 16:99–111, 1988.

38. M Schaer, KO Boernsen, E Gassmann. Fast protein sequence determination with matrix-assisted laser desorption ionization mass spectrometry. Rapid Commun Mass Spectrom 5:319–326, 1991.

39. C Dass. Mass spectrometry: instrumentation and techniques. In: DM Desiderio, ed. Mass Spectrometry, Clinical and Biochemical Applications. New York: Plenum Press, 1994, pp 1–52.

40. C Dass, DM Desiderio. Characterization of neuropeptides by fast atom bombardment and B/E linked-field scan techniques. Int J Mass Spectrom Ion Proc 92:267–287, 1989.

41. JA Loo, RD Smith. Tandem mass spectrometry of very large molecules: serum albumin sequence information from multiply charged ions formed by electrospray ionization. Anal Chem 63:2488–2499, 1991.

42. C Dass. High-performance liquid chromatography/electrospray ionization/mass spectrometry. In: FA Settle, ed. Handbook of Instrumental Techniques for Analytical Chemistry. Upper Saddle River, NJ: Prentice Hall PTR, 1997, pp. 647–664.

43. X Zhu, C Dass. Analysis of phosphoenkephalins by combined high-performance liquid chromatography and electrospray ionization mass spectrometry. J Chromatogr A (submitted).

11

Mass Spectrometry–Based Approaches Toward the Study of Regulated Biological Systems

Julian D. Watts, Axel Ducret*, Daniel Figeys, Ming Gu, Yanni Zhang, Paul A. Haynes, Rose Boyle, and Ruedi Aebersold
University of Washington, Seattle, Washington

---

*Current affiliation: Merck Frosst Canada, Inc., Montréal, Quebec, Canada.

I. INTRODUCTION

During the last two years, the complete genome sequences of a few selected species, among them the first genome sequence of a eukaryotic species, *Saccharomyces cerevisiae*, were completed and published. This represents a significant milestone in the history of biological science, since for the first time, the complete inherited information of biological species and with that, in principle, the composition of the organism have been determined. The comparative analysis of complete genome sequences encoded in the A,T,C,G four-letter code is providing new evolutionary insights into genome organization and dynamics, and has revealed a set of genes that appears to be close to the minimal number and composition required for a free-living organism [1]. However, the interpretation of the information encoded in the genome sequence in terms of biological function of the organism or particular systems thereof has proven significantly more difficult.

A first step in decoding the biological information contained in genome sequences usually involves the scanning of expressed gene sequences for the presence of motifs that correlate with an activity assigned to the translated product. Knowledge of the sequences and putative functions of all the elements involved in a biological system is, however, by itself insufficient to establish a model describing the system because neither the mechanisms that control individual activities nor the interdependence of the components is apparent.

In our group, we are working on the development and application of technologies for the comprehensive study of regulated biological systems and pathways. Ideally, the analysis of such systems would involve the determination of the function as well as the state of activity of each component involved in the system in a time-dependent manner. Due to technical limitations, such an approach is not currently feasible. We have therefore been focusing our efforts on the identification of those proteins that are part of a given biological system, and on the measurement of parameters that correlate with or indicate the state of activity of the protein. Such parameters include the level of expression, the degree of posttranslational processing and modification, and the association of proteins into functional protein:ligand complexes. The technology for the investigation of regulated biological systems at the level of the protein effector molecules has been revolutionized by both the rapid progress in genomic sequencing and the maturation of mass spectrometry (MS) of biomolecules with respect to generality, sensitivity, and affordability.

The general experimental strategy we have chosen is based on the integration of high-performance protein separation techniques with powerful MS-based analytical tools to determine both the amino acid sequence and the state of modification of the separated proteins at high sensitivity. Among the protein separation techniques available, high-resolution two-dimensional gel electrophoresis (2DE) is particularly powerful since it is highly resolving, sensitive, general, quantitative, and compatible with the procedures and reagents commonly used to extract proteins from biological sources. Furthermore, by performing specific experiments, 2DE is capable of displaying the subcellular location, the precise level of expression, the rate of synthesis and degradation, the presence of posttranslational modifications, and the state of association concurrently on hundreds to thousands of proteins. Such parameters, measured on key regulatory proteins, frequently reflect the state of activity of a biological system. By itself, 2DE is, however, strictly a descriptive technique. The full potential of the method is exploited if the identity of the observed proteins and the modifications that may reflect the state of activity of the system can be determined. Many of our efforts are therefore aimed at analyzing gel-separated proteins with respect to their covalent structure using tandem MS (MS/MS) techniques and at using this integrated approach to study regulated biological systems.

The following sections discuss how an analytical technology is being developed for the comprehensive analysis of the functional state of regulated biological systems at the protein level. In the first section work related to the mechanisms that lead to the activation of T cells is described. This complex biological system serves to crystallize the analytical challenges and acts as a test bed for newly developed technologies. The development of integrated systems for determining the identity and the amino acid sequence of gel separated proteins is described next. The two systems address the need for protein analysis at high throughput and high sensitivity, respectively. In the following two sections work related to the identification and localization within the polypeptide chain of posttranslational modifications is summarized. Section 5 focuses on the development of techniques for the identification of sites of protein phosphorylation, a key regulatory modification in T cells and signaling systems in general. In Section 6 an MS-based technique developed for screening protein samples for the presence of a recently discovered form of protein glycosylation that has been postulated to be reversible on intracellular proteins and to have regulatory function is described. The chapter is concluded by a section on developing a new approach to the determination of consensus recognition sequences for protein tyrosine kinases (PTKs). The project consists of the expression of mammalian T-cell PTKs in yeast and subsequent analysis of the sites of phosphorylation induced on endogenous yeast proteins by the expressed kinase. This project applies the high-sensitivity analytical techniques described in the other sections of this chapter to a genome-wide screen for the presence of induced tyrosine phosphate residues in yeast cells and is expected to yield significant insights into the function and control of PTKs in T-cell signaling.

II. T-CELL RECEPTOR SIGNALING: A COMPLEX, REGULATED SYSTEM*

This project is aimed at determining the sequence of molecular events that leads to the activation of lymphocytes. It integrates with the technology development projects described later on three levels. First, the application of more sensitive, more precise, and more general analytical technology enhances the understanding of the complex regulatory networks that control the state of activity of T lymphocytes. Second, the application of analytical technology to the T-cell receptor (TCR) signaling project highlights the boundaries of the established techniques and defines the objectives for further technology development. Third, successful application of newly developed techniques to a project like the TCR signaling problem demonstrates the maturity of the new technique and contributes to the understanding of the system. For the reasons just described, it is vital that any new and beneficial technology capable of enhancing the understanding in this field of research be generated in an atmosphere of close interaction between those who develop and those who apply the technology.

For some time it has been known that a cell's ability to respond to external stimuli, be they chemical, physical, or hormonal, is to a large part regulated by a wide range of cell surface receptor complexes. These, in turn, are linked to a variety of intracellular protein complexes and enzymes, which collectively regulate specific biological responses, a process often referred to as signal transduction. Of particular importance for the elucidation and regulation of such signaling pathways is the phosphorylation/dephosphorylation of key protein components of these signal transduction pathways, events performed by protein kinases and phosphatases, respectively, with many of the critical early phosphorylation events being carried out by PTKs. A second common event in signal transduction pathways is the induced reversible association of proteins into functional complexes and/or the induced dissociation of active or activating proteins from inactive complexes—events that themselves often involve reversible protein phosphorylation as a mechanism for biological control. To date, there have been two major obstacles to the unraveling of signal transduction pathways. First, each cell type expresses many different receptors, with each capable of eliciting a different set of biochemical responses. Thus the determination of the exact role played by individual pathway components and signaling events (such as specific phosphorylation events) is made difficult due to significant redundancy in these areas. Second, many of the signaling molecules of interest are present in very small quantities, thus severely limiting any direct

---

*Julian Watts.

biochemical analysis of in vivo–derived molecules of interest due to lack of material.

In T cells, the TCR similarly plays a role in the regulation of a wide range of biological responses via its interaction with a variety of co-receptor molecules, PTKs and protein tyrosine phosphatases, and the regulation of their enzymatic activities. A simplified representation of TCR structure, and some of the important co-receptors, PTKs, tyrosine phosphatases, and key phosphorylation events it has been linked with is shown in Fig. 1. Biological responses in which TCR, engagement plays a significant role include the induction of protein synthesis and transcriptional activation, the induction of both interleukin-2 (IL-2) and IL-2 receptor gene expression and subsequent IL-2 secretion, positive and negative selection of immature thymocytes in the thymus, cytotoxicity of mature thymocytes in the periphery, and the initiation of T-cell entry into either an active state of proliferation or an unresponsive (anergic) state. A number of PTKs are known to play a role in the regulation of these TCR-mediated biochemical and cellular responses (reviewed in ref. 2). These include members of the *src* family, in particular $p56^{lck}$ and $p59^{fyn}$, which interact with the CD4/CD8 co-receptors and the TCR, respectively, the *syk* family member ZAP-70, and $p50^{csk}$. These molecules are similar in structure in that they are comprised of various combinations of homologous functional subunits. They all contain a catalytic tyrosine kinase domain at their C-terminal end. They also contain combinations of two smaller subunits, known as SH2 and SH3 domains, which mediate protein:protein interactions via their preferential affinities for various phosphotyrosine-containing polypeptide sequences and certain proline-rich sequences, respectively (reviewed in ref. 3).

$p56^{lck}$ and $p59^{fyn}$ have been extensively studied in a number of model and transgenic cellular systems, and at least one or both have been shown to be vital components in multiple TCR-mediated signal transduction pathways, including positive and negative selection during thymocyte development, cell proliferation, IL-2 production, and TCR-mediated cytotoxicity. The activation of both $p56^{lck}$ and $p59^{fyn}$ is in turn regulated via reversible tyrosine phosphorylation at a conserved C-terminal tyrosine residue by the $p50^{csk}$ PTK and the CD45 transmembrane tyrosine phosphatase (reviewed in ref. 2).

Recent studies of immunodeficient individuals lacking ZAP-70 expression have shown that ZAP-70, like $p56^{lck}$ and $p59^{fyn}$, is vital to the induction of a full TCR-mediated response [4,5]. Following TCR engagement, a number of its subunits, in particular the ζ and CD3ϵ chains, become multiply tyrosine phosphorylated on a conserved motif, referred to as an immunoreceptor tyrosine-based activation motif (ITAM) [5–7]. The ZAP-70 PTK binds specifically to the doubly tyrosine phosphorylated ITAMs of the TCR ζ and CD3ϵ subunits (ζ having three such motifs and CD3ϵ having one), subsequently becoming itself both tyrosine phosphorylated and activated (reviewed in refs. 5 and 7).

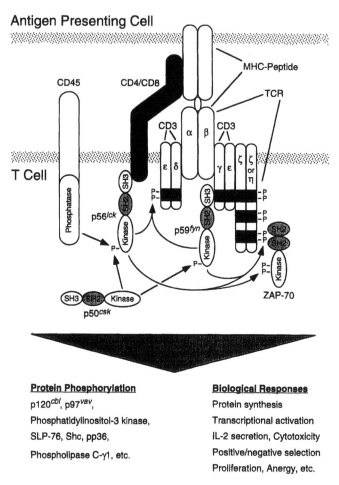

Figure 1 Schematic representation of an activated TCR and its associated PTKs and co-receptors, bound to the MHC-peptide complex on the surface of an antigen presenting cell. The subunits of a normal TCR (α, β, and ζ and the CD3 chains γ, δ, and ϵ) are labeled, with their respective ITAMs indicated (shaded boxes). The CD45 PTP, and the p56[lck], p59[fyn], p50[csk], and ZAP-70 PTKs are shown, along with their conserved domain structures: the catalytic domain (phosphatase and kinase, respectively), SH2 and SH3 domains, which recognize and bind, respectively, phosphotyrosine-containing and proline-rich polypeptide sequences. Inducibly phosphorylated tyrosine residues are indicated (P). Generally accepted phosphorylation/dephosphorylation events associated with TCR-mediated signaling are shown by solid arrows.

This ZAP-70/TCR ζ ITAM interaction requires both SH2 domains of ZAP-70 and both phosphorylated tyrosines of the ITAM. Indeed, it has been shown that if this ZAP-70/TCR ζ interaction is specifically blocked, TCR-mediated signaling is impaired [8].

The ZAP-70/TCR ζ interaction has also been implicated in the determination of which of two important biological responses occurs following TCR ligation by variant peptide ligands. Receptor engagement under conditions of optimal peptide presentation induces ζ phosphorylation, ZAP-70 recruitment, tyrosine phosphorylation and activation, and ultimately leads to cell proliferation and IL-2 secretion. Engagement under suboptimal conditions induces a (presumed to be) differentially or incompletely phosphorylated form of ζ [9,10], with the cells ultimately entering a state of anergy [9]. While this lower apparent molecular mass phospho-ζ isoform does appear to bind ZAP-70, the kinase does not become tyrosine phosphorylated or activated [10]. Thus it is clear that the identification of the sites of phosphorylation induced on molecules such as ZAP-70 and ζ and the characterization of other ligands complexed to these proteins are vital to the elucidation and understanding of the signal transduction pathways in which they are involved.

As previously stated, direct biochemical determination of in vivo phosphorylation sites inducible on known signaling molecules of interest and establishment of the identity of unknown phosphoproteins has been almost impossible to date. This has been due to the small quantities of such material recoverable from typical cell lysates and the often low stoichiometry of phosphorylation observed on such proteins. To circumvent these limitations, we adopted an approach based on comparative 2D phosphopeptide mapping (summarized in Fig. 2). Molecules of interest were obtained in expressed (or synthetic) form and were phosphorylated in vitro with a variety of available purified and/or expressed protein kinases. The same molecules were also isolated from suitable cellular sources (T cells), following in vivo ^{32}P-metabolic labeling and appropriate stimulation, usually via immunoprecipitation or a known association with another molecule. Following proteolytic fragmentation (usually with trypsin), the phosphorylation state of the molecule was determined by 2D phosphopeptide mapping on cellulose thin-layer chromatography (TLC) plates. Once conditions were identified capable of reproducing in vitro some or all of the phosphorylation events observed in vivo, the appropriate in vitro–derived phosphopeptide spots were recovered from the 2D TLC plate and analyzed using a selection of electrospray ionization (ESI) MS-based methodologies developed over the last few years in our laboratory and described in Section V of this chapter.

With on-line MS detection of peptides eluting from a narrow-bore HPLC column (LC-ESI-MS), we were able to determine that at least five, if not all six tyrosines in the three ITAMs of the TCR ζ cytoplasmic domain could be

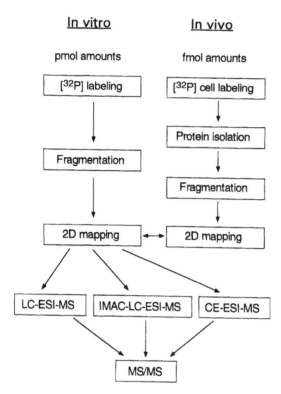

Figure 2 Schematic representation of experimental approach to the mapping of in vivo protein phosphorylation sites. Essentially, 2D phosphopeptide maps are derived from in vitro–labeled recombinant proteins of interest. These 2D maps are analyzed via one or more on-line separation with MS detection methodologies. Additional information such as limited peptide sequence can then be obtained by MS/MS techniques. These in vitro–derived maps are then compared with maps derived from the same protein of interest isolated (typically by immunoprecipitation) from stimulated and unstimulated ^{32}P-labeled cells. The identification of in vivo phosphorylation sites is made on the basis of comigration peptides in the two maps obtained for each molecule studied.

inducibly phosphorylated following TCR engagement of cultured T cells [11,12]. By coupling immobilized metal affinity chromatography (IMAC), which can be made to preferentially bind phosphopeptides [13] to the system just mentioned, we identified an unexpected autophosphorylation site for ZAP-70 (Tyr-292) and two other TCR-inducible phosphorylation sites, Tyr-492 and Tyr-493 [14], the latter of which has been subsequently shown to be critical for the in vivo regulation of ZAP-70 catalytic activity and biological function

[15,16]. This general approach was also shown to be valid for the identification of serine and threonine phosphorylation sites, since we demonstrated that the same serine residue of p56^{lck} is inducibly phosphorylated following either TCR or IL-2 stimulation of T cells [17,18] or antigen receptor stimulation of B cells [19]. Current work aimed at further increasing the sensitivity of phosphorylation site analysis is described in Section V of this chapter.

Numerous membrane, cytoplasmic, and nuclear proteins, downstream of TCR stimulation and its associated PTKs, also become phosphorylated following TCR engagement. Protein phosphorylation has been shown to regulate the activities and/or protein:protein interactions of many such proteins. These include phospholipase C-γ1, Shc, p76^{slp}, p120^{cbl}, p97^{vav} and phosphatidylinositol-3 kinase, to name but a few (reviewed in ref. 7). Some of these molecules may also be substrates for one or more of the PTKs discussed earlier. Indeed, it has also proven most difficult to determine whether a given protein is a direct substrate (or even a potential substrate for that matter) of a particular kinase in vivo. This problem is addressed in Section VII of this chapter, which is aimed at defining the consensus recognition sequences of T-cell PTKs. In addition to these known signaling phosphoproteins (i.e., their genomic and/or mRNA sequences have been determined), numerous potentially important but unidentified signaling proteins have been described in the literature that are also phosphorylated as a consequence of TCR stimulation. In general, such molecules have been characterized by size in polyacrylamide gels and immunoblots as coprecipitating phosphoproteins whose protein sequences have remained undetermined due to lack of material. For example, an unidentified 36–38 kD tyrosine phosphoprotein (pp36) associating with the Grb-2 adapter protein has been frequently reported in T cells following TCR activation [20–22]. A number of phosphoproteins of apparent molecular mass of 76, 90, 130 and 180 kD have been identified in ^{32}P-labeled B and T cells based on their affinity for expressed SH2 domains of the *blk*, *fyn(T)*, and *lyn* PTKs [23]. Unknown tyrosine phosphorylated proteins of 36, 62, and 130 kD have also been identified in T cells based on their affinity for p76^{slp} [24], which in turn associates with Grb-2.

There is therefore a clear need to be able to identify unknown signaling molecules. There is also a clear need to be able to identify sites of posttranslational modification on molecules such as those described, in particular the determination of sites of inducible phosphorylation, and to identify the kinase(s) phosphorylating a particular site. Again, the low abundance of such molecules in vivo is a major limiting factor. The difficulty in expressing and purifying many of these molecules (in particular larger proteins, such as p120^{cbl}, for example) can make the approach for the determination of phosphorylation sites outlined in Fig. 2 time-consuming, if not impossible. In addition, the inability to recreate certain in vivo phosphorylation events in vitro may be a significant limitation in some cases. For example, we observed two phosphopeptides in ZAP-70 precipitates following in vivo labeling and stimulation that we could not recreate in vitro [14]. Thus we were unable to determine if these phospho-peptides were indeed ZAP-70-derived and if so, what phosphorylation sites they represented.

The development of more sensitive, conclusive, robust, and simple to apply methodologies is thus necessary to address these limitations. The ability to identify proteins from complex mixtures and at high sensitivity is essential. It is also necessary to identify the sites of phosphorylation on proteins such as those discussed, and once such sites are determined, to identify the enzymes which induce a particular modification. Finally, technology to detect and char-acterize regulatory protein modifications other than phosphorylation is needed. In order to obtain the maximum relevant biochemical information, it will ulti-mately be necessary to eliminate the necessity for parallel in vitro experiments and directly analyze in vivo–derived samples. The work described next aims at meeting these analytical challenges. It appears that the recent developments in on-line capillary electrophoresis (CE) and microelectrospray MS technology of-fer the best promise for achieving these goals.

III. AUTOMATED, HIGH-THROUGHPUT IDENTIFICATION OF PROTEINS BY LC-ESI-MS/MS*

The objective of this project is the development of integrated instruments and procedures for the automated and high-throughput identification of proteins. Section II of this chapter has indicated the need for the conclusive and sensitive identification of a large number of proteins as a prerequisite for the analysis of complex regulated systems at the protein level. The advent of biological MS, with the introduction of matrix-assisted laser desorption/ionization [25] (MALDI) and ESI [26,27] sources, and the exponential growth of sequence databases have radically modified the strategies that were applied to the analy-sis of proteins only a few years ago. The growing sensitivity of MS and the possibility of obtaining structural information from peptides by MS/MS has rapidly reduced, by several orders of magnitude, the amounts of protein as well as the time required for analysis. Furthermore, rapid characterization of proteins for which sequences are represented in protein or DNA sequence databases has been made possible by the development of specialized software that com-pares the information obtained by MS with such databases. The integrated LC-ESI-MS/MS system we describe here accomplishes automated and rapid identi-

_____

*Axel Ducret.

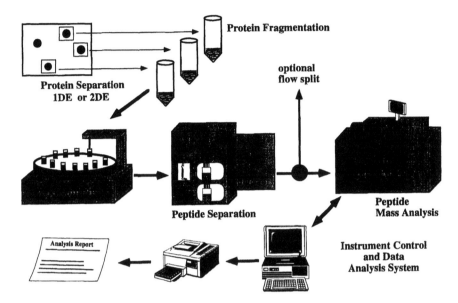

Figure 3 Design of automated protein workstation. After enzymatic cleavage, typically of gel-separated proteins, digests were loaded into an autosampler (Alcott model 738R, Norcross, GA) and sequentially injected onto a microbore HPLC system (Ultrafast Protein Analyzer, Michrom BioResources Inc, Auburn, CA) equipped with a Reli-asil C_{18} BDX column (0.5 × 150 mm) purchased from the same supplier. Peptides were separated using the following program (buffer A: 6% v/v acetonitrile, 0.025% v/v trifluoroacetic acid; buffer B: 80% v/v acetonitrile, 0.024% trifluoroacetic acid): 0–3 min, 0–20%B at 15 μl/min; 3–21 min, 20–50%B at 15 μl/min; 21–21.5 min, 50–100%B from 15 to 25 μl/min; 21.5–24.5 min, 100%B at 25 μl/min; 24.5–25 min, 100–0%B at 25 μl/min; 25–30.5 min, 0%B at 25 μl/min; 30.5–31 min, 0%B from 25 to 15 μl/min; 31–32 min, 0%B at 15 μl/min. The autosampler lines were washed between injections with buffer A at approximately 5 μl/min using a flow delivery module (Michrom BioResource, Inc.) to reduce memory effects. Peptides were detected by UV absorbance at 214 nm and analyzed by ESI-MS/MS. Alternatively, an optional flow split between the HPLC and the MS allowed the manual collection of 98% of the separated peptide. MS/MS spectra are analyzed by the Sequest and Sequest_Summary programs and the search results are automatically printed in the form of an one- to two-page report.

fication of proteins separated by gel electrophoresis by combining the opportunities provided by the sequence databases, advanced separation science, and MS.

In the late 1980s, sequence information from proteins and peptides was essentially obtained by the Edman degradation. While the automated protein sequencer generally produces unambiguous amino acid sequences from purified proteins or peptides, the method is slow and lacks the sensitivity required for the analysis of low abundance proteins. We therefore devised an integrated system for the automated, rapid, and conclusive identification of proteins based on MS determination of polypeptide sequences [28]. The protein workstation is schematically illustrated in Fig. 3. It consists of an autosampler, a microbore reverse-phase high-performance liquid chromatography (RP-HPLC) featuring a 500 μm ID column operated at 15 μl/min, and ESI-MS/MS on-line detection and analysis of the eluting peptides. Protein digests are loaded into the autosampler and consecutively injected onto the RP-HPLC column. The separated peptides are then analyzed by MS/MS. If the ion current reaches a threshold predetermined by the user, peptide ions are automatically subjected to collision-induced dissociation (CID). At the end of each LC-MS/MS run, the MS/MS spectra obtained are automatically analyzed by Sequest, a program that correlates the structural information of a CID spectrum with amino acid sequences contained in a database. Designed by John Yates and collaborators at the University of Washington in Seattle [29,30], Sequest first attempts to match an experimental MS/MS spectrum with the theoretical fragmentation pattern of all amino acid sequences contained in a database that are isobaric with the parent ion. After establishing a preliminary scoring table, the theoretical MS/MS spectra of the 500 best peptide candidates are generated and sequentially compared with the experimental data. The similarity between two spectra is expressed in the form of a correlation score. While good matches will generate high correlation scores, a correct answer is statistically obtained when the difference between the normalized correlation factors of the two best matches exceeds 0.1. This method does not presume any specificity of the enzyme used to generate the peptide and, because there is no formal interpretation of the MS/MS spectrum, the program can be run batchwise in a fully automated manner while the system acquires data from the next sample. Due to the high number of CID analyses that can be generated in a single LC-MS/MS run, we appended a second program, Sequest__Summary, to Sequest for rapid viewing of the combined results obtained from a sample. The program gives a score of 10 for every first-ranked peptide, a score of 8 for every second-ranked peptide, a score of 6 for a third-ranked peptide, a score of 4 for a fourth-ranked peptide, and a score of 2 for a fifth-ranked peptide. The scores of peptides originating from a common protein are added and the identified target(s) are listed according to their ranks. Although the characterization of one peptide is usually sufficient, we require a score of 20 or higher (that is, the identification of at least two first-ranked peptides originating from the same protein) to unambiguously identify a protein. Similarly, since each CID analysis gives rise to an independent identification by Sequest, protein mixtures are easily characterized by the independent analysis of peptides originating from different proteins.

We have used this workstation successfully for the investigation of 90 yeast proteins separated by 2D gel electrophoresis (Fig. 4) in a completely automated manner. Protein spots containing an estimated 50–1000 ng of protein were pooled from three blots, digested with trypsin, and analyzed in two batches of 40 samples each (10 samples of the second batch were pools of at least two proteins that were codigested) in unattended overnight or weekend runs. The samples were consecutively injected at a pace of one sample per 32 min by the autosampler, separated by RP-HPLC, and analyzed by MS/MS. At the end of each LC-MS/MS run, Sequest was automatically launched in the background as a postacquisition application and the program tentatively assigned a database entry for every MS/MS spectrum acquired during the run. Finally, Sequest_Summary identified the protein(s) by ranking the sequence entries that generated high correlation factors and summarized the results in a one- to two-page report. The analysis results from these experiments are summarized in Table 1. Proteins present in the gel at a level exceeding 100 ng were generally identified without difficulty by this strategy, while peptide maps generated by the digestion of proteins of lesser abundance were usually not sufficiently discriminating for identification purposes. Because the MS/MS spectrum of each peptide represents an independent analysis, unambiguous protein characterization was achieved, even when mixtures were present in a 2D gel spot. We actually took advantage of this feature to further reduce the analysis time to 16 min per protein by pooling the samples generated by the digestion of two different spots prior to analysis. Spots A–F (Fig. 4) consisted of two proteins that were jointly digested and analyzed in the protein workstation. Sequest and Sequest_Summary correctly identified the two proteins and, in one case, confirmed that one of the spots investigated was in fact a mixture of two protein isoforms.

The investigation by MS/MS of regulated systems in mammalian species, such as the TCR signaling pathway, is currently limited by the small fraction of the protein sequences that are contained in sequence databases. If a protein cannot be identified by the automated method just described, the primary objective usually aims at a de novo sequencing of peptides. Primers can be derived from peptide sequences for gene cloning by the polymerase chain reaction (PCR). Although some structural information can usually be manually derived from MS/MS data, unambiguous amino acid sequences to date have for the most part been obtained by the Edman degradation.

With the purpose of integrating LC-MS/MS into a general strategy for obtaining PCR primers for the cloning of an unknown gene, we have modified the protein workstation described as follows: Peptide mixtures are separated by micropreparative RP-chromatography using a 500 μm ID column at a flow rate of 15 μl/min. On-line, automated MS/MS analysis of the eluting peptides is

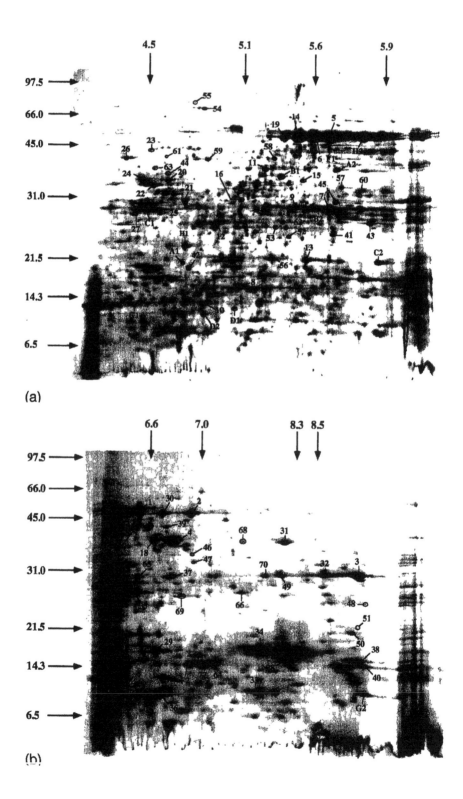

(a)

(b)

Table 1 Large-Scale Analysis of Yeast Proteins by Automated LC-ESI-MS/MS[a]

Spot intensity (estimated)	Total spots analyzed	Total proteins identified
$> 1 \mu g$	17 of 18[b]	21 Proteins 2 Samples: mixture of 2 proteins 1 Sample: mixture of 3 proteins
100 ng–1 μg	27 of 27	21 Proteins
< 100 ng	25 of 25	18 Proteins
Pooled (2 spots per analysis)	9 of 10[b]	19 Proteins
100 ng–1 ug		1 Sample: mixture of 3 proteins

[a] Ninety digests were analyzed in an unattended manner by the LC-MS/MS protein workstation in two batches. The MS/MS spectra were analyzed by Sequest using a yeast subset of the OWL database version 29.1 containing 8877 entries.
[b] Sample was not injected.

provided by a 2% flow split to the microelectrospray source, while the remainder of each eluting peptide is manually collected according to ultraviolet (UV) absorbance at 214 nm (Fig. 3). While relatively insensitive (the limit of the UV detection lies around 500 fmol of peptide loaded onto the column in these conditions), RP chromatography is tolerant to the most common biological contaminants (such as salts or detergents) and delivers the purified and concentrated peptides in a form directly suitable for Edman sequencing. On-line MS/MS detection fulfills several purposes. First, every fraction can be assessed with respect to the mass and the purity of the peptides collected. Second, the MS/MS spectra generated are useful to verify that the peptides in fact are derived from a protein of unknown sequence before they are submitted to Edman sequencing. Third, CID spectra very efficiently complement and confirm the sequence data obtained by chemical sequencing (Fig. 5).

Figure 4 2D (IEF/SDS-PAGE) separation of yeast cell lysate. First dimension was either a linear immobilized pH gradient of range 4–7 (a) or a linear immobilized pH gradient of 6–10 (b). Second dimension was 12% SDS-PAGE. Proteins were visualized by silver staining. For protein analysis, unstained gels were electroblotted onto nitrocellulose immediately after electrophresis and proteins were visualized by Amido Black staining. A total of 200 μg of protein was applied per gel. Indexed spots were identified using the automated LC-MS/MS system. Comigrating spots from three gels were pooled for analysis.

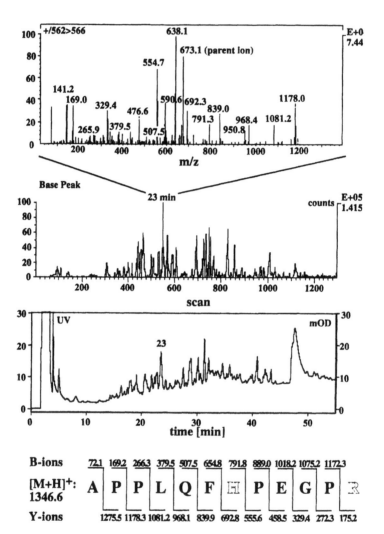

Figure 5 Complementarity of sequence information by Edman degradation and MS/ MS. The tryptic digest of an unknown protein was applied to an RP column and peptides were collected according to their UV absorbance at 214 nm. On-line MS analysis was performed on 2% of the sample split postcolumn to the micro-ESI source. Fraction 23 was subjected to automated Edman degradation and yielded the amino acid sequence APPLQFXPEGPX. The missing His and Arg residues were unambiguously deduced from the complementary CID spectrum.

In cases in which the purified peptide amount is below the sensitivity limit of the protein sequencer or when the overall protein amount is not likely to reach the picomole level, high-quality MS/MS analysis of selected fractions (or from the unfractionated sample) can nevertheless be attempted at unit resolution by on-line solid-phase extraction capillary electrophoresis MS/MS [31] (SPE-CE-MS/MS). This is described in more detail in Section IV of this chapter.

In summary, LC-MS/MS analysis of protein digests represents a sensitive and universal method for protein identification. In an automated mode, the protein workstation described here has proven to be a reasonably sensitive and extremely robust system. Tens of samples were reliably serially injected unattended, in overnight or weekend runs. The protein identification process was performed totally independently of the user and without the need for manual interpretation of CID spectra. Finally, since each MS/MS spectrum represents an independent analysis, redundant, unambiguous protein characterization was achieved by multiple identifications, even when mixtures were present in the digested protein sample.

While it is far less sensitive in its present configuration than the complementary SPE-CE separation method described in Section IV of this chapter, we anticipate that a further reduction in the diameter of RP columns, combined with a reduction in the flow rate, will increase the sensitivity limit of the protein workstation. A first design for automated peptide mapping featuring a 100 μm ID capillary RP column directly coupled to a microelectrospray tandem mass spectrometer has been presented [32], and the increase in sensitivity achieved by implementing capillary RP chromatography in a LC-MS/MS system is discussed in Section V of this chapter. While automated protein identification techniques will grow in importance as genome sequencing projects approach completion, the need for high-sensitivity amino acid sequencing by the Edman degradation will remain, because the method is robust, general, and the data are simple to interpret, even for large peptides. Our laboratory recently developed a modified Edman reagent, 311-PITC [33–35], that is designed for mass spectrometric detection of the resulting thiazolinones. Preliminary work performed on current commercial protein sequencers demonstrated that femtomole-level sensitivity is readily achieved using this reagent. The development of improved instrumentation combined with optimized degradation cycles for 311-PITC has the potential to increase the current sensitivity limit of chemical sequencing by several orders of magnitude. Finally, while the techniques described in this section are useful for the general determination of the amino acid sequence of proteins, the important analysis of posttranslational modifications will require adaptations of the strategies described here. Some approaches are summarized in Sections V and VI of this chapter.

IV. NOVEL TECHNOLOGIES FOR THE IDENTIFICATION OF PROTEINS AT THE LOW FEMTOMOLE SENSITIVITY LEVEL*

This project is aimed at establishing the identity of proteins at very high sensitivity. In contrast to DNA, which can be readily amplified, protein analysis is limited to the samples that can be isolated from the biological source. A high level of sensitivity is therefore essential for the study of regulatory proteins, which are frequently of low abundance. The approach described in this section builds on the general procedure described earlier. The two techniques share procedures for peptide sample preparation, automated generation of CID spectra in the ESI-MS/MS instrument, and the algorithms used for protein identification. The two techniques differ in the use of the peptide separation system that is integrated into the process. The system described here uses capillary electrophoresis (CE) for peptide separation. While CE offers superior sensitivity, it poses a unique set of challenges that have to be solved to make the technique useful for robust protein identification at very high sensitivity.

Recent developments in the field of MS have greatly increased the sensitivity achievable for protein and peptide analysis. For ESI-MS, major sensitivity improvements were obtained by the introduction of microspray and nanospray interfaces [36–43]. We and others have previously demonstrated the power of CE combined with micro ESI-MS/MS [39–42], in particular for the rapid identification of proteins. Specifically, we have demonstrated a sensitivity of detection of less than 300 amol for peptides and the ability to generate CID spectra of peptides of sufficient quality for automated database searching with less than 600 amol of peptide applied to the system [42].

In the last 2 years, we have designed an improved system that addresses the major limitation inherent in CE, namely, the incompatibility with the application of large sample volumes. By introducing an on-line solid-phase extraction (SPE) device, we were able to concentrate the analytes contained in large sample volumes into the small volumes compatible with CE. The SPE device consisted of a small RP chromatography column within the capillary, which concentrated the analytes by reversible adsorption [31,44,45]. The system is schematically illustrated in Fig. 6. Essentially, the protein of interest, typically separated by gel electrophoresis, is enzymatically cleaved off-line. Sample volumes of 10–30 μl of are typically obtained. The whole subsequent mixture of peptide fragments is injected onto the system and the peptides are accumulated on the SPE device. Once the system has been washed with an appropriate

---

*Daniel Figeys.

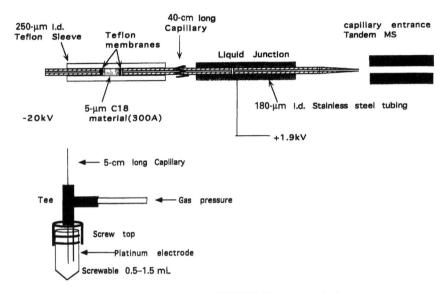

Figure 6   Schematic of SPE-CE-Micro ESI-MS/MS system. All the experiments were performed using 50 μm ID $\times$ 150 μm OD capillaries. The first step of the procedure consists of loading the sample to the SPE device by applying a pressure of 15 psi at the injection end of the capillary. Then the capillary is washed with the CE buffer [10 mM acetic acid/10% methanol (v/v), pH 3.2] for 5–10 min at 15 psi. The analytes are then eluted from the resin by injecting a plug of elution buffer of 100–200 nl (63% acetonitrile in 3 mM acetic acid; 6 psi for 6 sec). CE was performed by applying a potential of -20 kV and a pressure of 0.5 psi at the injection end of the capillary. The microelectrospray was generated by applying a potential of $+1.7$–1.9 kV at the liquid junction. The conditions used resulted in an approximate flow rate to the MS of 200 nl/min.

buffer, the peptides are eluted off the SPE device into the CE system and the CE experiment is started by applying a high voltage from the injection end of the capillary to the microelectrospray interface consisting of a liquid junction. The eluted peptides are electrophoretically stacked and separated. Using this approach, we have obtained concentration effects of up to 1000-fold. The peptides migrating out of the capillary are electrosprayed into the MS as described in Section III of this chapter for the LC-MS/MS system. The MS then automatically fragments each detected peptide by CID and separates the fragments according to their m/z ratio. The CID spectra obtained for all the peptides are searched against protein databases for possible matches using the Sequest program as previously described [29,30]. Each individual CID spectrum matched to a protein sequence in the database is, in principle, sufficient for the identification of the protein. Frequently, several peptides in the same sample are matched to the same protein. This redundancy dramatically enhances the level of confidence in the identification and makes the assignments conclusive. If the protein sequence is not represented in a database, de novo sequencing can be performed on the CID spectra.

The fabrication of the device is an important step to ensure success of the system. Since such devices are not commercially available, they were built in-house. For the operation of the SPE device it was critical to avoid a high back pressure that impeded the solvent flow through the capillary. Two principal potential causes for high back pressure were identified. The first was the resin packed in the extraction column, and the second was a misalignment of the capillaries in the liquid junction used to apply the spraying potential (Fig. 6). It is important to pack the C-18 RP material relatively loosely in the SPE device and to keep the amount of column material to a minimum. Excess material also resulted in band broadening. Misalignment of the capillaries in the stainless steel needle of the liquid junction could completely block the solvent flow. This mainly occurred if the spacing between the sprayer and separation capillary was so small that the CE capillary blocked the entrance to the microsprayer. Blockage in the liquid junction could be easily detected before use of the system. Provided that these parameters were controlled, the system could be built reproducibly in less than 30 min. Each device could be used for numerous analyses over several weeks. The cost of the fabrication of the device is negligible and the system is simple to operate.

The sensitivity of this system was determined using a triple-quadrupole mass spectrometer (model TSQ 7000, Finigan MAT). The absolute limit of detection of the system (LOD, defined as the amount of analyte required to generate a signal size equal to 3 times the standard deviation of the background signal if a volume of 20 μl of analyte was applied) was evaluated using a calibrated bovine serum albumin (BSA) tryptic digest. In the MS mode the absolute LOD was measured at 660 amol and in CID mode at 6 fmol (4 ng of BSA). Applying 20-μl aliquots of the sample to the system operated in the CID mode a concentration LOD of 300 amol/μl was calculated [31]. These detection limits clearly indicate that proteins can be identified with only a few fmol of material present in the digest.

The system was next applied to the identification of low nanogram amounts of proteins isolated by electroblotting of high-resolution 2DE gels, a standard protein isolation technique used in our laboratory. *Saccharomyces cerevisiae* was studied because the genome sequence for that species has recently been completed and every protein sequence should be represented in a sequence database [46,47]. For the preparation of the samples, 2DE gels similar to those used to evaluate the automated LC-based protein identification system were employed (Fig. 4). Aliquots of 200 μg of total yeast protein were separated by high-resolution 2DE. Separated proteins in duplicate gels were either elec-troblotted to nitrocellulose and stained with amido black or silver stained in the gel [48]. Protein spots of different intensities were cut out from the stained nitrocellulose membrane representing a single gel and digested in situ with trypsin [49]. One of the spots whose intensity was below the detection limit of protein staining on the 2D blot was located by alignment of the stained nitrocel-lulose membrane and the silver-stained gel, cut out, and digested on the mem-brane. In total, the proteins in 25 different spots were identified using this tech-nique (Fig. 7).

A typical electropherogram obtained for a spot containing approximately 10–20 ng of protein is shown in Fig. 8. The labeled peaks indicate those pep-tides which were identified as being from OYE2_yeast via Sequest search following CID. Peaks eluting before 10:30 min were derived from the elution buffer and nonpeptide contaminants present in the sample. Peaks that are not

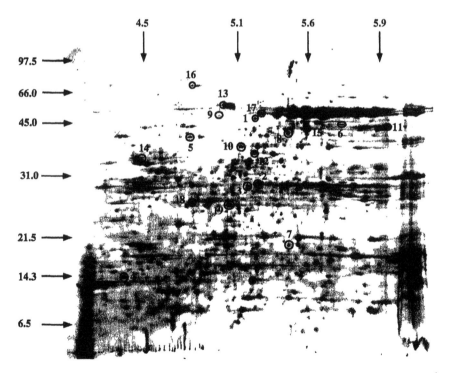

Figure 7   Identification of proteins by SPE-CE-MS/MS. Yeast proteins were separated by 2DE as described in Fig. 4 and selected proteins were identified using SPE-CE-MS/MS. Circled and numbered spots indicate identified proteins. Proteins from a single gel were analyzed.

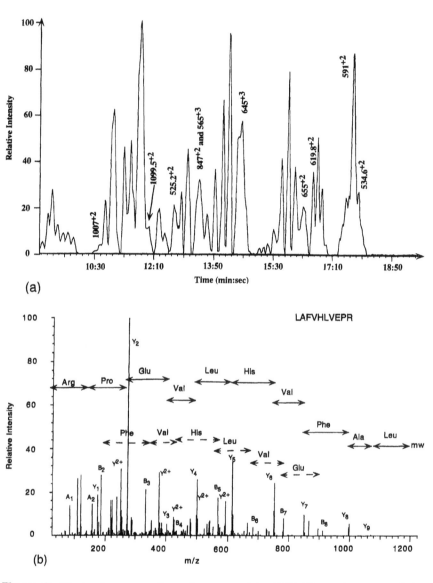

Figure 8 Electropherogram of protein separated by 2DE. Electropherogram of the protein spot identified as OYE2_yeast on the SPE-CZE-ESI-MS/MS system. Approximately 10 μl of digest at 12 fmol/μl was injected on the system for a total of 120 fmol of sample. Conditions as described in caption to Fig. 6. (a) Electropherogram indicating the m/z ratio and charge state of the detected peptides. (b) MS/MS spectrum obtained from the m/z = 591 peptide ion.

labeled in Fig. 8 were either duplicates of the peptides that were already identified, yielded poor-quality CID and were thus not identified, or peptides generated from the enzyme used for digestion or nonpeptide contaminants. From the staining intensity of the spot, we estimated that approximately 120 fmol of peptide was contained in the 10-μl aliquot that was applied to the system (12 fmol/μl; 10 μl of 15 μl applied). The intensity of the peaks was still well above the LOD achieved with this system. The CID spectra obtained from this amount of protein were very intense and of quality sufficient for de novo sequencing had the proteins not been represented in the database. This clearly illustrates that small amounts of proteins extracted from cells can be positively identified with this system. In fact, the system has been shown to be successful in identifying protein from different sources including cell lysates of mammalian cells, and protein isolated by a variety of different techniques including one- and two-dimensional gel electrophoresis and different modes of chromatography (data not shown).

The described system has proven robust and durable. Because the method is capable of identifying essentially any spot detectable by protein staining from a species with abundant genome sequence information, it should be a useful tool for the comprehensive analysis of biological processes such as signal transduction pathways. The level of sensitivity achieved could be further improved by a factor of 10 to 100 by better optimizing the system and incorporating isotachophoretic concentration of the samples eluted from the SPE device. Such a level of sensitivity would be compatible with the analysis of proteins that are only detectable after metabolic radiolabeling, for instance, with ^{35}S-Met or ^{32}P-orthophosphate. Obvious extensions of the method include automation of the whole process and the use of SPE devices with affinities for specific structures within peptides such as added phosphate or glycoforms. These are areas of development are being actively pursued.

V. DETERMINATION OF PHOSPHORYLATION SITES BY LC-MS/MS AND CE-MS/MS*

The objective of this project is the development of a generally applicable and integrated analytical technology for the determination of the sites of protein phosphorylation induced in vivo following cell stimulation. Phosphorylation of proteins is recognized as one of the most important posttranslational modifica-

---

*Ming Gu and Yanni Zhang.

tions of regulatory function, and plays an essential role in the control of cell growth, metabolism, and differentiation [50,51]. In order to understand the biological activities of many regulatory proteins it is critical to develop sensitive bioanalytical methods to identify the phosphorylation sites of the proteins, which are often among the least abundant cellular proteins.

The work described in this section builds on the LC-MS/MS and CE-MS/MS methods developed for the identification of proteins that were described in Sections III and IV of this chapter. The methods share the concept of connecting online the high-performance peptide separation techniques HPLC and CE with ESI-MS/MS for the analysis of the separated analytes and the use of computer algorithms to assist in the interpretation of the generated CID spectra. The methods differ in their requirements for the identification of a specific peptide within the peptide mixture generated by digestion of the protein under investigation. For establishing the protein identity a high-quality CID spectrum of any peptide is sufficient to yield a conclusive result. In contrast, for the determination of phosphorylation sites the phosphopeptide has to be first isolated from the peptide mixture before it can be analyzed. We are therefore using chromatographic and/or electrophoretic procedures with selectivity for phosphate to enrich for phosphopeptides prior to analysis of the sample by LC-MS/MS or CE-MS/MS. Hence we describe refinements of some of the techniques discussed in Sections II and III to determine essential phosphorylation events that control T-cell activation through the TCR.

During the past few years various MS-based techniques for the analysis of phosphoproteins have been developed. Particularly appealing has been the use of ESI triple-quadrupole MS because the characteristic fragments obtained in this type of instrument by CID of phosphopeptides, in principle, permit the identification of phosphopeptides in complex peptide mixtures.

Neutral loss scanning (NLS) in the positive-ion mode and the detection of specific negative reporter fragment ions in the negative-ion mode are two methods that have been used successfully for phosphopeptide analysis. NLS in combination with LC-MS/MS was demonstrated to detect phosphopeptides in tryptic digests of proteins [52]. Under the CID conditions used, only phosphopeptides eliminated a neutral fragment (H_3PO_4) of $m/z = 98$. Such a neutral loss was therefore diagnostic for the presence of a phosphopeptide. This technique was further improved by combining NLS with data-dependent CID, such that the phosphopeptides, once detected by NLS, could be automatically selected for product ion scanning to generate sequence information of the phosphopeptides for database searching or manual interpretation [53]. However, the NLS technique works only if the charge state of the ions is preselected. If the phosphopeptide has a charge state different from the one preselected, the phosphopeptide will inevitably go undetected.

Full MS scanning [54] with LC/MS and precursor ion scanning (PIS) with nanospray ionization [36,55] both employ characteristic negative ions to identify molecular ions of phosphopeptides within peptide mixtures and represent alternatives to NLS that are not dependent on the preselection of the phosphopeptide charge state. The former method uses a rapidly stepping voltage in the orifice region (between the atmospheric pressure inlet and the first skimmer) of the MS to induce low-mass, phosphopeptide-specific marker ions of $m/z = 79$ and 63. The latter method selectively detects phosphopeptides in nonseparated protein digests. Since these techniques rely on scanning in the negative-ion mode, which does not generate desired sequence information of phosphopeptides, it is often necessary to subsequently perform CID on corresponding positive peptide ions should additional sequence analysis be desired.

In addition to suffering from the idiotypic limitations indicated, the methods just described are in general not sufficiently sensitive to directly probe the in vivo state of phosphorylation of low-abundance proteins, can be difficult to interface with the biological source of the phosphoprotein, and do not yield quantitative information unless suitable internal standards are included in the experiments. To circumvent these limitations, an approach based on comparative 2D phosphopeptide mapping on cellulose TLC plates, which is summarized in Fig. 2 and described in Section II of this chapter, was adopted. Briefly, analytical 2D phosphopeptide maps are generated from phosphoproteins isolated from ^{32}P-labeled cells as well as from the same protein generated via in vitro kinase reactions using purified protein kinases. The phosphopeptide patterns are compared and the phosphopeptides from the in vitro generated maps are extracted from the cellulose plate and analyzed by MS. The identity of the in vivo sites of phosphorylation are therefore established by comigration of the phosphopeptides generated in vivo and in vitro. The main strengths of the 2D phosphopeptide mapping technique are its sensitivity and the ability to provide quantitative information about the stoichiometry of phosphorylation at individual sites.

An effective method for the analysis of phosphopeptides extracted from 2D phosphopeptide maps by combining IMAC with HPLC and ESI-MS/MS has been described [11,12,14] (Fig. 9). In Section II of this chapter the successful use of this approach for solving complex phosphorylation patterns in proteins controlling T-cell activation was discussed. Our current efforts are aimed at improving the sensitivity of the IMAC-LC-MS/MS method, making the results more conclusive and easier to interpret. To these ends we are incorporating into the system illustrated in Fig. 9 and described elsewhere [11,12,14] capillary-LC and nano-electrospray ion sources for high-sensitivity peptide separation and detection, and instrument control protocols for on-the-fly CID, and automated Sequest database searching for the identification of phosphopeptides [29].

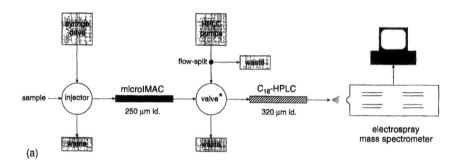

(a)

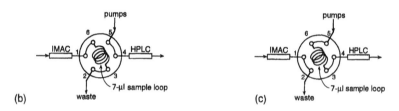

(b)   (c)

Figure 9   Schematic representation of IMAC-LC-MS system. (a) IMAC-LC-ESI-MS instrumentation employed for these studies. Microbore IMAC and HPLC columns were connected on-line to a triple-quadrupole ESI-MS. Expanded views of the switching valve from panel a (*) used to link the microIMAC column to the HPLC system are shown. (b) Valve position for IMAC column loading, washing and subsequent elution of bound phosphopeptides into the sample loop. (c) Valve position for loading contents of the sample loop onto the HPLC system and subsequent development of HPLC column.

Specifically, the system shown in Fig. 9 underwent the following modifications. First, Sepharose Fast Flow (Pharmacia Biotech), the IMAC resin initially used, was replaced with POROS IMAC resin (Perseptive Biosystems). The POROS resin can withstand the high pressures typically encountered in an HPLC system. Consequently, the IMAC column and the C-18 RP column could be operated in series at high pressure and phosphopeptides could be directly eluted from the IMAC column onto the RP column without the need for a low-pressure interface. Second, the 320 μm ID RP column was replaced with a 100 μm ID capillary column packed in-house with POROS C-18 by use of a home-made pressurized metal bomb. Third, a nano-electrospray ion source using a liquid junction similar to the one described in Section IV of this chapter was placed at the downstream end of the fused-silica capillary, which also contained the RP column. Fourth, a high-ratio precolumn flow splitter (Acurate from LC

Packings) was used to provide flow rates in the range of 150–200 nl/min compatible with nanospray ionization, and fifth, the instrument control protocols for automated CID and the capabilities for automated database searching described in Sections IV and V of this chapter were implemented.

The performance of this capillary chromatography-based ESI-MS/MS system was tested by applying 200-fmol aliquots of a tryptic digest of β-lactoglobulin and separating, detecting, and identifying the individual components. The results shown in Fig. 10 demonstrate a signal-to-noise ratio of about 1000:1 for the major peaks, indicating an LOD at the low femtomole sensitivity level. HPLC solvents containing 0.5% acetic acid provided superior sensitivity compared with separations performed in buffers containing trifluoroacetic acid (TFA); however, peak resolution was somewhat reduced in the former solvent.

An equimolar mixture of a peptide corresponding to residues 482–498 of human ZAP-70 [referred to as peptide ZAP-70 (482–498)] and the singly

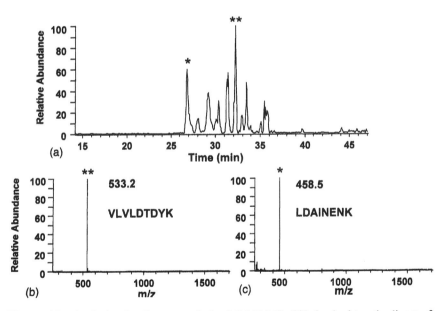

Figure 10 Analysis of a digest protein by LC-MS/MS: 200 fmol of tryptic digest of β-lactoglobulin was injected into a C-18 column (100 μm ID) and separated by gradient HPLC at a flow rate of 100 nl/min. The HPLC gradient ran from 0% of solvent A to 65% of solvent B in 35 min (solvent A, 0.5% acetic acid; solvent B, 80% acetonitrile and 0.5% acetic acid). (a) Base peak trace of the 200 fmol tryptic digest β-lactoglobulin. (b,c) Full-scan mass spectra of peptides of $m/z = 533.2$ and $m/z = 458.5$.

phosphorylated form of the same peptide that was phosphorylated by an in vitro kinase reaction at residue 493 [14] was used to evaluate the selectivity of the POROS IMAC resin for phosphopeptides. First, the peptide mixture was separated on the C-18 capillary chromatography system described earlier. The two peptides, although present in equimolar amounts, were detected in the MS at a 9:1 signal intensity ratio, with the nonphosphorylated peptide being detected with higher intensity. In contrast, when IMAC was used to enrich the peptide mixture for the phosphopeptide followed by separation of the IMAC eluate on the capillary chromatography system, the peptides were detected in the MS at a 4:1 intensity ratio, with the phosphorylated peptide being the dominant signal (data not shown). IMAC therefore demonstrated considerable selectivity for the ZAP-70 (482–498) phosphopeptide over its nonphosphorylated counterpart.

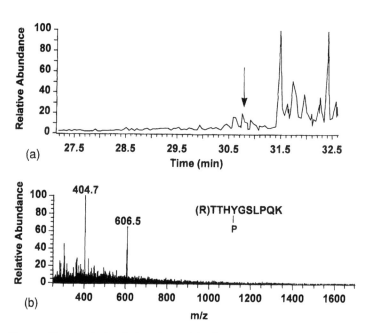

Figure 11 Analysis of phosphopeptide eluted from TLC plate, by IMAC/LC/MS/MS: 80-fmol aliquot of a 30-pmol phosphopeptide sample recovered from cellulose TLC plate was applied to an IMAC column (200 μm ID) and washed with 0.1% acetic acid. The peptide was eluted with 0.1% ammonium acetate, pH 8, 50 mM NaH$_2$PO$_4$ onto a C-18 column (100 μm ID) for LC-MS/MS analysis. The HPLC gradient ran from 0% to 65% of solvent B in 35 min (solvents A and B: 80% as in caption to Fig. 10). (a) Base peak trace and (b) full-scan spectrum.

To evaluate whether the method was compatible with the analysis of peptides isolated from cellulose phosphopeptide maps, an aliquot of 80 fmol of a phosphopeptide sample [sequence:TTH(p)YGSLPQK] extracted from a cellulose TLC plate was analyzed on the IMAC-LC-MS/MS system. The sample solvent was 20% aqueous acetonitrile and the applied volume was 1 μl. The phosphopeptide was derived by tryptic digestion of myelin basic protein that had been phosphorylated by p56^{lck}. The results shown in Fig. 11 indicate that the phosphopeptide ion was detected with sufficient intensity to permit automated CID and that the resulting spectrum was of sufficient quality to permit the identification of myelin basic protein as the source of the phosphopeptide.

While the technique as currently used is slow, mainly due to the time required to elute the phosphopeptides from the IMAC to the capillary RP column, the level of sensitivity achieved is approaching the LOD required to determine the sites of phosphorylation in phosphoproteins directly isolated from cellular sources [56].

The SPE-CE-MS/MS technique described in Section IV of this chapter is being adapted to the analysis of phosphopeptides. The construction of the device is the same as shown in Fig. 6 and described in Section IV, except that the RP chromatography resin was replaced with IMAC resin (HiTrap Chelate Sepharose, Pharmacia Biotech). At the time of this writing only very preliminary data are available. To test the system, aliquots of phosphorylated ZAP-70 (482–498) were applied to an IMAC-SPE device, washed with electrophoresis buffer, and eluted by injecting 100–200 nl of 10 mM phosphate buffer in methanol (90:10 v/v). The eluted peptides were separated by CE and detected by ESI-MS/MS as described in Section IV of this chapter. Results of an experiment in which 6 pmol of the phosphopeptide was applied are shown in Fig. 12. The ZAP-70 (482–498) phosphopeptide was unambiguously identified. However, the observed signal size was dramatically smaller, approximately by three orders of magnitude, than the signal expected for the picomole amounts of phosphopeptide applied to the CE-MS/MS system. There are several reasons for the observed poor detection limit. They relate to the difficulty of finding conditions that effectively elute the phosphopeptides from the IMAC resin and are at the same time compatible with peptide separation by CE and high-sensitivity analysis by ESI-MS. A number of problems have been identified in the current experimental design. The elution buffer used interfered with electrospraying, was ineffective in quantitatively eluting the phosphopeptide from the IMAC resin, and the pH induced a change in the electroosmotic flow, resulting in poor resolution and peak broadening. A systematic evaluation of IMAC elution conditions indicated that a column of the size of the SPE device required 1 μl of 50 mM sodium phosphate for complete phosphopeptide recovery. This amount of phosphate was completely incompatible with CE and ESI-MS. The elution solvent and conditions used therefore represent a relatively inefficient compromise between the requirements for phosphopeptide elution, CE, and ESI-MS. Work in progress focuses on finding suitable conditions for on-line phosphopeptide analysis by IMAC-SPE-CE-MS/MS to take advantage of the exquisite sensitivity this system offers in principle.

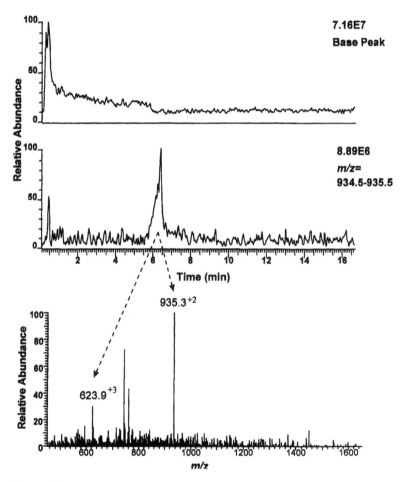

Figure 12 On-line IMAC-CE-MS/MS of phosphopeptide. Total amount of 6.3 pmol of phosphopeptide ZAP-70 (482–498) was loaded on the HiTrap (34 μm, Pharmacia Biotech) chelating column (154 μm ID, 280 μm OD × 1 mm). Elution buffer was 10 mM sodium phosphate and 0.02% ammonium acetate in 10% MeOH/water. All other conditions were similiar to the ones described in caption to Fig. 6.

VI. SEARCHING FOR AND ANALYSIS OF REGULATORY GLYCOSYLATED PROTEINS*

In addition to phosphoproteins, a specific class of glycoproteins has been impli-cated in cellular signaling. These are the proteins modified with *O*-linked *N*-acetylglucosamine (O-GlcNAc), which are usually found in the nucleus or cyto-plasm of a wide range of cell types. This project is aimed at the detection and identification of O-GlcNAc modified proteins. Although a number of such pro-teins have been identified, their role, if any, in signaling systems remains to be established. The development of a sensitive and rapid MS-based technique for the identification of such glycoproteins, and the further characterization of the exact sites of modification, will significantly accelerate research in this emerging field. The technique adapts the capillary chromatography, selective preenrichment strategies, and specific MS/MS scanning techniques that have been described in Sections III and IV for the identification of proteins and in Section V for the detection of sites of protein phosphorylation. Once estab-lished, the technique will be applied to study the involvement of O-GlcNAc protein modification in the T-cell activation models discussed in Section II, since it was in T cells that activation-induced O-GlcNAc modification was first described [57].

Advances in analytical methodology in recent years, particularly in the field of MS, have allowed the detailed structural analysis of oligosaccharide structures at ever-increasing levels of sensitivity. A myriad of novel glycopro-tein structures have been characterized, including those from newly purified molecules and also many from previously characterized glycoproteins. Many novel glycan structures have so far been identified only on a single type of molecule from a single organism. In recent years a new class of glycoproteins has been discovered, which has subsequently been found to be widely distrib-uted in nature [58]. These are the O-GlcNAc modified glycoproteins found as monosaccharides attached to serine or threonine residues. This modification is unusual when compared to most O-glycosylation, as it is a very simple struc-ture and is also highly dynamic and responsive to cellular stimuli in a manner somewhat analogous to phosphorylation [57]. It has also been shown, at least in one case, that O-GlcNAc modification occurs at the same sites as phosphory-lation, but in a mutually exclusive manner [59]. The function of these glycopro-teins is still uncertain, but one current hypothesis is that the O-GlcNAc modifi-cation acts as a counterpoint to phosphorylation, ensuring that the structural integrity of a protein is maintained even when phosphate groups are removed [60]. Since it was first reported [61], O-GlcNAc modification has been found

---

*Paul Haynes.

to reside almost exclusively in the nucleus and cytoplasm of eukaryotic cells ranging from trypanosomes to humans [60]. It seems certain that many more examples of this modification will be characterized in the future as analytical methods become increasingly sensitive and functional studies further elucidate its physiological role.

The most widely used method for detection of O-GlcNAc modified proteins currently involves labeling the O-GlcNAc residues using a galactosyl transferase. This enzyme transfers galactose residues onto a GlcNAc substrate to form a Gal(β1-4)GlcNAc, or N-acetyl-lactosamine group. By using UDP-[^{3}H]-Gal as the donor, a radiolabel can be incorporated into the glycoproteins and visualized by SDS-PAGE and autoradiography. This technique does not distinguish between O-GlcNAc linked to serine or threonine and GlcNAc attached to other groups such as the nonreducing terminus of an N-linked oligosaccharide. Thus, further analysis is needed after the galactosyl transferase labeling. Confirmatory procedures usually include the demonstration that the incorporated radiolabel is both resistant to peptide N-glycosidase F digestion and sensitive to alkaline β-elimination. The incorporation of a terminal galactose residue should also facilitate the isolation of glycoproteins carrying this modification, as they are then expected to bind to the *Ricinus communis* (RCA I) lectin [62].

Most published reports using the galactosyl transferase method have involved labeling known proteins that were previously purified by, for example, affinity chromatography. Our laboratory is investigating the use of serial lectin affinity chromatography to prepare fractions directly from lysates that are greatly enriched in the galactose labeled O-GlcNAc modified glycoproteins. It remains to be determined if these preparations will be of sufficient purity to allow the identification and characterization of the O-GlcNAc modified glycoproteins by LC-MS/MS.

After an O-GlcNAc-bearing glycoprotein has been visualized by SDS-PAGE or 2D gel electrophoresis and purified or enriched as much as possible by affinity chromatography, the protein is identified essentialy as described in Sections III and IV of this chapter. Using protease digests of bands excised from membranes containing SDS-PAGE-separated mixtures, many proteins can be identified by homology database searching using the Sequest program as explained in Sections III and IV of this chapter. This, of course, assumes that the protein sequence of the isolated glycoprotein is available in a database. Initially the yeast *Saccharomyces cerevisiae* is being used because the genome sequence of this species has been completed. A schematic diagram of the whole process from cell lysis to identification of O-GlcNAc modified proteins is shown in Fig. 13. It is worth noting that there are several problems inherent in isolating O-GlcNAc-bearing glycoproteins from yeast, not the least of which is that the chitinous yeast cell walls contain many terminal O-GlcNAc residues,

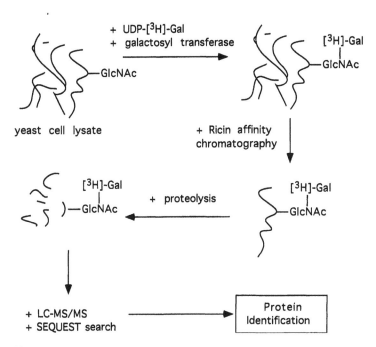

Figure 13 Schematic diagram of techniques used in the detection, isolation, and identification of O-GlcNAc modified glycoproteins.

which can act as substrates for the galactosyl transferase used in the assay and sequester the radiolabel.

The principal approach to determine the identity of glycoproteins that were previously established as being modified by O-GlcNAc is microbore LC-MS/MS. It is possible to selectively detect O-GlcNAc-containing glycoproteins via MS/MS on the basis of the loss of mass caused by the highly labile N-acetylglucosamine side chain [63]. O-GlcNac-containing glycopeptides can be detected in highly complex mixtures using neutral loss scanning (NLS) for the $m/z = 204$ ion generated by loss of an N-acetylglucosamine during relatively low energy collision. This is shown in Fig. 14b, where the peak eluting at 21:09 min is a synthetic O-GlcNAc peptide that can be detected by NLS even though there is insufficient peptide present to be detected by UV absorbance. There are several other obvious peaks detected using the NLS, most of which were also present in a reagent-only blank digestion. The peak eluting at time 21:09 is the only one in which the spectra collected clearly indicate the $m/z = 204$ ion is generated from a single peptide. A similar NLS procedure can

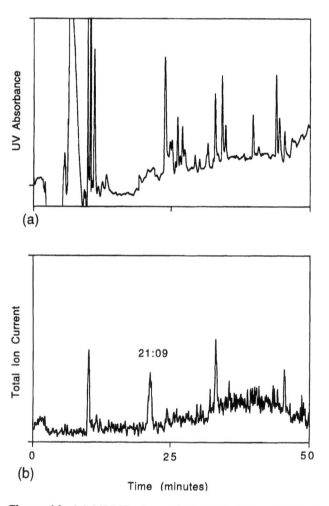

Figure 14 LC-MS/MS of a synthetic O-GlcNAc modified glycopeptide in the presence of a complex mixture of peptides produced by trypsin digestion of bovine carbonic anhydrase. (a) UV absorbance trace at 214 nm; (b) NLS for loss of m/z = 204. The synthetic peptide with the sequence (PSVPVS(GlcNAc)GSAPGR) was a generous gift from Professor G.W. Hart, University of Alabama, Birmingham.

be used for the $m/z = 366$ ion generated for O-GlcNAc modified glycoproteins that have been labeled with galactosyl transferase. This approach may well prove to be useful in confirming that a particular peptide seen in LC-MS/MS was originally associated with the O-GlcNAc modification, rather than purified inadvertently as a contaminant. The other available means of confirmation would be to demonstrate that the introduced radiolabel comigrates with the peptide in question.

The expected continuous advances in sensitivity of the analytical techniques will be essential for the identification of additional physiologically relevant O-GlcNAc modified proteins, even those expressed at extremely low levels. In the future, our laboratory will use this approach to study the biology of O-GlcNAc modified proteins in T cells in the specific states of activation described in Section II of this chapter.

VII. A GENOMIC APPROACH TO SEARCHING FOR PROTEIN TYROSINE KINASE SUBSTRATES*

The objective of this project is the development of a general approach for the identification of the consensus recognition sequence of PTKs and potentially their substrates. The identification of the physiological substrates of protein kinases is technically difficult but important for the investigation of regulatory networks. The approach described here uses the MS/MS-based analytical techniques described in Sections III–V of this chapter to determine the sites of phosphorylation in tyrosine-phosphorylated proteins isolated from yeast cells that were engineered to express mammalian T-cell PTKs. In particular we are interested in determining the consensus recognition sequence of the T-cell PTK ZAP-70, which is a key regulator in the early activation of T cells and for which to date no physiological substrates have been identified. The identification of potential ZAP-70 kinase substrates will therefore greatly assist in the further elucidation of the sequence of early activation events in T cells.

Identification of the physiological substrate(s) of a protein kinase is difficult due to several factors, including the low abundance of signaling proteins, the high background of phosphorylated proteins in a cell, the redundancy of phosphorylation of the same substrate under different stimulatory conditions, the perceived promiscuity of protein kinases, and the absence of the gene sequences of many putative kinase substrates from sequence databases.

Two established methods for identifying a kinase consensus recognition sequence in vitro consist of testing the kinase against a library of peptides with candidate recognition sequences. In one method, a vast library of randomly generated peptides, each containing tyrosine, is presented to the kinase in an in vitro assay [64]. Relative kinase activity is taken to indicate specificity for a peptide sequence. Another method uses a "designed" library. Starting from a candidate sequence, typically derived from a genuine kinase substrate, peptides

---

*Rose Boyle.

are constructed varying by one mutation at a time and tested in in vitro kinase reactions [65]. The relative activities of the kinase toward the tested peptide sequences are then used to derive a consensus recognition sequence.

These in vitro studies suffer from a number of technical and conceptual limitations. First, kinase reaction conditions in vitro may be sufficiently different from those in vivo so as to inhibit kinase activity or to affect substrate recognition. Second, in vitro methods test peptides, not folded proteins. A determinant that is conformational rather than merely sequential will not be detected this way. Third, in vitro kinase reactions do not take into account the location of the substrate within the cell. Fourth, the random generation of peptides representing the library will result in a nonuniform distribution of peptide sequences, thus skewing measured kinase activities by the relative abundance of each peptide in the library. It is also possible that a given sequence will not be represented at all. Finally, degenerate peptide libraries are very complex, even if only a few residues are randomized.

To alleviate some of these limitations we are exploring a strategy, schematically illustrated in Fig. 15. The method consists of expressing a catalytically active PTK in yeast via a plasmid. The cells are then grown under conditions that allow the PTK to phosphorylate yeast proteins. The tyrosine-phosphorylated proteins are detected by immunoblotting and identified by the LC-MS/MS or CE-MS/MS methods described in Sections IV and V of this chapter. Finally, the precise site of phosphorylation and therefore the amino acid sequence recognized by the PTK are determined by MS/MS analysis of the phosphopeptide(s). This approach combines the strengths of the automated protein analysis technology with the ease with which yeast cells can be genetically manipulated to express foreign proteins. Several indicators suggest that this strategy might be successful. First, yeast has no endogenous PTKs, so there is low background of tyrosine-phosphorylated proteins. Second, yeast provides a library of several thousand proteins in native conformation under the conditions provided by a eukaryotic cell. We can therefore anticipate that potential recognition sequences are presented to the kinase in an optimal or near-optimal configuration. Third, knowledge of the complete yeast genome sequence significantly aids in the task of identifying the phosphorylated proteins, even if they are of low abundance.

We have started to test the approach and to establish the parameters for the determination of the induced phosphorylation site(s) using a yeast strain that constitutively expresses a catalytically active form of the PTK Src, kindly provided by Dr. Larry Rohrschneider of the Fred Hutchinson Cancer Research Center, Seattle, WA. Yeast expressing the PTK and a control wild-type strain were grown under standard conditions and lysed. The presence of active Src PTK was observed to not adversely affect the viability or growth characteristics of the cells. The protein lysates were separated by SDS-PAGE, electroblotted

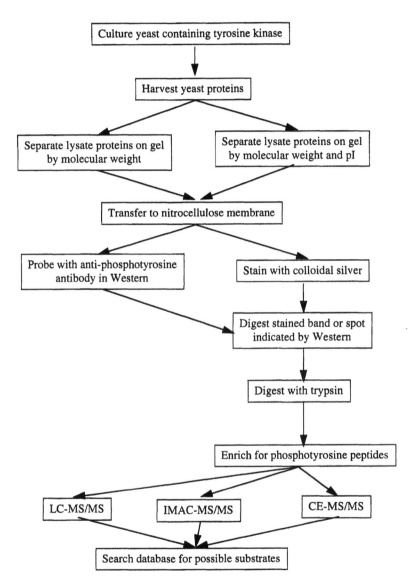

Figure 15 Schematic diagram of procedure used to identify consensus recognition sequence of PTK. The kinase gene is inserted into yeast so that the kinase will be expressed in an active form. After culturing the yeast, the lysate proteins are harvested and separated by either SDS-PAGE or 2DE. Western blotting indicates bands or spots of interest, which are digested by trypsin. Phosphotyrosine peptides are purified from the digest by antiphosphotyrosine antibody precipitation. These peptides are analyzed by MS/MS, and the results are examined and compared to the yeast sequence database for the identification of a consensus recognition sequence for the expressed kinase.

onto nitrocellulose, and the blots were probed with the phosphotyrosine-specific antibody 4G10. Figure 16 indicates that in the cells expressing the mammalian Src PTK several proteins were phosphorylated on tyrosine residues that were not detected in the wild-type lysate. The proteins that did react with the antibody in the wild-type cells are likely the substrates of dual specificity (Ser/Thr and Tyr) protein kinases expressed in yeast. Parallel membranes were stained with colloidal silver [66], and the silver-stained bands with identical electrophoretic mobilities to the bands detected in the tyrosine-phosphate specific immunoblot were cut out and digested with trypsin. The LC-MS/MS and CE-MS/

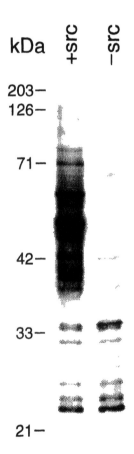

Figure 16 Western blot indicating phosphorylation by Src PTK. Yeast strain L40 +Src (lane 2) and L40 −Src (lane 2) (60 μg total lysate protein each) were loaded onto 12.5% acrylamide gel, transferred to 0.45 μm nitrocellulose, and blotted with antiphosphotyrosine monoclonal antibody (4G10).

MS methods described in Sections III and IV are being used to identify the phosphoproteins and determine the respective site(s) of phosphorylation.

This analysis may be complicated by the fact that several proteins may comigrate in the gel band and because the tyrosine-phosphorylated protein may not necessarily be the most abundant species. 2DE may in this case be a better approach for protein separation. In addition, enrichment for tyrosine-phosphory-lated peptides by subjecting the protein digest to anti-phosphotyrosine affinity chromatography prior to analysis by MS may prove valuable.

The analysis of phosphotyrosine peptides from the yeast strain expressing the PTK Src is expected to confirm the sequence pattern believed to be recognized by the Src family of kinases and to indicate whether this approach can be developed into a general strategy for the determination of consensus recognition sequences of PTKs and possibly other classes of enzymes. If the strategy appears successful we will focus on determining the recognition sequence of the T-cell PTK ZAP-70 by expressing this molecule in the same strain of yeast.

VIII. CONCLUSIONS

This chapter describes several projects being carried out in our laboratory with the aim of developing an integrated approach to comprehensively study the biochemistry of regulated biological systems. The application of these techniques to study the regulation of signal transduction pathways in T cells is also described. The projects focus on the rapid and sensitive identification of key regulatory proteins and on the mechanisms that control their respective levels of activity. The rapid recent progress in MS techniques, in particular ESI-MS/MS, now makes MS an essential element of any protein analytical technology. Modern MS instruments clearly have the sensitivity required for the analysis of low-abundance proteins, and the rapidly growing sequence databases significantly complement MS-based protein analysis. The major challenge remains in the development of suitable tools to interface the MS with the biological protein source. Techniques similar to the ones described here will make the study of complex, regulated systems by direct analysis of the key regulatory proteins an exciting, feasible, and powerful approach.

ACKNOWLEDGMENTS

Projects described in this chapter were supported by the National Science Foundation (NSF) Science and Technology Center for Molecular Biotechnology. D. Figeys and A. Ducret were supported in part by fellowships from the National

Science and Engineering Research Council (NSERC, Canada) and from Oxford Glyco Sciences, respectively.

REFERENCES

1. CM Fraser, JD Gocayne, O White, MD Adams, RA Clayton, RD Fleischmann, CJ Bult, AR Kerlavage, G Sutton, JM Kelley, et al. The minimal gene complement of *Mycoplasma genitalium*. Science 270:397–403, 1995.
2. KG Peri, A Veillette. Tyrosine protein kinases in T lymphocytes. Chem Immunol 59:19–39, 1994.
3. T Pawson. SH2 and SH3 domains in signal transduction. Adv Cancer Res 64:87–110, 1994.
4. C Hirvoz, A Fischer. Immunodeficiency diseases. Multiple roles for ZAP-70. Curr Biol 4:731–733, 1994.
5. NSC van Oers, A Weiss. The Syk/ZAP-70 protein tyrosine kinase connection to antigen receptor signalling processes. Semin Immunol 7:227–236, 1995.
6. AS Shaw, LK Timson Gauen, Y Zhu. Interactions of TCR based activation motifs with tyrosine kinases. Semin Immunol 7:13–20, 1995.
7. RL Wange, LE Samelson. Complex complexes: signaling at the TCR. Immunity 5:197–205, 1996.
8. RL Wange, N Isakov, TR Burke Jr., A Otaka, PP Roller, JD Watts, R Aebersold, LE Samelson. $F_2(Pmp)2$-TAMζ_3, a novel competitive inhibitor of the binding of ZAP-70 to the T cell antigen receptor, blocks early T cell signaling. J Biol Chem 270:944–948, 1995.
9. J Sloan-Lancaster, AS Shaw, JB Rothbard, PM Allen. Partial T cell signaling: altered phospho-ζ and lack of Zap-70 recruitment in APL-induced T cell anergy. Cell 79:913–922, 1994.
10. J Madrenas, RL Wange, JL Wang, N Isakov, LE Samelson, RN Germain. ζ phosphorylation without ZAP-70 activation induced by TCR antagonists or partial agonists. Science 267:515–518, 1995.
11. M Affolter, JD Watts, D Krebs, R Aebersold. Evaluation of two-dimensional phosphopeptide maps by electrospray mass spectrometry of recovered peptides. Anal Biochem 223:74–81, 1994.
12. JD Watts, M Affolter, DL Krebs, RL Wange, LE Samelson, R Aebersold. Electrospray ionization mass spectrometric investigation of signal transduction pathways: determination of sites of inducible protein phosphorylation in activated T cells. In: F Snyder, ed. Biological and Biotechnological Applications of Electrospray Ionization Mass Spectrometry. Washington, DC: ACS Press, 1996, pp 381–407.
13. L Andersson, J Porath. Isolation of phosphoproteins by immobilized metal (Fe^{3+}) affinity chromatography. Anal Biochem 154:250–254, 1986.
14. JD Watts, M Affolter, DL Krebs, RL Wange, LE Samelson, R Aebersold. Identification by electrospray ionization mass spectrometry of the sites of tyrosine phosphorylation induced in activated Jurkat T cells on the protein tyrosine kinase ZAP-70. J Biol Chem 269:29520–29529, 1994.

15. RL Wange, R Guitián, N Isakov, JD Watts, R Aebersold, LE Samelson. Activating and inhibitory mutations in adjacent tyrosines in the kinase domain of ZAP-70. J Biol Chem 270:18730–18733, 1995.
16. AC Chan, M Dalton, R Johnson, G Kong, T Wang, T Thoma, T Kurosaki. Activation of ZAP-70 kinase activity by phosphorylation of tyrosine 493 is required for lymphocyte antigen receptor function. EMBO J 14:2499–2508, 1995.
17. JD Watts, JS Sanghera, SL Pelech, R Aebersold. Phosphorylation of serine 59 of p56^{lck} in activated T cells. J Biol Chem 268:23275–23282, 1993.
18. JD Watts, MJ Welham, L Kalt, JW Schrader, R Aebersold. IL-2 stimulation of T lymphocytes induces sequential activation of mitogen-activated protein kinases and phosphorylation of p56^{lck} at serine-59. J Immunol 151:6862–6871, 1993.
19. MR Gold, R Chiu, RJ Ingham, TM Saxton, I Van Oostveen, JD Watts, M Affolter, R Aebersold. Activation and serine phosphorylation of the p56^{lck} protein tyrosine kinase in response to antigen receptor cross-linking in B lymphocytes. J Immunol 153:2369–2380, 1994.
20. M Sieh, A Batzer, J Schlessinger, A Weiss. GRB2 and phospholipase C-γ1 associate with a 36- to 38-kilodalton phosphotyrosine protein after T-cell receptor stimulation. Mol Cell Biol 14:4435–4442, 1994.
21. DG Motto, SE Ross, JK Jackman, Q Sun, AL Olson, PR Findell, GA Koretzky. In vivo association of Grb2 with pp116, a substrate of the T cell antigen receptor-activated protein tyrosine kinase. J Biol Chem 269:21608–21613, 1994.
22. T Fukazawa, KA Reedquist, G Panchamoorthy, S Soltoff, T Trub, B Druker, L Cantley, SE Shoelson, H Band. T cell activation-dependent association between the p85 subunit of the phosphatidylinositol 3-kinase and Grb2/phospholipase C-γ1-binding phosphotyrosyl protein pp36/38. J Biol Chem 270:20177–20182, 1995.
23. SN Malek, S Desiderio. SH2 domains of the protein-tyrosine kinases Blk, Lyn, and Fyn(T) bind distinct sets of phosphoproteins from B lymphocytes. J Biol Chem 268:22557–22565, 1993.
24. DG Motto, SE Ross, J Wu, LR Hendricks-Taylor, GA Koretzky. Implication of the GRB2-associated phosphoprotein SLP-76 in T cell receptor-mediated interleukin 2 production. J Exp Med 183:1937–1943, 1996.
25. M Karas, F Hillenkamp. Laser desorption ionization of proteins with molecular masses exceeding 10,000 daltons. Anal Chem 60:2299–3001, 1988.
26. TR Covey, RF Bonner, BI Shushan, J Henion. The determination of protein, oligonucleotide and peptide molecular weights by ion-spray mass spectrometry. Rapid Commun Mass Spec 2:249–256, 1988.
27. JB Fenn, M Mann, CK Meng, SF Wong, CM Whitehouse. Electrospray ionization for mass spectrometry of large biomolecules. Science 246:64–71, 1989.
28. A Ducret, EJ Bures, B Fisher, R Aebersold. Towards routine, femtomole level protein sequencing. Proceeding of MPSA '96: XIth international conference on methods in protein structure analysis, Annecy, 1996, p 31.
29. J Eng, AL McCormack, JR Yates. An approach to correlate tandem mass spectral data of peptides with amino acid sequences in a protein database. J Am Soc Mass Spectrom 5:976–989, 1994.

30. JR Yates, J Eng, AL McCormack, D Schieltz. A method to correlate tandem mass spectra of modified peptides to amino acid sequences in the protein database. Anal Chem 67:1426, 1995.

31. D Figeys, A Ducret, JR Yates, R Aebersold. Protein identification by solid phase microextraction-capillary zone electrophoresis-microelectrospray-tandem mass spectrometry. Nature Biotech 14:1579–1583, 1996.

32. A Ducret, M Gu, PA Haynes, JR Yates, R Aebersold. Simple design for a capillary liquid chromatography-microelectrospray-tandem mass spectrometry system for peptide mapping at the low femtomole sensitivity range. Proceedings of ABRF '97: International Symposium: Techniques at the Genome/Proteome Interface, Baltimore, MD, 1997, p 69.

33. R Aebersold, EJ Bures, M Namchuk, MH Goghari, B Shushan, TC Covey. Design, synthesis, and characterization of a protein sequencing reagent yielding amino acid derivatives with enhanced detectability by mass spectrometry. Protein Sci 1:494–503, 1992.

34. D Hess, H Nika, DT Chow, EJ Bures, HD Morrison, R Aebersold. Liquid chromatography-electrospray ionization mass spectrometry of 4-(3-pyridinylmethylamino-carboxypropyl) phenylthiohydantoins. Anal Biochem 224:373–381, 1995.

35. EJ Bures, H Nika, DT Chow, HD Morrison, D Hess, R Aebersold. Synthesis of the protein sequencing reagent 4-(3-pyridinylmethyl-aminocarboxypropyl) phenyl isothiocyanate and characterization of 4-(3-pyridinylmethyl-aminocarboxypropyl) phenylthiohydantoins. Anal Biochem 224:364–372, 1994.

36. M Wilm, M Mann. Analytical properties of the nanoelectrospray ion source. Anal Chem 68:1–8, 1996.

37. M Wilm, A Shevchenko, T Houthaeve, S Breit, L Schweigerer, T Fotsis, M Mann. Femtomole sequencing of proteins from polyacrylamide gels by nano-electrospray mass spectrometry. Nature 379:466–469, 1996.

38. MR Emmett, RM Caprioli. Micro-electrospray mass spectrometry: Ultra-high-sensitivity analysis of peptides and proteins. J Am Soc Mass Spectrom 5:605–613, 1994.

39. J Wahl, DC Gale, RD Smith. Sheathless capillary electrophoresis—electrospray ionization mass spectrometry using 10 μm I.D. capillaries: analyses of tryptic digests of cytochrome C. J Chromatogr A 659:217–222, 1994.

40. J Wahl, RD Smith. Comparison of buffer systems and interface designs for capillary electrophoresis-mass spectrometry. J Capillary Electrophoresis 62–71, 1994.

41. J Wahl, SA Hofstadtler, RD Smith. Direct electrospray ion current monitoring detection and its use with on-line capillary electrophoresis mass spectrometry. Anal Chem 67:462, 1995.

42. D Figeys, I van Oostveen, A Ducret, R Aebersold. Protein identification by capillary zone electrophoresis/microelectrospray ionization-tandem mass spectrometry at the subfemtomole level. Anal Chem 68:1822–1828, 1996.

43. PE Andren, MR Emmett, RM Caprioli. Micro-electrospray: zeptomole/attomole per microliter sensitivity for peptides. J Am Soc Mass Spectrom 5:867–869, 1994.

44. D Figeys, A Ducret, R Aebersold. Identification of proteins by capillary electrophoresis-tandem mass spectrometry. Evaluation of an on-line solid-phase extraction device. J Chromatogr A 1997 (in press).

45. D Figeys, R Aebersold. High sensitivity identification of proteins by electrospray ionization tandem mass spectrometry: initial comparison between an ion trap mass spectrometer and a triple quadrupole mass spectrometer. Electrophoresis 1997 (in press).

46. National Center for Human Genome Research. Press Release. April 24, 1996.

47. A Goffeau, BG Barrell, H Bussey, RW Davis, B Dujon, H Feldmann, F Galibert, JD Hoheisel, C Jacq, M Johnston, EJ Louis, HW Mewes, Y Murakami, P Philippsen, H Tettelin, SG Oliver. Life with 6000 genes. Science 274:546–567, 1996.

48. T Rabilloud. Two-dimensional electrophoresis of basic proteins with equilibrium isoelectric focusing in carrier ampholyte-pH gradients. Electrophoresis 15:278–282, 1994.

49. RA Aebersold, J Leavitt, RA Saavedra, LE Hood, SBH Kent. Internal amino acid sequence analysis of proteins separated by one- or two-dimensional gel electrophoresis after in situ protease digestion on nitrocellulose. Proc Natl Acad Sci USA 84:6970–6974, 1987.

50. P Cohen. Signal integration at the level of protein kinases, protein phosphatases and their substrates. Trends Biochem Sci 17:408–413, 1992.

51. MJ Hubbard, P Cohen. On target with a new mechanism for the regulation of protein phosphorylation. Trends Biochem Sci 18:172–177, 1993.

52. T Covey. LC/MS and LC/MS/MS screening for the sites of post-translational modification. In: H Jornvall, ed. Methods in Protein Sequence Analysis. Birkhauser: ALS, 1991, pp 249–256.

53. JR Yates, JK Eng, AL McCormack. Mining genomes: correlating tandem mass spectra of modified and unmodified peptides to sequences in nucleotide databases. Anal Chem 67:3202–3210, 1995.

54. MJ Huddleston, RS Annan, MF Bean, SA Carr. Selective detection of phosphopeptides in complex mixtures by electrospray liquid chromatography/mass spectrometry. J Am Soc Mass Spectrom 4:710–717, 1993.

55. SA Carr, MJ Huddleston, RS Annan. Selective detection and sequencing of phosphopeptides at the femtomole level by mass spectrometry. Anal Biochem 239:180–192, 1996.

56. SG Whalen, AC Gingras, L Amankwa, S Mader, PE Branton, R Aebersold, N Sonenberg. Phosphorylation of eIF-4E on serine 209 by protein kinase C is inhibited by the translational repressors, 4E-binding proteins. J Biol Chem 271:11831–11837, 1996.

57. RS Haltiwanger, WG Kelly, EP Roquemore, MA Blomberg, L-YD Dong, L Kreppel, T-Y Chou, GW Hart. Glycosylation of nuclear and cytoplasmic proteins is ubiquitous and dynamic. Biochem Soc Trans 20:264–269, 1992.

58. GW Hart. Glycosylation. Curr Opin Cell Biol 4:1017–1023, 1992.

59. T-Y Chou, GW Hart, CV Dang. c-Myc is glycosylated at threonine 58, a known phosphorylation site and a mutational hot spot in lymphomas. J Biol Chem 270:18962–18965, 1996.

60. GW Hart, KD Greis, L-YD Dong, MA Blomberg, T-Y Chou, M-S Jiang, EP Roquemore, DM Snow, LK Kreppel, RN Cole, FI Comer, CS Arnold, BK Hayes. O-Linked N-acetylglucosamine: the yin-yang of Ser/Thr phosphorylation? Adv Exp Med Biol 376:115–123, 1995.

61. C-R Torres, GW Hart. Topography and polypeptide distribution of terminal *N*-acetylglucosamine residues on the surfaces of intact lymphocytes. Evidence for *O*-linked GlcNAc. J Biol Chem 259:3308–3317, 1984.

62. BK Hayes, KD Greis, GW Hart. Specific isolation of *O*-linked *N*-acetylglucos-amine glycopeptides from complex mixtures. Anal Biochem 228:115–122, 1995.

63. KD Greis, BK Hayes, Fl Comer, M Kirk, S Barnes, TL Lowary, GW Hart. Sensi-tive detection and site-analysis of O-GlcNAc-modified glycopeptides by B-elimi-nation and tandem electrospray mass spectrometry. Anal Biochem 234:38–49, 1996.

64. J Wu, QN Ma, KS Lam. Identifying substrate motifs of protein kinases by a ran-dom library approach. Biochemistry 33:14825–14833, 1994.

65. BE Kemp, RB Pearson. Protein kinase recognition sequence motifs. Trends Bio-chem Sci 15:342–346, 1990.

66. I van Oostveen, A Ducret, R Aebersold. Colloidal silver staining of electroblotted proteins for high sensitivity peptide mapping by liquid chromatography-electro-spray ionization tandem mass spectrometry. Anal Biochem 1997 (in press).

12

Gels in Vacuo? A Minimalist Approach for Combining Mass Spectrometry and Polyacrylamide Gel Electrophoresis

Rachel R. Ogorzalek Loo and Philip C. Andrews
University of Michigan, Ann Arbor, Michigan

Joseph A. Loo
Parke-Davis Pharmaceutical Research, Division of Warner-Lambert Company, Ann Arbor, Michigan

I. INTRODUCTION

For protein analysis, both molecular biologists and mass spectrometrists desire an effective meld between gel electrophoresis and mass spectrometry (MS)

325

[1,2]. Often, protein samples cannot be analyzed directly because they are impure or are insoluble (e.g., intrinsic membrane proteins or extracellular matrix proteins). Beyond its separation capabilities and ability to overcome the solubility problems of a broad range of proteins (e.g., by using sodium dodecyl sulfate [SDS] or urea gels), polyacrylamide gel electrophoresis (PAGE) allows parallel processing of either a large number of samples (by using one-dimensional [1-D] slab gels) or of a single, ultracomplex mixture (by using two-dimensional [2-D] gels). With the gargantuan volume of structural information being unearthed by genome and proteome projects, rapid methods must be developed to characterize the very large numbers of predicted protein products if the full benefits of these massive sequencing efforts are to be realized. Two-dimensional gel electrophoresis/mass spectrometry is poised to play an important role in this endeavor.

Most reported mass spectrometric approaches to PAGE analysis of proteins involve proteolysis, mass analysis of the resulting fragments, and database searches to match product peptide masses [3–9], although the analysis of intact proteins has been performed from electroblotted membranes [10–18]. We describe methods that allow high-resolution gel electrophoresis to be linked directly to the high throughput and sensitivity of matrix-assisted laser desorption/ ionization (MALDI) without electroelution or electroblotting and with a methodology uniquely suited to applications requiring the measurement of intact protein masses [19–21]. The key to success with these methods is the use of ultrathin polyacrylamide gels compatible with 1-D or 2-D separations and that dry to thicknesses of 10 μm or less and have the additional advantages of rapid preparation and run times. Spectra have been acquired from isoelectric focusing (IEF), native, and SDS gels. The IEF-based approach also makes it possible to run virtual 2-D gels in which proteins are separated in the first dimension on the basis of their charge (i.e., IEF gels) while the second dimension is molecular weight, measured by MALDI-MS instead of the traditional SDS gel electrophoresis method.

II. EXPERIMENTAL

A. Gel Electrophoresis

1. Isoelectric Focusing Gels

Electrophoresis is carried out on a PhastSystem automated gel electrophoresis assembly (Pharmacia, Uppsala, Sweden) using Pharmacia's precast thin polyacrylamide gels marketed for this equipment. Isoelectric focusing employs 5% polyacrylamide gels that are 0.35 mm thick when hydrated. These gels are cast onto a polyester film backing, making them easier to handle. The sample is dissolved in a buffer prepared by diluting Novex (San Diego, CA) pH 3–7 sample buffer (lysine, glycerol, water, bromophenol blue) 1:10 with water and adding 0.5 mg methyl red (Aldrich, Milwaukee, WI) to 10 ml of diluted buffer. The methyl red serves as a tracking dye, turning red when it reaches its isoelectric point (pI), indicating that the gel has run to completion. Samples are loaded onto the pH 3–9 gradient gel by dipping a comb (0.5, 1, or 4 μl capacity/lane) into the sample solutions and placing the comb on top of the gel. Because the samples are not loaded in wells, the proteins migrate primarily along the surface of the gel, making them more accessible for laser desorption or diffusive transfer to membranes than proteins run on thicker gels. The IEF procedure detailed by the manufacturer is followed [22], except that the cooling temperature is reduced to 10°C. Gels to be stained are fixed in 20% trichloroacetic acid, washed in 30% methanol/10% acetic acid, stained in 0.02% PhastGel Blue R solution (Pharmacia) or Coomassie brilliant blue R-250 or G-250 (Sigma) in 30% methanol/10% acetic acid, 0.1% $CuSO_4$, and destained in 30% methanol/ 10% acetic acid [23]. Gels to be examined unstained are soaked in 20% trichloroacetic acid/H_2O for about 20 min and then washed in two changes of 10% acetic acid/30% methanol for 15 and 30 min each, respectively. The gels are allowed to dry in a lightly covered petri dish at room temperature. Sinapinic acid matrix is spotted onto dried gels prior to mass analysis. Alternatively, gels can be soaked in matrix solution for 10 min immediately following the last methanol/acetic acid wash, and allowed to dry as already described.

2. Native Gels

Native gel electrophoresis employs gels with 12.5% polyacrylamide that are 0.45 mm thick when hydrated but that also dry to less than 10 μm. Gradient gels (8–25%) have a similar thickness. The commercial gels are precast in 0.112 M Tris acetate, pH 6.5, and use 3% agarose IEF grade buffer strips containing 0.88 M L-alanine and 0.25 M Tris, pH 8.8. Samples dissolved in Novex native sample buffer 1:10 are loaded onto the gel as described earlier. Electrophoresis conditions [24] and gel staining conditions [23] recommended by the manufacturer are followed. Gels to be analyzed unstained are washed with two changes of 30% methanol/10% acetic acid for at least 30 min followed by a 15-min soak in 0.1 g sinapinic acid in 5 ml acetic acid and 5 ml acetone. The gels are allowed to dry at room temperature and are spotted with matrix when dry. The higher cross-linking of these gels renders them more brittle after drying, which causes them to pull away from the polyester film backing. This is observed after a few days or while under vacuum, unless the gel has been specially treated with the sinapinic acid/acetic acid/acetone solution. A soaking step with 10% glycerol is included in the protocol for stained gels [23], substituting for the sinapinic acid/acetone/glacial acetic acid soaking step applied to unstained native gels. This treatment enables them to adhere to the backing after drying. Washes in 5–10% glycerol can be used to treat gels for MALDI analysis in lieu of the sinapinic acid/acetic acid/acetone soak, although the quality of the subsequently deposited matrix crystals tends to be highly variable.

3. Sodium Dodecyl Sulfate Gels

Sodium dodecyl sulfate electrophoresis employs gels with the same characteristics as those used for native gel electrophoresis (0.112 M Tris acetate, pH 6.5). Unlike other commercial precast SDS gels, these are not poured with SDS in the gel. The SDS is present only in the 3% agarose buffer strips (0.2 M tricine, 0.2 M Tris, 0.55% SDS, pH 8.1). Samples are dissolved in 10 mM Tris/HCl, 1 mM ethylenediamine tetraacetic acid (EDTA) pH 8.0, 2.5% SDS, 0.01% bromophenol blue, and 0.55% β-mercaptoethanol and heated at 80°C for 5 min. The samples are loaded onto the gel as described earlier, and electrophoresis is carried out according to the manufacturer's directions [25]. Unstained and stained gels are washed and treated as described earlier for native gels. Dried gels are spotted with matrix prior to analysis.

4. General Comments on Polyacrylamide Gels

The long-term stability of the gels is a concern because cracking of gels and pulling away from the polyester backing is more likely as they age and is certainly affected by the composition of the gel (i.e., higher percent acrylamide gels are more prone to cracking). The treatment of the gel (e.g., how it is washed, whether it is soaked in matrix or whether matrix is spotted onto the surface) is also important to the quality of the gels. We usually examine gels by MALDI within a few days of preparation, although isoelectric focusing gels are stable for weeks. The methods described here apply to Pharmacia Phast gels (gels from other manufacturers or even from the same manufacturer, but sold for other systems, may be more or less robust).

    For desorption directly from polyacrylamide, isoelectric focusing gels have the preferred properties. IEF gels rarely crack inside the mass spectrometer vacuum and they have survived several cycles (pump-downs) from atmospheric pressure to high vacuum. The SDS gels have been more difficult to handle. They often crack after about an hour in the mass spectrometer, although we have not had problems with cracked gels completely detaching from their backing under vacuum. Certainly all edges of the gel should be carefully secured to the sample stage to minimize problems. We have been able to continue acquiring spectra from cracked gels, but mass accuracy is compromised if the gel is lifting from its backing. It is recommended that SDS and native gels be mass analyzed soon after preparation (but after complete drying) and that transfers in and out of the vacuum chamber be limited. In general, the higher the percentage of acrylamide, the more apt a gel is to crack or lift off from the backing in vacuo. With 20% acrylamide gels under vacuum, we have observed that the dried gel often detaches from its backing. We have then attached the gel without backing to the sample stage for analysis. This method is limited in that detached gels deform following subsequent in-gel chemistry, although they are still compatible with methods for extracting the products from the gel.

B. Materials

SDS molecular weight marker proteins and precast polyacrylamide gels are supplied by Pharmacia (Uppsala, Sweden), while IEF markers are purchased from BioRad (Hercules, CA). All other proteins and $CuSO_4$ are obtained from Sigma Chemical (St. Louis, MO). All materials are used without further purification. Stock solutions of standard proteins are prepared by weight.

C. Mass Spectrometry

Some of the data were acquired with a Vestec LaserTec Research reflectron time-of-flight (TOF) mass spectrometer (PerSeptive Biosystems, Framingham, MA) operated in the linear mode (1.3 m pathlength) with continuous ion extraction. The instrument, fitted with a 100-position sample plate modified by milling a 4 cm $\times$ 4 cm square to a depth of 0.5 mm to better accommodate gels, has been previously described [19–21]. Additional work was performed with PerSeptive Elite (with time-lag focusing) [19–21] and Elite-XL (with continuous extraction) TOF mass spectrometers using milled or unmilled sample plates. The Elite (2.0 m pathlength) and Elite-XL (4.2 m pathlength) mass spectrometers were operated in the linear mode. All of the instruments employ nitrogen laser (337 nm) irradiation.

Gels with the polyester backing intact are trimmed at the edges to fit the MALDI sample plate and are mounted with adhesive tape. Because of the good match between the gel and sample plate size, all of the lanes from a gel are mounted on a single plate. Generally 25–100 laser shots are averaged for each spectrum.

Matrix solution, prepared from saturated sinapinic acid (Aldrich) in 1:1 $CH_3CN:0.1\%$ trifluoroacetic acid/H_2O or 1:2 $CH_3CN:0.1\%$ trifluoroacetic acid/H_2O is spotted (0.5 μl) onto the bands of interest for native, SDS, and IEF gels. Alternatively, matrix is applied to IEF gels by soaking, as described earlier. Most gels are examined unstained; bands are located by comparing a duplicate, stained gel. Unstained gels required 5–10 min to pump down in the mass spectrometer, while Coomassie blue–stained gels needed up to 0.3 hr.

III. RESULTS

A. Isoelectric Focusing Gels

Isoelectric focusing gels, which separate proteins on the basis of pI, are valued in electrophoresis because of their ability to separate proteins that have subtle differences, (e.g., isoforms that differ by only one lysine). These gels also provide sharper bands (higher spatial resolution) than do SDS gels, and they can be run under non-denaturing conditions, preserving protein–protein interactions. Increasing use of these gels has probably been limited only by their inability to provide molecular weight information, a problem that we are now poised to overcome.

Gels are traditionally visualized by staining. However, when Coomassie blue–stained gels are examined by MALDI, dye-adducted protein ions are often observed. The ion signal is distributed over several peaks, which reduces the overall sensitivity, although the number of dye adducts is not as extensive as those reported from membranes in infrared MALDI-MS using a succinic acid matrix [14]. For relatively low-molecular-weight proteins, as shown in Fig. 1 for bovine ubiquitin (M_r 8565), the 800-Da adducts are easily resolved from both ubiquitin and its inactive des-GlyGly form [26], ubiquitin t. The degree of Coomassie adduction is variable, reflecting the extent of destaining and the fact that protein staining is sequence dependent. To eliminate dye adducts, we prefer to work with unstained gels or to use alternative staining methods such as a colloidal gold stain [21].

Polyacrylamide gel electrophoresis-MALDI-MS is well suited to mixture analysis, as illustrated by loading a protein mixture spanning a pI range from 8.7 to 4.5 and containing bovine carbonic anhydrase (700 fmol), bovine hemoglobin (700 fmol each of α- and β-chain), bovine trypsinogen (2 pmol), soybean trypsin inhibitor (2 pmol), and bovine serum albumin (2 pmol) onto a single lane of an isoelectric focusing gel. Matrix is applied to an unstained gel by soaking in sinapinic acid. An image of a similar Coomassie blue–stained gel prepared at higher loadings is illustrated in Fig. 2. Figure 3 shows spectra (25–50 shot averages) obtained for various vertical positions on the gel. Bands are located by acquiring mass spectra while moving the laser spot vertically down a lane. The approximate position can also be inferred from a stained duplicate gel. When a protein band is encountered it is apparent from the high-m/z ion signals produced. This method can be used to characterize proteins isolated on polyacrylamide gels quickly and sensitively. It can also be combined with in-

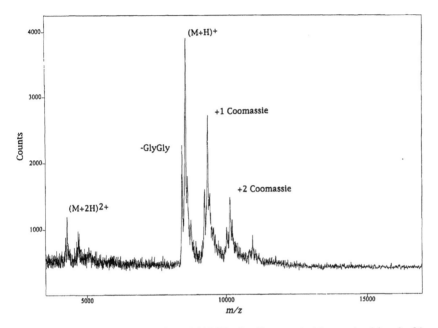

Figure 1   Spectrum obtained from MALDI of a Coomassie blue–stained band of bovine ubiquitin. The entire IEF gel was adhered to the probe. (Reprinted with permission from ref. 20. Copyright 1996 American Chemical Society.)

gel cleavage methods, such as reaction with CNBr (cleavage at Met residues), to further characterize gel-isolated proteins [19,20].

Mass spectra, illustrated in Figs. 3e and 3f, of two very similarly sized proteins partially separated under these IEF conditions were obtained at laser positions separated by a few tenths of a millimeter at the pI of trypsin inhibitor (approximately 4.5). The two 20-kD proteins differ in mass by 350 Da (Fig. 3e). These spectra underscore the superior spatial resolution available under these analysis conditions, a capability that may be lost with analytical methods relying on excised bands. Moreover, the presence of more than one component is not evident on the Coomassie-stained gel.

Many biologists and biochemists would like to use MALDI as an alternative densitometer, particularly if sensitivities better than those for silver staining can be achieved. Knowing the difficulties involved in quantitation by MALDI (ion suppression, and the variations in response for different proteins), a MALDI densitometer is likely to be problematic. However, "semiquantitative" MALDI visualization is attainable. Figure 4 illustrates the three-dimensional (3-

a)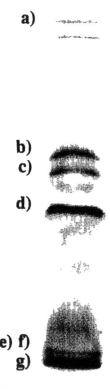

b)
c)

d)

e) f)
g)

Figure 2 Coomassie blue–stained duplicate gel illustrating the migration positions for (a) bovine trypsinogen, (b and c) bovine hemoglobin, (d) carbonic anhydrase, (e and f) soybean trypsin inhibitor, and (g) bovine albumin in isoelectric focusing. The top of the gel corresponds to the most basic proteins. The letters also indicate positions where mass spectra (shown in Fig. 3) were acquired from a duplicate, unstained gel. (From ref. 21.)

D) plot assembled from mass spectra acquired at discrete positions spanning 1 cm of a lane on a 1-D IEF gel. This 1 cm contains cytochrome c, bovine trypsinogen, lentil lectin, bovine hemoglobin, and horse heart myoglobin bands. The x, y, and z figure axes correspond to m/z, ion intensity, and laser position down the gel lane. At the most basic pI (smallest position on the z axis) cytochrome c is observed at 12.4 kD. At less basic pI (higher z-axis position), three bands containing the lentil lectin α-chain (monomer not shown, dimer is the heterogeneous species at 11–12 kD) and β-chain (20 kD) appeared as singly and doubly charged ions, followed by α- and β-chain hemoglobin (1+ and 2+) and myoglobin (1+ and 2+) at 15–17 kD in mass. Because of mass accuracy

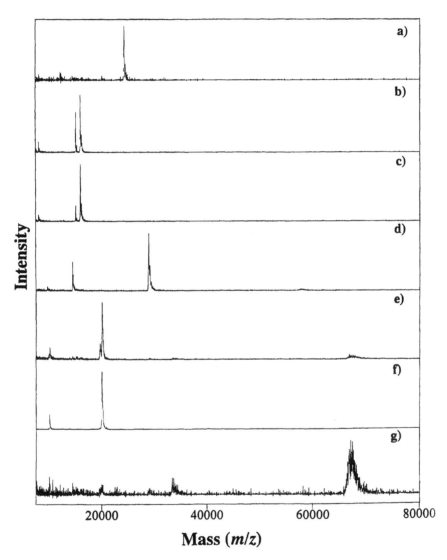

Figure 3 Composite MALDI spectrum from a mixture of proteins on an IEF gel similar to that shown in Fig. 2, but unstained and at a lower loading. The proteins are (a) bovine trypsinogen (2 pmol distributed over two bands), (b and c) bovine hemoglobin (700 fmol each α-chain and β-chain), (d) carbonic anhydrase (700 fmol), (e and f) soybean trypsin inhibitor (2 pmol), and (g) bovine albumin (2 pmol). (From ref. 21.)

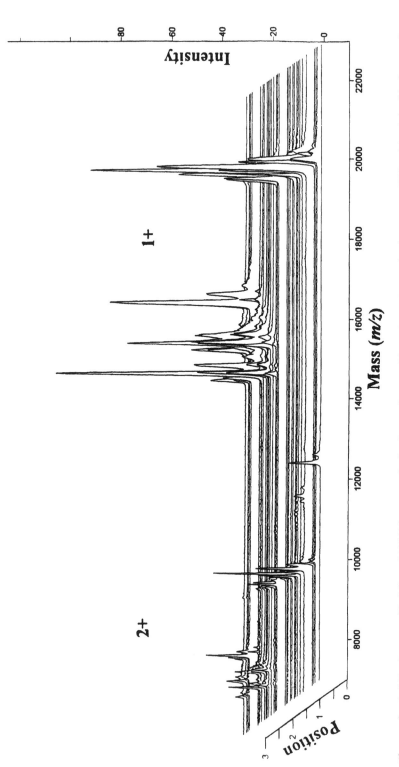

Figure 4 A 3-D plot assembled from MALDI spectra obtained at various positions down 1 cm of a lane on an IEF gel loaded with the proteins of Fig. 1 at a 10–20 pmol level in addition to myoglobin, cytochrome *c*, and lentil lectin.

limitations, the proteins do not line up perfectly on the 3-D plot. This problem (and the solution) are discussed later in the section on mass accuracy. These results show IEF's ability to detect noncovalent complexes—the α- and β-subunits of lentil lectin appear together in the spectra (same pI on the gel) because they migrate as complexes under the non-denaturing conditions employed. Hemoglobin shows similar behavior. Clearly, scanning entire 1-D gels is achievable with this technique.

B. Native and SDS Gels

Much of our work is performed with isoelectric focusing gels where mass spectrometry is particularly valuable, because IEF gels do not provide molecular weight information. Moreover, SDS and native gels are not nearly as rugged as IEF gels when handled under conditions suitable for mass spectrometry [19–21]. We encourage users to exercise caution when examining higher percentage acrylamide gels in the mass spectrometer; cracking of high-percentage acrylamide gels under vacuum is not unusual. Nevertheless, the analytical methodology is successful with ultrathin SDS gels, as illustrated in Fig. 5 for bovine albumin [M_r 66,430], and has been extended to proteins up to 85 kD [21]. For many proteins, multiple adducts of about 72 Da, suggestive of acrylamide, are observed. These adducts are well known from previous mass spectrometry per-

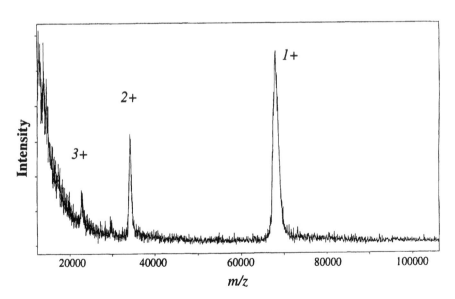

Figure 5   Mass spectrum of 30 pmol bovine albumin isolated on an SDS gel. (From ref. 21.)

formed on gel-isolated proteins [27–31] and from Edman sequencing of electroblotted proteins. They are not observed in our mass spectrometry experiments from isoelectric focusing gels, possibly because the gels were run under
non-denaturing conditions (without β-mercaptoethanol). The extent of acrylamide modification will also depend on the characteristics of the polyacrylamide gel, and may reflect differences in the quality of the gels we use for IEF
versus SDS electrophoresis. Precast gels are superior to freshly poured gels in
minimizing acrylamide adducts and are valued for applications employing Edman sequencing because they minimize artifactual NH_2-terminal blockage.
These same properties make them valuable for mass spectrometry. Acrylamide
adducts can be overcome by the commonly employed practice of reducing and
alkylating proteins prior to electrophoresis; well-established protocols are presented in many texts on protein modification.

Mass spectrometry should be as important to native gels as it is for isoelectric focusing. In SDS gels, the protein is heavily complexed to negatively
charged SDS molecules, and protein migration is governed by the amount of
SDS that the protein binds, which is proportional primarily to the size of the
protein. These properties allow a molecular mass to be estimated from protein
migration. In native gels, separation is based on the mass-to-charge ratio of the
protein itself; estimating mass from native gel migration can be difficult without prior information about protein charge. Also, native gels are frequently run
under non-denaturing conditions to elucidate noncovalent interactions. Mass
spectrometry can assist analyses greatly by rapidly confirming which species
migrate to particular positions on native gels, and can be helpful in uncovering
various participants in protein–protein interactions. Figure 6 illustrates the compatibility of MALDI directly from native gels in an analysis of horse heart
myoglobin. Although the non-covalently associated heme is not observed in the
MALDI spectrum of the protein, its association can still be determined visually
by the red band present at that position on the gel.

C. Mass Accuracy

Mass accuracy is an important issue. Concepts that are frequently taken for
granted in desorption/ionization of samples spotted onto probes must be re-
thought carefully; for example, the suitability of internal mass standards will be
more limited. In order to obtain good mass values when desorbing from gels
and membranes in continuous ion extraction TOF instruments, considerable attention must be paid to the method of mass calibration. For our purposes, *external calibration* refers to spotting a standard protein *near* a band of interest on
the gel, while *internal calibration* refers to spotting the standard directly on *top*
of the band of interest. In one case, a 1% drift in calibration was observed
without time-lag focusing (or delayed extraction) when we scanned down an

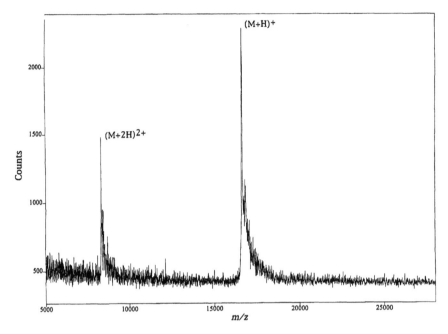

Figure 6 MALDI mass spectrum of horse heart myoglobin from a native gel. (Reprinted with permission from ref. 20. Copyright 1996 American Chemical Society.)

entire lane after calibrating with a single spot at the top of the gel. The size of the drift reflected the distance from the standard spot. Problems arise from several sources: uneven gel thicknesses, difficulty in mounting gels flat, and surface charging. Problems with charging are particularly bothersome in experiments looking at entire gels or blots, as opposed to excised bands, because one can spend an hour or more examining a few lanes, constantly building up charge on the insulator surface.

Accuracies for 1-D gels are improved by placing multiple spots of internal mass standards down a blank lane bordering the sample lane and calibrating with nearby standard spots, thus improving mass accuracy to 0.1–0.2% in the linear mode, without using delayed extraction. The extension of this technique to 2-D gels is limited without delayed extraction because one does not have the luxury of applying molecular weight calibration spots anywhere on the gel. The density of protein spots may be particularly high in a given region of the 2-D gel. Also, a "blank" region indicated by a stained 2-D gel may, in fact, contain several protein spots that are below the detection limits of staining. MALDI-MS appears to have superior sensitivity to silver staining.

In principle, internal calibration would be superior, but it is particularly difficult with gels, because nonoptimal standard loadings can overwhelm sample signals. In practice, we have found that internal calibration with a continuous ion extraction source design yields mass accuracies of $\pm 0.1\%$ using a single stage ion source. However, because of ion suppression effects, sometimes several different ratios of sample to internal standard concentration must be tried before a spectrum yielding ions for both can be obtained. This behavior is usually of no consequence when samples are spotted onto probes, because multiple mixtures can be spotted and examined until the desired result is obtained. On a gel, however, the size of the sample spot or band limits how many attempts to spot the proper amount of internal standard can be made. Estimating the proper amount of internal standard to add is more difficult because spotted proteins desorb from gels more readily than proteins entrained in the gel. Moreover, unless a duplicate gel has been stained, there may not be any estimate of the quantity of protein in that spot or band. Once a standard has been spotted onto a band, it must be accounted for in subsequent chemistry and analyses performed on that band. Armed with the sequence of the spotted internal standard, one could digest or chemically cleave the gel-entrained protein (and the spotted internal standard) and ignore the standard-related products. In practice, the standard's products could suppress sample-related peaks in subsequent MALDI analyses, necessitating additional extraction and separation steps.

These issues make spotting of external standards desirable, particularly because the quality of calibration cannot be assumed when insulating surfaces are employed. It is also useful to evaluate the mass accuracy that can be achieved for gels prepared for positional scanning, with matrix deposited across the entire gel, rather than in discrete spots. Moreover, there is the issue that proteins entrained in gels desorb quite differently from proteins spotted onto surfaces—for example, higher laser powers are required. The ions desorbed from proteins embedded in gels may possess very different initial velocities from those arising from spotted samples.

In light of these concerns, we chose to explore mass accuracy and precision in desorption from gels for TOF mass spectrometers with and without time-lag focusing. Since the seminal publication by Wiley and McLaren [32] detailing time-lag focusing and resolution in TOF mass spectrometry, a number of groups have applied the technique to improve resolution by focusing ions [33–37] or to measure velocity and angular distributions of photodissociation products more precisely by defocusing ions [38–41]. Several laboratories have added this capability to MALDI mass spectrometry [42–49]. Our interest in applying time-lag focusing to gels arose because of the superior space-focusing capabilities of the two-stage ion sources employed on these instruments; insensitivity to spatial distortions is desirable for desorption from uneven gel surfaces. Moreover, the pulsed electric field offered promise for reducing problems due to charging at the insulator surface. Compensation for differences in initial velocity could also be helpful in obtaining better precision in mass calibration.

Delayed extraction shows promise for overcoming some of the limitations due to both charging and slightly distorted gels, provided that the matrix crystals are not too large (they should be at or below the size usually achieved with the dried drop method for sinapinic acid). In the linear operating mode, mass accuracies better than 0.1% were readily obtained for individual measurements on the 5-kD α-subunit proteins of lentil lectin measured over the course of an hour and referenced to a single calibration. Based on 11 measurements over that hour, the gel lectin bands yielded 5875.0 ± 1.5 Da for one component, a standard deviation of 0.025%, while electrospray mass spectrometry yielded 5876.7 ± 0.4 Da. This series of measurements employed multiple IEF bands for the noncovalent complex [20,50–52], spanning a distance of approximately 0.6 cm (about 18% of the length of the lane). Similarly, a series of seven measurements on the 15–16 kD hemoglobin subunits showed a precision of $\pm 0.02\%$. When the hemoglobin bands were measured, based on a calibration obtained 30% of the lane's length away, near the top of the gel and close to where it was secured to the probe, the mass accuracy was $\pm 0.5\%$. However, while this suggests that calibration spots can be spaced more widely with delayed extraction, they cannot be eliminated.

IV. CONCLUSIONS

The direct mass analysis of proteins separated by PAGE is a viable approach. Subpicomole amounts of protein entrained in polyacrylamide gels can be routinely mass measured. To answer the question asked by the title, gel surfaces are compatible with the high vacuum requirements of mass spectrometry. This should not be too surprising, as it was not long ago that fast atom bombardment (FAB) was the ionization method of choice for nonvolatile biomolecules such as peptides and small proteins. The analyte was dissolved or suspended in a FAB matrix such as glycerol and placed into the source vacuum region of the mass spectrometer. An early attempt to link gel electrophoresis with mass spectrometry used FAB as the desorption/ionization approach [53].

With delayed ion extraction on TOF instruments, mass accuracy is much improved. This factor is extremely important for the PAGE-MALDI-MS marriage if the method is to have wide utility. A potential application of the technique is proteome mapping. The protein components encoded by an organism's genome is defined as the proteome. Each organism has its unique genome, whereas it may have several proteomes because the environment of the cell can influence which proteins are produced and to what levels. Different posttranslational modifications may occur under various environments. The identification of proteins coded by currently undefined genes is a goal of proteome projects. Further improvements and refinements in all aspects of PAGE-MS, including electrophoresis and mass spectrometry, and its combination with other tools such as gel image analysis and protein database searching should prove to be a high-throughput method for determining the identity and function of cell proteins.

ACKNOWLEDGMENTS

Encouragement from J. E. Hoover was greatly appreciated by PCA. Also, we thank PerSeptive Biosystems for use of their instrumentation, and Tracy Stevenson (Parke-Davis) and Charles Mitchell (University of Michigan) for their technical expertise.

REFERENCES

1. SD Patterson, R Aebersold. Mass spectrometric approaches for the identification of gel-separated proteins. Electrophoresis 16:1791–1814, 1995.
2. SD Patterson. From electrophoretically separated protein to identification: Strategies for sequence and mass analysis. Anal Biochem 221:1–15, 1994.
3. JR Yates III, S Speicher, PR Griffin, T Hunkapiller. Peptide mass maps: A highly informative approach to protein identification. Anal Biochem 214:397–408, 1993.
4. WJ Henzel, TM Billeci, JT Stults, SC Wong, C Grimley, C Watanabe. Identifying proteins from 2-dimensional gels by molecular mass searching of peptide fragments in protein sequence databases. Proc Natl Acad Sci USA 90:5011–5015, 1993.
5. DJC Pappin, P Hojrup, AJ Bleasby. Rapid identification of proteins by peptide-mass fingerprinting. Curr Biol 3:327–332, 1993.
6. M Mann, P Hojrup, P Roepstorff. Use of mass spectrometric molecular weight information to identify proteins in sequence databases. Biol Mass Spectrom 22:338–345, 1993.
7. J Eng, AL McCormack, JR Yates III. An approach to correlate tandem mass spectral data of peptides with amino acid sequences in a protein database. J Am Soc Mass Spectrom 5:976–989, 1994.
8. JR Yates III, J Eng, AL McCormack, D Schieltz. Anal Chem 67:1426–1436, 1994.
9. M Wilm, A Shevchenko, T Houthaeve, S Breit, L Schweigerer, T Fotsis, M Mann. Femtomole sequencing of proteins from polyacrylamide gels by nano-electrospray mass spectrometry. Nature 379:466–469, 1996.
10. SD Patterson. Matrix-assisted laser desorption/ionization mass spectrometric approaches for the identification of gel-separated proteins in the 5–50 pmol range. Electrophoresis 16:1104–1114, 1995.
11. JC Blais, P Nagnan-Le-Meillour, G Bolbach, JC Tabet. MALDI-TOFMS identifica-

tion of 'odorant binding proteins' (OBPs) electroblotted onto poly(vinylidene difluoride) membranes. Rapid Commun Mass Spectrom 10:1–4, 1996.

12. C Eckerskorn, K Strupat, M Karas, F Hillenkamp, F Lottspeich. Mass spectrometric analysis of blotted proteins after gel electrophoretic separation by matrix-assisted laser desorption/ionization. Electrophoresis 13:664–665, 1992.

13. X Liang, J Bai, Y-H Liu, DM Lubman. Characterization of SDS-PAGE-separated proteins by matrix-assisted laser desorption/ionization mass spectrometry. Anal Chem 68:1012–1018, 1996.

14. K Strupat, M Karas, F Hillenkamp, C Eckerskorn, F Lottspeich. Matrix-assisted laser desorption ionization mass spectrometry of proteins electroblotted after polyacrylamide gel electrophoresis. Anal Chem 66:464–470, 1994.

15. K Strupat, C Eckerskorn, M Karas, F Hillenkamp. Infrared-matrix-assisted laser desorption/ionization mass spectrometry (IR-MALDI-MS) of proteins electroblotted onto polymer membranes after SDS-PAGE separation. In: AL Burlingame, SA Carr, eds. Mass Spectrometry in the Biological Sciences. Totowa, NJ: Humana Press, 1996, pp 203–216.

16. MM Vestling, C Fenselau. Poly(vinylidene difluoride) membranes as the interface between laser desorption mass spectrometry, gel electrophoresis, and in situ proteolysis. Anal Chem 66:471–477, 1994.

17. MM Vestling, C Fenselau. Polyvinylidene difluoride (PVDF): An interface for gel electrophoresis and matrix-assisted laser desorption/ionization mass spectrometry. Biochem Soc Trans 22:547–551, 1994.

18. M Schreiner, K Strupat, F Lottspeich, C Eckerskorn. Ultraviolet matrix assisted laser desorption ionization-mass spectrometry of electroblotted proteins. Electrophoresis 17:954–961, 1996.

19. RR Ogorzalek Loo, C Mitchell, T Stevenson, JA Loo, PC Andrews. Interfacing polyacrylamide gel electrophoresis with mass spectrometry. In: D Marshak, ed. Techniques in Protein Chemistry. San Diego: Academic Press, 1996, pp 305–313.

20. RR Ogorzalek Loo, TI Stevenson, C Mitchell, JA Loo, and PC Andrews. Mass spectrometry of proteins directly from polyacrylamide gels. Anal Chem 68:1910–1917, 1996.

21. RR Ogorzalek Loo, C Mitchell, TI Stevenson, SA Martin, W Hines, P Juhasz, D Patterson, J Peltier, JA Loo, PC Andrews. Sensitivity and mass accuracy for proteins analyzed directly from polyacrylamide gels: Implications for proteome mapping. Electrophoresis 18:382–390, 1997.

22. Pharmacia LKB Biotechnology AB. Separation Technique File No. 100. Uppsala, Sweden.

23. Pharmacia LKB Biotechnology AB. Separation Technique File No. 200. Uppsala, Sweden.

24. Pharmacia LKB Biotechnology AB. Separation Technique File No. 121. Uppsala, Sweden.

25. Pharmacia LKB Biotechnology AB. Separation Technique File No. 111. Uppsala, Sweden.

26. KD Wilkinson, TK Audhya. Stimulation of ATP-dependent proteolysis requires ubiquitin with the carboxy-terminal sequence Arg-Gly-Gly. J Biol Chem 256:9235–9241, 1981.

27. E Mortz, O Vorm, M Mann, P Roepstorff. Identification of proteins in polyacryl-amide gels by mass spectrometric peptide mapping combined with database search. Biol Mass Spectrom 23:249–261, 1994.

28. WZ Zhang, AJ Czernik, T Yungwirth, R Aebersold, BT Chait. Matrix-assisted laser desorption mass spectrometric peptide mapping of proteins separated by two-dimensional gel electrophoresis—Determination of phosphorylation in synapsin I. Protein Sci 3:677–686, 1994.

29. C Bonaventura, J Bonaventura, R Stevens, D Millington. Acrylamide in polyacryl-amide gels can modify proteins during electrophoresis. Anal Biochem 222:44–48, 1994.

30. SC Hall, DM Smith, FR Masiarz, VM Soo, HM Tran, LB Epstein, AL Burlingame. Mass spectrometric and edman sequencing of lipocortin I isolated by two-dimensional SDS/PAGE of human melanoma lysates. Proc Natl Acad Sci USA 90:1927–1931, 1993.

31. M le Maire, S Deschamps, J Moller, JP Le Caer, J Rossier. Electrospray ionization mass spectrometry on hydrophobic peptides electroeluted from sodium dodecyl sulfate-polyacrylamide gel electrophoresis. Application to the topology of the sarcoplasmic reticulum Ca^{2+} ATPase. Anal Biochem 214:50–57, 1993.

32. WC Wiley, IH McLaren. Time-of-flight mass spectrometer with improved resolution. Rev Sci Instrum 26:1150, 1955.

33. JA Browder, RL Miller, WA Thomas, G Sanzone. High-resolution TOF mass spectrometry. II. Experimental confirmation of impulse-field focusing theory. Int J Mass Spectrom Ion Phys 37:99–108, 1981.

34. NL Marable, G Sanzone. High resolution time-of-flight mass spectrometry theory of the impulse-focused time-of-flight mass spectrometer. Int J Mass Spectrom Ion Phys 13:185, 1974.

35. RL Miller, JA Browder, G Sanzone. High-resolution time-of-flight mass spectrometry impulse field focusing pulse generators. Rev Sci Instrum 57:1523, 1986.

36. ML Muga. Velocity compaction time-of-flight mass spectrometer for mass range 1000–10000 u. Biomed Environ Mass Spectrom 16:131–132, 1987.

37. RB Opsal, KG Owens, JP Reilly. Resolution in the linear time-of-flight mass spectrometer. Anal Chem 57:1884–1889, 1985.

38. R Ogorzalek Loo, H-P Haerri, GE Hall, and PL Houston. Methyl rotation, vibration, and alignment from a multiphoton ionization study of the 266 nm photodissociation of methyl iodide. J Chem Phys 90:4222–4236, 1989.

39. R Ogorzalek Loo, GE Hall, H-P Haerri, PL Houston. State-resolved photo fragment velocity distributions by pulsed extraction time-of-flight mass spectrometry. J Phys Chem 92:5–8, 1988.

40. HJ Hwang, J Griffiths, MA El-Sayed. The one dimensional photofragment translational spectroscopic technique: Intramolecular clocking of energy redistribution for molecules falling apart. Int J Mass Spectrom Ion Proc 131:265–282, 1994.

41. JA Syage. Photofragment imaging by sections for measuring state-resolved angle-velocity differential cross sections. J Chem Phys 105:1007–1022, 1996.

42. ML Vestal, P Juhasz, P Martin. Delayed extraction matrix-assisted laser desorption time-of-flight mass spectrometry. Rapid Commun Mass Spectrom 9:1044–1050, 1995.

43. RS Brown, JJ Lennon. Mass resolution improvement by incorporation of pulsed ion extraction in a matrix-assisted laser desorption/ionization linear time-of-flight mass spectrometer. Anal Chem 67:1998–2003, 1995.

44. SM Colby, TB King, JP Reilly. Improving the resolution of matrix-assisted laser desorption/ionization time-of-flight mass spectrometry by exploiting the correlation between ion position and velocity. Rapid Commun Mass Spectrom 8:865, 1994.

45. RM Whittal, L Li. High-resolution matrix-assisted laser desorption/ionization in a linear time-of-flight mass spectrometer. Anal Chem 67:1950–1954, 1995.

46. SM Colby, JP Reilly. Space-velocity correlation focusing. Anal Chem 68:1419–1428, 1996.

47. P Juhasz, MT Roskey, IP Smirnov, LA Haff, ML Vestal, SA Martin. Applications of delayed extraction matrix-assisted laser desorption ionization time-of-flight mass spectrometry to oligonucleotide analysis. Anal Chem 68:941–946, 1996.

48. B Spengler, Y Pan, RJ Cotter, L-S Kan. Molecular weight determination of underivatized oligodeoxyribonucleotides by positive-ion matrix-assisted ultraviolet laser-desorption mass spectrometry. Rapid Commun Mass Spectrom 4:99–102, 1990.

49. TB King, SM Colby, JP Reilly. High resolution MALDI-TOF mass spectra of three proteins obtained using space-velocity correlation focusing. Int J Mass Spectrom Ion Proc 145:L1–L7, 1995.

50. L Bhattacharyya, CF Brewer. Isoelectric focusing studies of concanavalin A and the lentil lectin. J Chromatogr 502:131–142, 1990.

51. IJ Goldstein, RD Poretz. Isolation, physicochemical characterization, and carbohydrate-binding specificity of lectins. In: IE Liener, N Sharon, IJ Goldstein, eds. The Lectins. Properties, Functions, and Applications in Biology and Medicine. Orlando, FL: Academic Press, 1986, pp 70–73.

52. IK Howard, HJ Sage, MD Stein, NM Young, MA Leon, DF Dyckes. Studies on a phytohemagglutinin from the lentil. J Biol Chem 246:1590–1595, 1971.

53. P Camilleri, NJ Haskins, AJ Hill, AP New. A coordinated approach towards the molecular weight determination of peptides by gel electrophoresis and fast-atom bombardment mass spectrometry. Rapid Commun Mass Spectrom 3:440–442, 1989.

13

Studying Noncovalent Interactions by Electrospray Ionization Mass Spectrometry

Joseph A. Loo and Kristin A. Sannes-Lowery*
Parke-Davis Pharmaceutical Research, Division of Warner-Lambert Company, Ann Arbor, Michigan

I. INTRODUCTION

Mass spectrometry (MS) is playing an increasingly important role in biomedical research. The development of electrospray ionization (ESI) [1] and matrix-

---

*Current affiliation: Wacker Silicones Corporation, Adrian, Michigan.

assisted laser desorption/ionization (MALDI) [2] is making sensitive and accurate molecular mass determinations available for a variety of large biological molecules. In addition, ESI is also used to study the complexation of molecules [3,4]. Biological processes are a result of a cascade of intra- and intermolecular interactions. Gas-phase multiply charged ions representing specific noncovalently bound complexes can be detected with ESI-MS. During the early development of ESI, it was suggested that weakly bound complexes could be detected. ESI is a gentle ionization method, yielding little, if any, molecular fragmentation. Adducts attributed to salt and solvent species (e.g., isopropanol and other organic cosolvents) on peptide and protein ions were observed in some of the early examples reported [5,6]. Aggregates (e.g., dimers, trimers, etc.) of polypeptide analytes attributed to gas-phase processes were observed, especially at higher protein concentrations [7]. Even a complex between protoporphyrin IX (heme) and myoglobin protein was evident in some of the earlier ESI-MS spectra [6,8]. At the time, it was not recognized that such gas-phase complexes may reflect the solution-phase characteristics of the biological molecule. (Indeed, the denaturing conditions such as acidic pH and high organic solvent content used in these early studies most likely prevented the specific interaction between heme and apomyoglobin.) Not until the report of the globin–heme interaction of myoglobin [9] and an example of a receptor–ligand complex, FKBP and FK506 [10], were obtained from aqueous solution near neutral pH was it suggested that complexes that associate through specific noncovalent forces could be detected by ESI-MS. Since these initial reports, several other types of noncovalent binding systems have been studied, including protein–cofactor [11], enzyme–substrate pairings [12], protein subunit complexes [13–23], antibody–antigen [24,25], inclusion complexation [26–28], oligonucleotide duplex assemblies [29–35], and protein–DNA complexes [36,37].

The purpose of this report is to discuss the experimental variables that affect the direct observation of ions due to noncovalent complexes by ESI-MS. Many other reviews of the subject have appeared in the literature [3,4,38–41]. Over 100 examples of noncovalent interactions studied by ESI-MS have been published. However, as we and others have found in the laboratory, the use of ESI in this endeavor is not without its share of complications and is not considered a "routine" experiment. Each biological and experimental system has its unique characteristics. Several examples of noncovalently bound complexes will be discussed that highlight some of these "features."

II. THE UTILITY OF ESI-MS FOR STUDYING NONCOVALENT COMPLEXES

There are several established instrumental methods that have been applied for the study of macromolecular interactions, such as spectroscopic approaches (e.g., circular dichroism, light scattering, and fluorescence), differential scanning calorimetry, isothermal titration calorimetry, analytical ultracentrifugation, and surface plasmon resonance (SPR). Each biophysical method has strengths and weaknesses, as recently reviewed by Hensley [42]. For example, association and dissociation rate constants can be measured by several methods, such as SPR. Microcalorimetry provides thermodynamic information. On the other hand, large quantities of material are typically required for techniques such as NMR. Often, proteins may precipitate at the high concentrations necessary for NMR. X-ray crystallography can provide high-resolution structures, but the sample crystallization process can be problematic for many samples. The accuracy of the molecular weight measurement from ultracentrifugation and non-denaturing gel electrophoresis also can be problematic, leading to questionable stoichiometry results.

By using ESI-MS to study noncovalent complexes, several important features of MS can be exploited over the other more traditional methods. Speed and sensitivity are the most obvious advantages of a MS-based method. Results using Fourier transform MS (FTMS) [16] and time-of-flight (TOF) analyzers [22] show low picomole to femtomole sensitivity for such experiments. Submicromolar concentration solutions are normally required for such experiments. The recently demonstrated advantages of low-flow ESI sources (or commonly referred to as "nanoelectrospray" [43]) offers additional sensitivity enhancements.

Specificity is a critical advantage of ESI-MS. Complexes come together by specific interactions, based on key structural and/or energetic features, and nonspecific interactions (e.g., aggregation). Small structural changes in a ligand can dramatically affect solution binding. The binding of $pp60^{v-Src}$ SH2 (Src homology 2) domain protein is specific toward phosphopeptides and is variable in its amino acid sequence immediately on the C-terminal side of the phosphotyrosine residue. An ESI-MS study showed the expected greater affinity of phosphopeptides to Src SH2 relative to an unphosphorylated peptide of the same sequence [44].

The number of ligands that form a unique and biologically relevant complex is an important issue for many systems. Stoichiometry of the complex can be easily obtained from the resulting mass spectrum because the molecular weight of the ligands and the complex are directly and accurately measured. Many enzyme systems are composed of identical and nonidentical subunits that associate together to form the fully active species. Mass spectrometry experiments with protein oligomers can yield information on both the stoichiometry and molecular nature of subunit interactions. Whether a protein's quaternary structure involves formation of a monomer, dimer, trimer, tetramer, and so on can be determined by ESI-MS. For example, the oligomeric structure of the enzyme 4-oxalocrotonate tautomerase (4-OT) was studied by ESI with a TOF mass analyzer [22]. By gel permeation chromatography and ultracentrifugation,

4-OT was estimated to be a pentamer. However, the ESI-TOF-MS studies clearly demonstrated that 4-OT (62-residue monomer) exists as a hexamer in solution, consistent with x-ray crystallography experiments. Moreover, analogues of 4-OT were used to demonstrate the importance of specific residues for maintaining the hexameric state. The analysis of (oxidized-Met[45])-4-OT and (des-Pro[1])-4-OT showed predominantly monomeric ions, consistent with structural studies by circular dichroism spectroscopy.

Reports on the determination of the relative and absolute strength of these solution-phase interactions by ESI-MS provide a gauge on whether solution-phase binding events are being monitored by this gas-phase method. From mixtures of various peptide inhibitors, where the total peptide concentration is much greater than the protein concentration (competitive binding conditions), the relative abundances of the Src SH2 protein–phosphopeptide complexes observed in the ESI mass spectrum were consistent with their measured solution-phase binding constants [44]. The relative affinities for even D/L-isomeric peptides could be determined. One of the most popular graphical methods for obtaining the intrinsic dissociation constant, K_D, from equilibrium receptor–ligand binding systems is the Scatchard plot [45,46]. For the simple reaction:

$$[R] + [L] \leftrightarrow [RL]$$

the association constant K_a is:

$$K_a = [RL]/[R][L] = 1/K_D$$

At equilibrium, the total number of binding sites, B_{max}, is the number of unbound sites plus the bound sites, or the unbound sites (R) will be equal to:

$$R = B_{max} - RL$$

Thus, the equilibrium equation becomes:

$$K_a = [RL]/\{(B_{max} - [RL])[L]\}$$
$$\text{Bound/free} = [RL]/[L] = K_a[R] = K_a(B_{max} - [RL]) = K_D^{-1}(B_{max} - [RL])$$

A Scatchard plot of the moles of bound ligand per total moles of receptor (i.e., bound/free) on the ordinate axis versus the moles of bound receptor (i.e., bound) on the abscissa should produce a straight line where the slope is $-1/K_D$ and the abscissa intercept is equal to B_{max} (or stoichiometry). The first demonstration that data from ESI-MS experiments can be used to construct conventional Scatchard plots was reported for the binding of vancomycin antibiotics with tripeptide ligands [47]. The MS gas-phase measurements were in reasonable agreement with previously reported solution-phase values and verified the validity of the overall methodology. Likewise, the association of albumin protein with oligonucleotides was measured to have dissociation constants in the micromolar range by Scatchard analysis of ESI-MS titration data, and independently verified by capillary electrophoresis [36].

III. GENERAL EXPERIMENTAL CONSIDERATIONS

The solution conditions needed to maintain an intact, noncovalently bound complex do not match those typically used for routine ESI operation. Solution pH of 2–4 for positive ionization and pH 8–10 for negative ion ESI are typical for maximum sensitivity. The addition of an organic cosolvent such as methanol or acetonitrile also enhances sensitivity and ion signal stability. However, these conditions are not generally tolerated for studying noncovalent complexes. Denaturation of the complex can occur in solution at pH ranges outside of the pH 6–8 range, and the addition of a high concentration of an organic cosolvent can likewise disrupt the stability of the complex. This can act as a simple control experiment for an ESI-MS study. Systems that are known to denature under these conditions should only yield ions for the unbound species.

Although quadrupole mass analyzers were the first types of systems used to detect noncovalent complexes by ESI, all of the other common types of mass spectrometers, such as magnetic-sector instruments, ion traps and FTMS systems, and TOF instruments, have been successfully used. Analyzers with high mass-to-charge range (greater than m/z 4000) have advantages for studying the full range of biochemical complexes because of the tendency for many noncovalent complexes to exhibit relative low charge states. Even quadrupole analyzers modified for high mass-to-charge ion transmission and focusing (m/z range > 10,000) have been utilized to study low-charged complexes [13].

Most versions of ESI–atmospheric pressure interfaces (e.g., differentially pumped nozzle-skimmer interface, heated glass or metal capillary inlet with or without a countercurrent bath gas) have been employed to observe complexes. Careful control of the various instrumental settings associated with each interface has proven to be important. Variables that affect ion desolvation and ion activation can effect the success of the experiment. Solvent molecules need to be removed from the ion prior to detection. This is accomplished through a variety of means, including the use of a counterflow of nitrogen gas, heat (either a warm countercurrent gas, a heated capillary inlet, or a warm interface chamber), and collisional activation downstream of the initial interface. Often, it is difficult to find the balance between sufficient desolvation of the gas-phase complex and dissociation of the complex. The amount of ion activation can be controlled by the voltage difference between the capillary-skimmer or nozzle-skimmer lens elements and more often needs to be reduced to maintain the intact complex.

Most gas-phase complexes are relatively fragile—that is, it is normally very easy to dissociate a complex once it is in the gas phase. ESI interface conditions need to be as gentle as possible to maintain the intact complex. However, protein–DNA complexes [37,48] and protein–RNA complexes [49] are examples where dissociation of the gas-phase complex is very difficult; that is, the multiply charged ions for the complex are stable at high interface energies. Protein complexes with oligonucleotides usually involve extensive electrostatic forces that can be very strong in the gas phase. At the other extreme, the gas-phase complexes between acyl coenzyme A (CoA) binding protein and acyl CoA derivatives were found to be sensitive to the ESI source temperature; increasing the temperature from 20 to 80°C reduced the proportion of complexed species to zero [50]. Cooling the ESI nebulizer gas and the analyte solution produced the best results [50]. In this regard, the use of low-flow electrospray sources may prove to have advantages for aqueous solutions and studying complexes. The nanoelectrospray data for the 4-OT hexameric enzyme complex suggests that the nanoESI-MS combination may be more gentle than higher flow rate sources [22].

In most published reports, the observed stoichiometry is consistent with the expected result. Sometimes, more ligand than expected is observed to bind to the protein receptor due to nonspecific gas-phase aggregation. The prevalence of nonspecific aggregation can be reduced by reducing the solution concentration of the analytes [3,7]. Control experiments can be used to rule out ubiquitous nonspecific interactions. In addition, some solution-phase interactions such as hydrophobic and nonpolar stacking may not be maintained in the gas phase. The acyl CoA-binding protein example illustrates a case where the ESI-MS data are not perfectly faithful to the solution-phase characteristics [50]. Acyl CoA ligands with different solution dissociation constants to the protein were not differentiated by the ESI-MS experiment. A combination of hydrophobic, electrostatic, and nonpolar stacking interactions maintain protein–ligand binding in this example. Changes in the length of the hydrocarbon acyl chain, while greatly affecting solution binding, did not appear to affect the stability of the gas-phase complex.

IV. SELECTED EXAMPLES

There are several types of noncovalent interactions that have been studied in our laboratory in recent years. A few selected examples that illustrate the types of problems that can be solved will be discussed. The important experimental procedures necessary for each case will be highlighted.

The following examples used either one of the following mass spectrometers: a single-quadrupole system with a differentially pumped nozzle-skimmer atmosphere–vacuum interface and a countercurrent desolvation gas flow or a commercial double-focusing hybrid mass spectrometer (EBqQ geometry, Finnigan MAT 900Q, Bremen, Germany) with a mass-to-charge range of 10,000 at 5 kV full acceleration potential. Two slightly different ESI interfaces on the magnetic-sector system have been used: an interface based on a heated glass capillary inlet with a nitrogen gas flow countercurrent to the electrospray axis for droplet and ion desolvation [1], and a heated metal capillary inlet with no countercurrent gas [51]. In addition, gas-phase collisions, controlled by adjustment of the voltage difference between the tube lens at the exit of the capillary and the first skimmer element (ΔV_{TS}) [19,52,53], were used to augment the desolvation of the ESI-produced droplets and ions for both magnetic sector interfaces. A stream of SF_6 coaxial to the spray suppressed corona discharges in the ESI source, especially important for ESI of aqueous solutions.

The solution conditions necessary for maintaining the intact complex is dependent on the biochemical system of interest. Some protein complexes do not survive many repeated freeze–thaw cycles. Ligand binding conditions may vary; incubation for a defined period of time may be required. The list of volatile buffers that are compatible with ESI is relatively short: ammonium acetate or ammonium bicarbonate, with buffer concentrations typically between 5 and 25 mM with adjustment of the pH (pH 5.5–9.0) with acetic acid or ammonium hydroxide. For ESI systems that cannot tolerate even a low-concentration, volatile buffer, unbuffered 100% aqueous solutions are often used. However, the pH of the analyte solution (not just the solvent) should be measured. The stability of the complexes may be very pH sensitive. The use of an organic modifier also depends on the system of interest. Even addition of 10% (v/v) methanol can improve the stability of the electrospray signal. Up to 50% (v/v) methanol can be tolerated for some metal-binding proteins without significant denaturation. However, results without using organic cosolvents should be acquired to determine if the addition of organic solutions alters the results of the MS experiment.

Conditioning the solution transfer line (e.g., fused silica, PEEK, or metal capillary) is important, much like the ritual employed by experimentalists performing capillary electrophoresis. In our laboratory, we generally rinse the capillary with equal volumes of a 65/32.5/2.5 methanol/water/acetic acid solution, followed by a 1% ammonium hydroxide solution, and then the ammonium acetate/bicarbonate buffer solution.

A. Myoglobin-Heme Binding

Myoglobin is composed of a polypeptide chain (16.9 kD for equine myoglobin) and a heme prosthetic group (Fe-protoporphyrin IX) noncovalently bonded in a hydrophobic crevice. The myoglobin–heme system has been well studied by

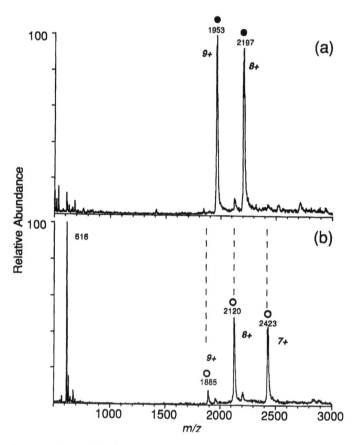

Figure 1   Positive-ion ESI mass spectra of horse heart myoglobin (1.9 μM) in 10 mM ammonium acetate, pH 6.9. The spectra were acquired on a magnetic-sector mass spectrometer (Finnigan MAT 900Q, Bremen, Germany) with a heated metal capillary ESI interface (capillary temperature 150°C). The potential between the tube lens and the skimmer of the interface (ΔV_{TS}) was varied from (a) +60 V to (b) +125 V. The closed circles (●) represent ions for the myoglobin-heme complex and the open circles (○) designate ions for the apo-protein.

ESI-MS, largely because of the relative ease with which experimental conditions for observing the complex can be established on most instruments [9,54–57]. Organic solvents and low pH conditions are known to denature myoglobin, expelling the heme from the pocket. The mass spectrum of myoglobin taken from solutions of acidic pH reflects a molecular mass for the apo-protein only (and a peak representative of a singly charged ion for the heme). ESI mass spectra taken from pH 6 aqueous solutions shows only ions for the myoglobin protein–heme complex. The sensitivity for observing the noncovalent myoglobin complex has been demonstrated to be as low as 100 fmol [56]. Increasing the voltage difference between the tube lens and first skimmer element of the electrospray interface promotes collisional dissociation of the gas-phase complex to the separated polypeptide and Fe-porphyrin species.

The mass spectra in Fig. 1 illustrate these points. Myoglobin (horse heart, apo-protein mass 16,951, holo-protein mass 17,567) in 10 mM ammonium acetate solution, pH 6.9, shows ions for the 8+ and 9+ charge states of the noncovalent myoglobin–heme complex (Fig. 1a). Collisional dissociation of the complex can be induced in the ESI interface by increasing ΔV_{TS} from a +60 V to +125 V, yielding ions for the polypeptide chain (7+ to 9+ charges) and a singly charged ion at m/z 616 for heme (Fig. 1b).

The myoglobin complex is a good test case for developing experimental conditions for observing noncovalent complexes on all mass spectrometers. The protein is relatively inexpensive and readily available. The goal should be to determine the conditions necessary to observe the holo-protein with good sensitivity and signal stability with a minimum amount of fragmentation or complex denaturation. In general, electrospraying 100% aqueous solutions is not easy compared to the more typical organic, low pH solutions. Adjustment of the spray voltage and maybe even the spray direction is usually necessary. Using weakly buffered solutions (e.g., 5–10 mM ammonium acetate) is preferable. Although complexes such as myoglobin can be observed from unbuffered water solutions, the charge distributions tend not to be reproducible on a daily basis and from laboratory to laboratory. A variable amount of denaturation may also be evident because of the variability of the pH, depending on the water supply.

B. Ribonuclease S Protein–Peptide Interactions

The ribonuclease S (RNase S) system is an example of a peptide–protein interaction studied by ESI-MS [58,59]. RNase S is a modified form of 124 residue RNase A in which limited proteolysis by subtilisin yields the S-peptide (residues 1–20) and S-protein (residues 21–124). The noncovalent association of S-peptide and S-protein to form the active species has a dissociation constant (K_D) of approximately 10^{-10} M in solution (pH 7, 0°C).

RNase S represents a particularly challenging complex for ESI-MS study because of its thermal instability. The solution-phase and gas-phase stability of the RNase S complex has been extensively investigated [58,59]. Different ESI interfaces on a quadrupole mass spectrometer were utilized, including a conventional differentially pumped nozzle-skimmer interface and a heated metal capillary inlet for ESI-MS and capillary electrophoresis-ESI-MS.

Figures 2 and 3 show ESI mass spectra of ribonuclease S under a variety of experimental conditions. With the temperature of the ESI interface of the

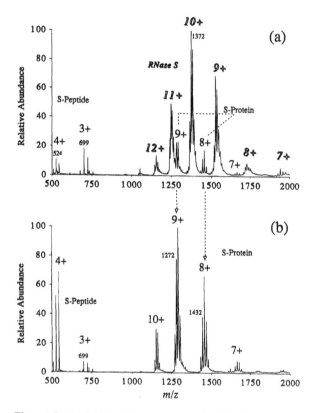

Figure 2 ESI-MS of ribonuclease S in deionized water (a) without (pH 5.3) and with (b) 5% acetic acid (v/v, pH 2.4). The mass spectra were acquired with a single-quadrupole mass spectrometer with a differentially pumped nozzle-skimmer ESI interface. The voltage difference between the nozzle and skimmer lens elements (ΔV_{NS}) was set to 0 V. (From Ref. 58. Copyright © 1993 American Chemical Society.)

quadrupole mass spectrometer at room temperature and the voltage difference between the nozzle and skimmer (ΔV_{NS}) at zero volts, multiply charged ions for the protein–peptide complex can be observed from an aqueous solution of RNase S (Fig. 2a) [58]. Adducts of mass 98 Da are also observed, most likely due to residual phosphate or sulfate buffers used in the protein preparation. Increasing ΔV_{NS} to +250 V induces collisional dissociation of the complex to the free components (Fig. 3). Moreover, acidifying the solution pH by addition of acetic acid to 5% (v/v) also eliminated the presence of the complex and enhanced the abundance of the ions for the free-protein and free-peptide (Fig. 2b).

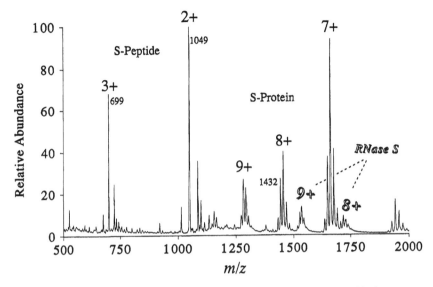

Figure 3 ESI-MS of ribonuclease S in deionized water (pH 5.3) with ΔV_{NS} was set to $+250$ V, showing collisional dissociation of the protein complex to the S-protein and S-peptide species. The mass spectra were acquired with a single quadrupole mass spectrometer with a differentially pumped nozzle-skimmer ESI interface.

Temperature of the interface and countercurrent gas and the nozzle-skimmer potential greatly affected the relative abundance of the RNase S complex ions [59]. The "softest" conditions were necessary to maintain stability of the complex (i.e., low nozzle-skimmer bias, low-temperature gas and/or capillary). However, some activation is necessary to desolvate the ions prior to MS detection, creating a balancing act that needs to be maintained. Even the temperature of the RNase S solution affected the results; incubation of RNase S at 60°C prior to ESI-MS resulted in very low abundance RNase S complex ions, indicating thermal denaturation of the protein complex in the liquid phase.

C. Nucleocapsid Protein–Zinc Interactions

Metal ions are essential for the function and structural stability of many metalloproteins. For example, zinc finger proteins contain zinc-binding domains composed of Cys and/or His ligands that participate in nucleic acid interactions [60]. Many transcription factors and gene-regulatory proteins such as transcription factor IIIA (TFIIIA) include zinc finger structures that appear to be well suited for specific DNA recognition. A few zinc finger proteins have been studied using ESI-MS, such as the DNA binding domain of the glucocorticoid receptor [61] and nucleocapsid protein NCp7 [62–66]. In human immunodeficiency virus (HIV), the first 55 amino acids of the mature form of NCp7 have been shown to contains two zinc fingers that are involved in encapsulation of genomic RNA prior to or during HIV viral assembly.

The binding of zinc to NCp7 results in the ESI mass spectra shown in Fig. 4. The spectrum acquired at pH 2.5 showed a measured mass of 6369.5 ± 0.4, consistent for the apo-protein (theory 6369.5). The predominant

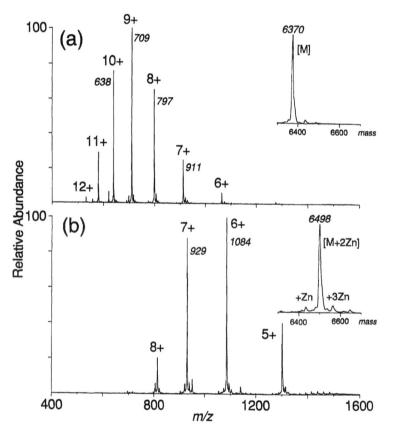

Figure 4 Effect of pH on zinc binding measured by ESI-MS. A 15-μM NCp7 solution containing 30 μM ZnCl$_2$ was prepared in (a) 65.0/32.5/2.5 methanol/water/acetic acid (v/v/v) (pH 2.5) or (b) 25 mM ammonium acetate, pH 6.9 (aqueous), and analyzed by ESI-MS. Both the ESI mass spectra and the deconvoluted spectra (converted to the mass domain, inset) are shown. The mass spectra were acquired on a magnetic-sector mass spectrometer (Finnigan MAT 900Q, Bremen, Germany) with a heated glass capillary ESI interface. The spectrum acquired at pH 2.5 is consistent for the apo-protein. At pH 6.9, the dominant species is consistent for two zinc ions bound to the protein. (From Ref. 65. Copyright © 1996 American Chemical Society.)

species at pH 6.9 is the holo-protein where two zinc ions are bound to the protein. From accurate mass measurements using ESI-MS and chemical modification experiments, Fenselau's lab determined that two thiol groups are deprotonated upon zinc complexation by Cys-Cys-His-Cys-type zinc finger structures such as nucleocapsid protein (i.e., two protons are lost for each zinc ion complexed) [66]. Loo and co-workers followed the time course of the zinc ejection induced by covalent addition of several anti-HIV dithiobis(benzamides) and benzisothiazolones [65]. From further proteolytic degradation and LC-MS experiments, it was determined that the C-terminal zinc finger was attacked preferentially by the disulfide compounds to subsequently eject its coordinated zinc.

The solution conditions that are compatible with ESI-MS for detection of metal-binding ions can vary with the type of biochemical system. For NCp7, up to 25% methanol (v/v) can be tolerated (5–10 mM ammonium acetate, pH 6.9) for observation of the multiply charged ions for the holo-protein and no ions for the apo-protein. However, the order of addition to the protein solution may make a difference. For example, only apo-NCp7 ions were observed when the following conditions were used to prepare the protein solution: 10 μl of methanol was added to a 20-μl solution of NCp7/Zn$_2$; the 30-μl solution was then further diluted with 70 μl of 10 mM ammonium acetate, pH 6.9. Reversing the order of addition of the methanol and the aqueous ammonium acetate solutions yielded only zinc-bound protein ions. Apparently, in the first procedure, the presence of methanol at 33% (v/v) was sufficient to denature the protein. Protein folding back to the native state upon dilution with the high aqueous buffer solution was slow compared to the time scale of the experiment.

D. Streptavidin Quaternary Structure

Noncovalent interactions between protein subunits are responsible for folding the polypeptide chains into multimers, which defines the protein's quaternary structure. Protein complexes can be homocomplexes or heterocomplexes. Determining the number of subunits in the quaternary ensemble is a task that is well suited to ESI-MS.

Streptavidin is composed of four 159-amino-acid-residue polypeptides (*Streptomyces avidinii*) interacting to form a tetrameric complex. Each subunit chain has one high-affinity biotin binding site. Several reports of observing the tetrameric streptavidin complex as well as the biotin binding characteristics using ESI-MS have appeared in the literature [16,17,23]. The ESI mass spectrum shown in Fig. 5 is similar to those previously published. ESI-MS of streptavidin from an acidic pH solution (65.0/32.5/2.5 methanol/water/acetic acid) yields polyprotonated ions for a mixture of monomer subunits of molecular mass 12,971 and 13,042 Da, consistent for polypeptides composed of residues 14–136 and 13–136, respectively. Streptavidin undergoes degradation during fermentation [23]. The

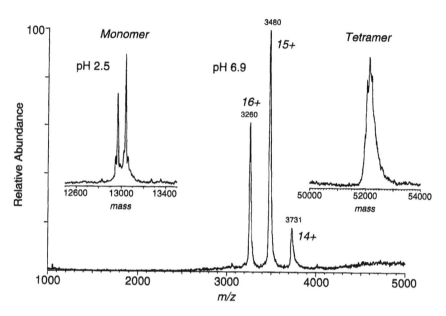

Figure 5 ESI mass spectrum of streptavidin (Boehringer Mannheim, Indianapolis, IN) in 10 mM ammonium acetate, pH 6.9, and a protein monomer concentration of 5 μM. The insets show mass deconvoluted spectra from strepavidin at pH 2.5 and pH 6.9, showing the monomer(s) and tetramer molecular weights. A magnetic-sector mass spectrometer (Finnigan MAT 900Q, Bremen, Germany) with a heated metal capillary ESI inlet was used. The metal capillary temperature was set to 140°C and $\Delta V_{TS} = +56$ V.

mass spectrum obtained from a pH 6.9 solution (10 mM ammonium acetate) shows only multiply charged ions for the 52-kD tetramer at high mass-to-charge ratio (Fig. 5). The peaks are relatively wide (200 Da in the mass deconvoluted spectrum) because of all the possible combinations for two different mass protein chains coming together to form a tetramer complex. The relatively low charge states and the narrow charge distribution is consistent with previous observations of protein quaternary complexes. The stability of the gas-phase complex is somewhat sensitive to the ESI atmosphere/vacuum conditions. Increasing the energy of the ions as they traverse the interface by increasing ΔV_{TS} from $+56$ V to $+79$ V produces abundant ions for the monomer and trimer species.

E. Protein–RNA Complexes

The expression of the genetic information found in nucleic acids is dependent upon the specificity of their interaction with proteins. Proteins serve as the regulators of the genetic information provided by nucleic acids. Thus, the development of techniques to study and understand the molecular details and function of protein–DNA/RNA interactions is an important endeavor. The observation of intact noncovalent protein–DNA complexes by ESI-MS has been recently reported. Both positive ion and negative ion ESI-MS have been used to observe the association of a polyanionic oligonucleotide and a polycationic protein.

We have studied a variety of protein–oligonucleotide systems using ESI-MS. Sample purity is an important consideration for the success of the experiment. Adducts from alkali salts can degrade the ESI mass spectra. For peptides, purification by reversed-phase high-performance liquid chromatography (HPLC) is usually required. In the case of RNAs and DNAs synthesized using phosphoramidite chemistry, the samples are first purified by polyacrylamide gel electrophoresis and then further desalted by cold ethanol precipitation as an ammonium salt. During the cold ethanol precipitation, sodium cations (Na^+) and potassium cations (K^+) bound to the RNA backbone are exchanged for ammonium ions (NH_4^+), which do not interfere with obtaining high-quality ESI mass spectra. For between 10 and 100 nmol of RNA, the ethanol precipitation procedure we use is as follows:

1. Dry down the RNA solution (in an Eppendorf tube) using either a centrifugal dryer or a lyophilizer.
2. Dissolve the RNA in 80 μl of water that was deionized using a Millipore Milli-Q system (water that is not deionized is a source for Na^+ and K^+ ions).
3. Add 16 μl of 10 M ammonium acetate (NH_4OAc). When adding the NH_4OAc solution, the pipet tip should be below the surface of the water and the NH_4OAc solution should fall to the bottom of the tube.
4. Vortex and centrifuge the solution several times, making sure that all of the RNA has dissolved.
5. Add 240 μl of cold ethanol (100%). Let the ethanol run down the side of the tube into the solution. (Note that ethanol is difficult to pipet because of its tendency to leak out of the pipet tip.)
6. Gently mix the ethanol-containing solution by inverting the tube so that all surfaces of the tube are coated. Then briefly centrifuge the sample.
7. Place the tube in a dry ice/isopropanol bath for 2 hr.
8. Centrifuge the solution for 30 min in a refrigerated unit. Immediately remove the sample tubes from the centrifuge. Remove the liquid with a pipet by placing the pipet tip as far away from the precipitated RNA pellet. This must be done quickly because the RNA can redissolve in the solution if allowed to sit.

9. Repeat steps 2–8 two more times.
10. Wash the precipitated RNA with 300 μl of refrigerated 80% ethanol and pipet the liquid immediately. This final wash step helps remove any remaining salts.
11. Dry the RNA sample in a centrifugal dryer for 10–15 min to remove any remaining ethanol.
12. Dissolve in 1 ml of Milli-Q water and quantitate using ultraviolet (UV) spectroscopy by measuring the absorbance at 260 nm [67].

For RNA samples, the secondary structure of the RNA is often important in the binding of peptides. In order to insure a proper secondary structure, the RNA sample should be annealed for 4 min at 95°C in a 10 mM NH$_4$OAc solution (pH 6.9). The solution is then allowed to cool slowly to room temperature before use. For the ESI-MS experiments, 10% (v/v) methanol is added to the 10 mM NH$_4$OAc solution to increase signal stability without altering the resulting spectra to a significant extent for these systems. Higher amounts of organic solvents may inhibit the formation of noncovalent complexes. The solution also contains 0.2 mM 1,2-diaminocyclohexane-N,N,N',N'-tetraacetic acid (CTDA), a chelating agent, to help eliminate any remaining sodium or potassium ions [68].

For highly basic peptides, aggregation may be a problem. Dissolving the peptide samples in 10 mM NH$_4$OAc (pH 6.9) containing 0.01% (v/v) Nonidet P-40, a nonionic detergent, and heating for 15 min at 37°C followed by vortexing can help minimize the aggregation. At a 0.01% level, spectral interference by Nonidet P-40 is a problem. Reducing the amount of Nonidet P-40 by a factor of 100 minimized this interference. The dilution of the peptide sample usually occurs when it is added to the RNA sample to form the noncovalent complex.

Noncovalent peptide–RNA complexes can be observed by both positive- and negative-ion mode ESI-MS. Care must be taken to insure that the noncovalent complexes observed are due to specific interactions between the peptide and RNA and are not the result of nonspecific gas-phase aggregation. From our experiences, specific noncovalent complexes that are formed through electrostatic interactions between the peptide and RNA are stable to dissociation attempts. On the other hand, nonspecific gas-phase complexes dissociate easily. Another way to confirm if the complexes are specific is to add acid to the solution. The acid will denature the RNA and thus destroy the noncovalent complex. In the case of positive-ion mode ESI, the free peptide will be observed upon the addition of acid to the solution.

ESI-MS has proved to be useful for studying protein–ribonucleic acid recognition important in the replication cycle of HIV-1 and the effect of targeted drug inhibitors on these complexes [49]. Tat protein from HIV is a viral trans-activator that is essential for viral replication. Tat is required to increase the rate of transcription from the HIV long terminal repeat (LTR) and its action is dependant on the region near the start of transcription in the viral LTR called the trans-activation responsive (TAR) element. TAR RNA contains a three-nucleotide pyrimidine bulge that is essential for Tat binding and activity. ESI-MS was used to study the complex formation between Tat protein (9.8 kD) and TAR RNA (31-mer, 9.2 kD) [69]. The Tat protein–TAR RNA complex high-

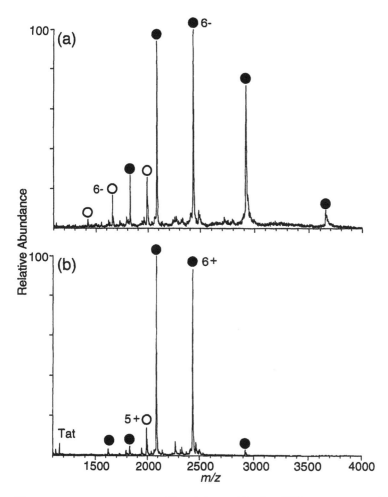

Figure 6 (a) Negative-ion ESI and (b) positive-ion ESI mass spectra of the 1:1 Tat peptide (40 residues, M_r 4644)–TAR RNA (31-mer, M_r 9941) complex. Tat peptide was added to a solution containing previously annealed TAR (13 pmol μl^{-1}), 0.3 mM CDTA, 10% (v/v) methanol, and 10 mM ammonium acetate, pH 6.9. A magnetic sector mass spectrometer with a heated metal capillary ESI inlet was used. The closed circles ($\bullet$) represent ions for the peptide–RNA complex and the open circles ($\bigcirc$) designate ions for the unbound RNA.

lights the importance of RNA structure in protein recognition of RNA. Tat protein contains an arginine-rich region that is essential for RNA binding. A 40-residue peptide containing this basic region has very similar binding characteristics to TAR RNA compared to the protein. Figure 6 shows the positive ion and negative ion ESI mass spectra of Tat peptide binding to TAR as the 1:1 complex. Under competitive binding conditions, ESI-MS spectra show that Tat peptide affinity for TAR RNA is greatly reduced for the bulgeless 28-mer RNA, consistent with solution-phase measurements.

V. CONCLUSIONS

Even though our laboratory has used ESI-MS to study many types of biochemical systems involving some type of noncovalent interaction, these experiments are still not considered "routine." Not all experiments lead to positive results. Each system has its unique experimental variables that may not apply to another system. However, as more examples are studied and the experience levels increase, these types of experiments may be considered more routine in the future.

The application of ESI-MS for investigating weakly bound macromolecular complexes has grown in utility, as more researchers attempt such experiments. Mass spectrometry has its place in this field, especially when sample quantities are limited and time is also precious. The information gained by this methodology, ligand stoichiometry, covalent or noncovalent binding, and possibly relative binding affinities for a mixture of ligands can be invaluable to the researcher. ESI-MS should be used as a complementary method to other biophysical techniques for the study of macromolecular assemblies.

ACKNOWLEDGMENTS

We gratefully acknowledge the following for their experimental and intellectual contributions to several of the examples we describe: Peifeng Hu (Baxter Health Care), David P. Mack (Parke-Davis), Houng-Yau Mei (Parke-Davis), Rachel R. Ogorzalek Loo (University of Michigan), and Richard D. Smith (Pacific Northwest National Laboratory).

REFERENCES

1. JB Fenn, M Mann, CK Meng, SF Wong, CM Whitehouse. Electrospray ionization for mass spectrometry of large biomolecules. Science 246:64–71, 1989.

2. F Hillenkamp, M Karas, RC Beavis, BT Chait. Matrix-assisted laser desorption/ionization mass spectrometry of biopolymers. Anal Chem 63:1193A–1203A, 1991.
3. RD Smith, KJ Light-Wahl. Perspectives—The observation of non-covalent interactions in solution by electrospray ionization mass spectrometry—Promise, pitfalls and prognosis. Biol Mass Spectrom 22:493–501, 1993.
4. JA Loo. Bioanalytical mass spectrometry: Many flavors to choose. Bioconj Chem 6:644–665, 1995.
5. CM Whitehouse, RN Dreyer, M Yamashita, JB Fenn. Electrospray interface for liquid chromatographs and mass spectrometers. Anal Chem 57:675–679, 1985.
6. CG Edmonds, JA Loo, CJ Barinaga, HR Udseth, RD Smith. Capillary electrophoresis-electrospray ionization-mass spectrometry. J Chromatogr 474:21–37, 1989.
7. RD Smith, KJ Light-Wahl, BE Winger, JA Loo. Preservation of noncovalent associations in electrospray ionization-mass spectrometry: Multiply charged polypeptide and protein dimers. Org Mass Spectrom 27:811–821, 1992.
8. TR Covey, RF Bonner, BI Shushan, J Henion. The determination of protein, oligonucleotide and peptide molecular weights by ion-spray mass spectrometry. Rapid Commun Mass Spectrom 2:249–256, 1988.
9. V Katta, BT Chait. Observation of the heme-globin complex in native myoglobin by electrospray-ionization mass spectrometry. J Am Chem Soc 113:8534–8535, 1991.
10. B Ganem, Y-T Li, JD Henion. Detection of noncovalent receptor-ligand complexes by mass spectrometry. J Am Chem Soc 113:6294–6296, 1991.
11. JT Drummond, RR Ogorzalek Loo, RG Matthews. Electrospray mass spectrometric analysis of the domains of a large enzyme: observation of the occupied cobalamin-binding domain and redefinition of the carboxyl terminus of methionine synthase. Biochemistry 32:9282–9289, 1993.
12. B Ganem, Y-T Li, JD Henion. Observation of noncovalent enzyme–substrate and enzyme–product complexes by ion-spray mass spectrometry. J Am Chem Soc 113:7818–7819, 1991.
13. KJ Light-Wahl, BE Winger, RD Smith. Observation of the multimeric forms of concanavalin-A by electrospray ionization mass spectrometry. J Am Chem Soc 115:5869–5870, 1993.
14. KJ Light-Wahl, BL Schwartz, RD Smith. Observation of the noncovalent quaternary associations of proteins by electrospray ionization mass spectrometry. J Am Chem Soc 116:5271–5278, 1994.
15. BL Schwartz, KJ Light-Wahl, RD Smith. Observation of noncovalent complexes to the avidin tetramer by electrospray ionization mass spectrometry. J Am Soc Mass Spectrom 5:201–204, 1994.
16. BL Schwartz, JE Bruce, GA Anderson, SA Hofstadler, AL Rockwood, RD Smith, A Chilkoti, PS Stayton. Dissociation of tetrameric ions of noncovalent streptavidin complexes formed by electrospray ionization. J Am Soc Mass Spectrom 6:459–465, 1995.
17. BL Schwartz, DC Gale, RD Smith, A Chilkoti, PS Stayton. Investigation of noncovalent ligand binding to the intact streptavidin tetramer by electrospray ionization mass spectrometry. J Mass Spectrom 30:1095–1102, 1995.

18. VF Smith, BL Schwartz, LL Randall, RD Smith. Electrospray mass spectrometric investigation of the chaperone SecB. Protein Sci 5:488–494, 1996.
19. JA Loo, RR Ogorzalek Loo, PC Andrews. Primary to quaternary protein structure determination with electrospray ionization and magnetic sector mass spectrometry. Org Mass Spectrom 28:1640–1649, 1993.
20. JA Loo. Observation of large subunit protein complexes by electrospray ionization mass spectrometry. J Mass Spectrom 30:180–183, 1995.
21. X-J Tang, CF Brewer, S Saha, I Chernushevich, W Ens, KG Standing. Investigation of protein-protein noncovalent interactions in soybean agglutinin by electrospray ionization time-of-flight mass spectrometry. Rapid Commun Mass Spectrom 8:750–754, 1994.
22. MC Fitzgerald, I Chernushevich, KG Standing, CP Whitman, SBH Kent. Probing the oligomeric structure of an enzyme by electrospray ionization time-of-flight mass spectrometry. Proc Natl Acad Sci USA 93:6851–6856, 1996.
23. K Eckart, J Spiess. Electrospray ionization mass spectrometry of biotin binding to streptavidin. J Am Soc Mass Spectrom 6:912–919, 1995.
24. IV Chernushevich, W Ens, KG Standing, PC Loewen, MC Fitzgerald, SBH Kent, RC Werlen, M Lankinen, X-J Tang, CF Brewer, S Saha. Studies of non-covalent interactions by time-of-flight mass spectrometry. Proceedings of the 43rd ASMS Conference on Mass Spectrometry and Allied Topics, Atlanta, GA, 1995, p 1327.
25. IV Chernushevich, W Ens, KG Standing. Electrospray ionization time-of-flight mass spectrometry. In: RB Cole, ed. Electrospray ionization mass spectrometry: Fundamentals, Instrumentation, and Applications. New York: John Wiley & Sons, 1997, pp 203–234.
26. P Camilleri, NJ Haskins, AP New, MR Saunders. Analysing the complexation of amino acids and peptides with β-cyclodextrin using electrospray ionization mass spectrometry. Rapid Commun Mass Spectrom 7:949–952, 1993.
27. P Camilleri, NJ Haskins, DR Howlett. β-Cyclodextrin interacts with the alzheimer amyloid β-A4 peptide. FEBS Lett 341:256–258, 1994.
28. A Selva, E Redenti, M Zanol, P Ventura, B Casetta. A study of β-cyclodextrin and its inclusion complexes with piroxicam and terfenadine by ionspray mass spectrometry. Org Mass Spectrom 28:983–986, 1993.
29. KJ Light-Wahl, DL Springer, BE Winger, CG Edmonds, DG Camp, III, BD Thrall, RD Smith. Observation of a small oligonucleotide duplex by electrospray ionization mass spectrometry. J Am Chem Soc 115:803–804, 1993.
30. DC Gale, DR Goodlett, KJ Light-Wahl, RD Smith. Observation of duplex DNA–drug noncovalent complexes by electrospray ionization mass spectrometry. J Am Chem Soc 116:6027–6028, 1994.
31. DC Gale, RD Smith. Characterization of noncovalent complexes formed between minor groove binding molecules and duplex DNA by electrospray ionization-mass spectrometry. J Am Soc Mass Spectrom 6:1154–1164, 1995.
32. B Ganem, YT Li, JD Henion. Detection of oligonucleotide duplex forms by ionspray mass spectrometry. Tetrahedron Lett 34:1445–1448, 1993.
33. MJ Doktycz, S Habibigoudarzi, SA McLuckey. Accumulation and storage of ionized duplex DNA molecules in a quadrupole ion trap. Anal Chem 66:3416–3422, 1994.

34. E Bayer, T Bauer, K Schmeer, K Bleicher, M Maler, HJ Gaus. Analysis of double-stranded oligonucleotides by electrospray mass spectrometry. Anal Chem 66:3858–3863, 1994.

35. X Cheng, Q Gao, RD Smith, K-E Jung, C Switzer. Comparison of 3',5'- and 2',5'-linked DNA duplex stabilities by electrospray ionization mass spectrometry. J Chem Soc, Chem Commun 747–748, 1996.

36. MJ Greig, H Gaus, LL Cummins, H Sasmor, RH Griffey. Measurement of macromolecular binding using electrospray mass spectrometry. Determination of dissociation constants for oligonucleotide-serum albumin complexes. J Am Chem Soc 117:10765–10766, 1995.

37. XH Cheng, AC Harms, PN Goudreau, TC Terwilliger, RD Smith. Direct measurement of oligonucleotide binding stoichiometry of gene V protein by mass spectrometry. Proc Natl Acad Sci USA 93:7022–7027, 1996.

38. DL Smith, Z Zhang. Probing noncovalent structural features of proteins by mass spectrometry. Mass Spectrom Rev 13:411–429, 1994.

39. RD Smith, X Cheng, BL Schwartz, R Chen, SA Hofstadler. Noncovalent complexes of nucleic acids and proteins studied by electrospray ionization mass spectrometry. In: AP Snyder, ed. Biochemical and Biotechnological Applications of Electrospray Ionization Mass Spectrometry (ACS Symposium Series 619). Washington, DC: American Chemical Society, 1996, pp 294–314.

40. M Przybylski. Mass spectrometric approaches to the characterization of tertiary and supramolecular structures of biomacromolecules. Adv Mass Spectrom 13:257–283, 1995.

41. M Przybylski, MO Glocker. Electrospray mass spectrometry of biomacromolecular complexes with noncovalent interactions—New analytical perspectives for supramolecular chemistry and molecular recognition processes. Angew Chem Int Ed Engl 35:807–826, 1996.

42. P Hensley. Defining the structure and stability of macromolecular assemblies in solution: the re-emergence of analytical ultracentrifugation as a practical tool. Structure 4:367–373, 1996.

43. M Wilm, M Mann. Analytical properties of the nanoelectrospray ion source. Anal Chem 68:1–8, 1996.

44. JA Loo, P Hu, P McConnell, WT Mueller, TK Sawyer, V Thanabal. A study of Src SH2 domain protein–phosphopeptide binding interactions by electrospray ionization mass spectrometry. J Am Soc Mass Spectrom 8:234–243, 1997.

45. G Scatchard. The attractions of proteins for small molecules and ions. Ann NY Acad Sci 51:660–672, 1949.

46. DJ Winzor, WH Sawyer. Quantitative Characterization of Ligand Binding. New York: Wiley-Liss, 1995.

47. H-K Lim, YL Hsieh, B Ganem, J Henion. Recognition of cell-wall peptide ligands by vancomycin group antibiotics: studies using ion spray mass spectrometry. J Mass Spectrom 30:708–714, 1995.

48. X Cheng, PE Morin, AC Harms, JE Bruce, Y Ben-David, RD Smith. Mass spectrometric characterization of sequence-specific complexes of DNA and transcription factor PU.1 DNA binding domain. Anal Biochem 239:35–40, 1996.

49. KA Sannes, JA Loo, P Hu, H-Y Mei, D Mack. Studying drug binding to the nonco-

valent Tat peptide-TAR RNA complex by ESI-MS. Proceedings of the 44th ASMS Conference on Mass Spectrometry and Allied Topics, Portland, OR, 1996, p 1405.

50. CV Robinson, EW Chung, BB Kragelund, J Knudsen, RT Aplin, FM Poulsen, CM Dobson. Probing the nature of noncovalent interactions by mass spectrometry. A study of protein-CoA ligand binding and assembly. J Am Chem Soc 118:8646–8653, 1996.

51. SK Chowdhury, V Katta, BT Chait. An electrospray-ionization mass spectrometer with new features. Rapid Commun Mass Spectrom 4:81–87, 1990.

52. P Dobberstein, E Schroeder. Accurate mass determination of a high molecular weight protein using electrospray ionization with a magnetic sector instrument. Rapid Commun Mass Spectrom 7:861–864, 1993.

53. P Dobberstein, H Muenster. Application of a new atmospheric pressure ionization source for double focusing sector instruments. J Chromatogr, A 712:3–15, 1995.

54. Y-T Li, Y-L Hsieh, JD Henion, B Ganem. Studies on heme binding in myoglobin, hemoglobin, and cytochrome c by ion spray mass spectrometry. J Am Soc Mass Spectrom 4:631–637, 1993.

55. Y Konishi, R Feng. Conformational stability of heme proteins *in vacuo*. Biochemistry 33:9706–9711, 1994.

56. JA Loo, AG Giordani, H Muenster. Observation of intact (heme-bound) myoglobin by electrospray ionization on a double-focusing mass spectrometer. Rapid Commun Mass Spectrom 7:186–189, 1993.

57. SA McLuckey, RS Ramsey. Gaseous myoglobin ions stored at greater than 300 K. J Am Soc Mass Spectrom 5:324–327, 1994.

58. RR Ogorzalek Loo, DR Goodlett, RD Smith, JA Loo. Observation of a noncovalent ribonuclease S-protein S-peptide complex by electrospray ionization mass spectrometry. J Am Chem Soc 115:4391–4392, 1993.

59. DR Goodlett, RR Ogorzalek Loo, JA Loo, JH Wahl, HR Udseth, RD Smith. A study of the thermal denaturation of ribonuclease S by electrospray ionization mass spectrometry. J Am Soc Mass Spectrom 5:614–622, 1994.

60. A Klug, JWR Schwabe. Protein motifs 5—Zinc fingers. FASEB J 9:597–604, 1995.

61. HE Witkowska, CHL Shackleton, K Dahlman-Wright, JY Kim, J-A Gustafsson. Mass spectrometric analysis of a native zinc-finger structure: the glucocorticoid receptor DNA binding domain. J Am Chem Soc 117:3319–3324, 1995.

62. A Surovoy, D Waidelich, G Jung. Nucleocapsid protein of HIV-1 and its Zn^{2+} complex formation analysis with electrospray mass spectrometry. FEBS Lett 311:259–262, 1992.

63. A Surovoy, D Waidelich, G Jung. Electrospray mass spectroscopic analysis of metal-peptide complexes. In: CH Schneider, AN Eberle, eds. Peptides 1992 (Proceedings of the 22nd European Peptides Symposium). Leiden, Netherlands: ESCOM Science, 1993, pp 563–564.

64. C Fenselau, X Yu, D Bryant, MA Bowers, RC Sowder, II, LE Henderson. Characterization of processed gag proteins from highly replicating HIV-1MN. In: C Fenselau, ed. Mass Spectrometry for the Characterization of Microorganisms (ACS Symposium Series 541). Washington, DC: American Chemical Society, 1994, pp 159–172.

65. JA Loo, TP Holler, J Sanchez, R Gogliotti, L Maloney, MD Reily. Biophysical characterization of zinc ejection from HIV nucleocapsid protein by anti-HIV 2,2'-dithiobis[benzamides] and benzisothiazolones. J Med Chem 39:4313–4320, 1996.

66. D Fabris, J Zaia, Y Hathout, C Fenselau. Retention of thiol protons in two classes of protein zinc ion coordination centers. J Am Chem Soc 118:12242–12243, 1996.

67. PF Crain. Preparation and enzymatic hydrolysis of DNA and RNA for mass spectrometry. In: JA McCloskey, ed. Methods in Enzymology. San Diego, CA: Academic Press, 1990, pp 782–790.

68. PA Limbach, PF Crain, JA McCloskey. Molecular mass measurement of intact ribonucleic acids via electrospray ionization quadrupole mass spectrometry. J Am Soc Mass Spectrom 6:27–39, 1995.

69. JA Loo, KA Sannes-Lowery, P Hu, DP Mack, H-Y Mei. Studying noncovalent protein–RNA interactions and drug binding by electrospray ionization mass spectrometry. In: KG Standing, ed. New Methods for the Study of Molecular Aggregates. Dordrecht, the Netherlands: Kluwer, 1997, in press.

14

Protein–Protein and Protein–Ligand Interactions Studied by Hydrogen/Deuterium Exchange Mass Spectrometry

Carol V. Robinson
Oxford Center for Molecular Sciences, New Chemistry Laboratory, Oxford, England

I. INTRODUCTION

The advent of mass spectrometers capable of analyzing proteins from conditions under which the protein remains close to its native state initiated studies of higher order protein structure. These studies have included both protein–protein and protein–ligand interactions as well studies of the relative stability of protein variants. To observe these interactions and stability differences using mass spectrometry (MS) it is necessary to preserve higher order protein structure in the gas phase. The higher order structure of a protein is defined as the spatial arrangement of amino acid residues into elements of secondary structure such as α-helices and β-sheets. These elements of secondary structure arise from local hydrogen bonding interactions, while the spatial interactions of amino acids, far apart in the linear sequence, impose conformations on the protein and give rise to the tertiary structure. For application of MS to higher order protein structure a method is required in which differences in conformations of proteins may be monitored. In the early days of electrospray ionization (ESI) MS, many studies reported changes in the charge state distribution arising from different conformations of the protein. These changes in charge state distribution are observed for proteins in the presence and absence of cofactors or ligands. For example, Fig. 1 shows the ESI mass spectra obtained from bovine α-lactalbumin in the presence and absence of Ca^{2+}. The protein with Ca^{2+} present shows the expected increase in mass and in addition a lower charge state distribution. However, comparison of charge states to show differences in tertiary structure relies on keeping all instrument parameters and solution conditions identical. In particular, changes in solvent composition, often necessary to effect changes in protein conformation (discussed later), also affect the charge state distribution [1]. While it is possible to use changes in charge state distribution as a "fingerprint" for studying three-dimensional (3D) structure of proteins, a more reliable method is required. One such method, successfully applied in nuclear magnetic resonance (NMR), is that of hydrogen deuterium exchange labeling (HX) [2].

The labeling process exploits the fact that proteins contain amides and certain exchangeable side chains that are involved in secondary structure and/or buried from solvent and therefore exchange more slowly, due to H-bonding interactions and sequestering from solvent. ^2H on the surface of the protein and those not involved in secondary structure will exchange more rapidly. Thus, these differential exchange rates provide valuable insight into the extent of secondary structure and solvent accessibility within proteins [3,4]. The technique has found widespread application in the study of protein stability and conformational change by NMR [5] and, since hydrogen and deuterium have different relative masses, has successfully been monitored by MS [6].

Using hydrogen exchange monitored by mass spectrometry (HX-MS), it has been shown that a number of ^2H are protected from exchange in the native state of cytochrome c [6], ubiquitin, and lysozyme [7]. HX-MS has also been

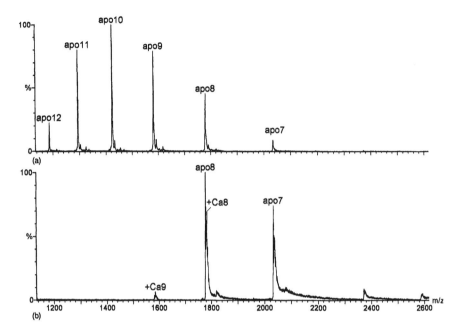

Figure 1 Typical ES mass spectra recorded for (a) *apo*- and (b) *holo*-bovine α-lactalbumin, illustrating changes in the charge state distribution observed between these two
protein conformations.

used to study lysozyme during protein folding experiments [8] and to compare
the relative stabilities of ferrocytochrome c_2 and site-directed mutants [9]. Different exchange rates are observed for myoglobin in its *apo* and *holo* forms
[10], and the solution dynamics of β-sheet and α-helical peptides have also
been measured [11]. These initial studies have demonstrated the potential of
HX-MS. Further applications from our laboratory are presented here together
with the practical techniques necessary to carry out this methodology.

II. METAL BINDING PROTEINS

Bovine α-lactalbumin (BLA) is a Ca^{2+}-binding milk protein, and extensive study
by other biophysical techniques has provided a precise picture of the change in
conformation under various pH and solution conditions [12]. Under aqueous conditions at pH 3.8 in the presence of Ca^{2+} the protein is in its fully folded native
state; however, removal of Ca^{2+} and lowering the pH changes the conformation
to that of a partially folded form. To measure the HX properties of the protein in
these two different conformations the protein is diluted into D_2O, either with or

without Ca^{2+} present [12]. The number of sites remaining protected from exchange is calculated from the change in mass with time and the number of exchange-labile sites is determined from the protein sequence. BLA has 220 labile protons and NMR studies have shown that, under the conditions of these experiments, all the side-chain N-H and O-H groups exchange within the first few minutes of diluting into D_2O. The slowly exchanging backbone amides, hydrogen bonded in elements of secondary structure, remain protected over a much longer time scale. Hydrogen exchange of these backbone amides takes place at different rates depending on solvent penetration of the core of the protein. The hydrogen exchange kinetics of *apo* BLA (in the absence of Ca^{2+}) and *holo* BLA (Ca^{2+}-bound state) are plotted as a function of mass against time in Fig. 2. Comparison of these two kinetic profiles shows that the partially folded *apo* form of BLA exchanges more readily than fully folded *holo* BLA. After 150 min, 55 sites remain in the native state compared with only 23 in the partially folded form. These results are in accord with solution measurements made by NMR [12] and demonstrate that the more tightly folded native state of the protein, with its solvent-excluded core, remains more protected to exchange than the partially folded state in which solvent is able to penetrate.

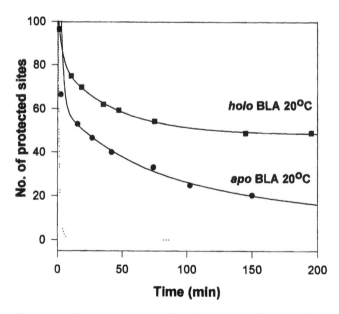

Figure 2   Hydrogen exchange kinetics for *holo* (■), *apo* (●), and completely unstructured (· ·) bovine α-lactalbumin at pH 3.8 and 20°C. Each point on the graph represents a mass spectrum obtained after a given time for exchange. The exchange of unstructured bovine α-lactalbumin is a simulation calculated from near neighbor inductive effects [27].

The variation in peak width for a protein undergoing hydrogen exchange also provides valuable insight into the population of molecules undergoing exchange. For example, Fig. 3 shows the variation in peak width throughout the time course for exchange of *holo* Ca^{2+}-bound BLA. At the start of exchange the peak width of the protein is dictated by its natural-abundance isotopes, principally ^{13}C, and the resolution capabilities of the mass spectrometer. Under the conditions used on a quadrupole mass spectrometer this peak width approaches that calculated from a binomial distribution of isotopes, assuming a resolving power of 1000 [12]. After 1.5 min of exposure to D_2O the peak is considerably broader. The increased peak width occurs since there is a distribution of deuterium throughout the population of protein molecules. This distribution arises since hydrogen exchange occurs more rapidly in some molecules, either as a result of global unfolding or local perturbations, and hence a distribution of different states, carrying a range of deuterium atoms, is observed. As hydrogen exchange proceeds, the peak width narrows, since the number of sites available for exchange decreases with time. At the plateau for exchange, after 180 min, the protein retains the protective core of 55 hydrogens and the peak width is closely similar to that of the protein before exchange. This information, from peak width comparisons, can provide valuable insight into populations of molecules and hydrogen exchange mechanisms. For example, experimental peak widths for exchange of BLA were compared with a random exchange model in which all 220 sites are exposed for exchange via global unfolding and a core protection model in which 158 labile sites are exposed via local fluctuations (Fig. 3). Close agreement is found with the core protection model. This analysis of peak widths highlights the potential for distinguishing between global unfolding, involving ^{1}H in the core of the protein, and local unfolding, in which a protected core of ^{1}H remains unaffected by exchange. Peak width analysis has been applied to protein folding studies [8], delineating hydrogen exchange mechanisms [12] and macromolecular complexes (discussed later).

III. CONFORMATION AND DYNAMICS OF PROTEINS

HX-MS has provided valuable insight into the relative stability of site-directed mutant proteins [9,13]. Comparison of the HX kinetic profiles of cytochrome c_2 from *Rhodobacter capsulatus* with site-directed mutants has allowed detailed investigation of the effects of individual amino acid substitutions on the overall conformation and stability of the protein [9].

In our laboratory two naturally occurring variants of human lysozyme, implicated in the formation of amyloid fibrils in liver [14], have been compared with wild-type human lysozyme. X-ray crystallography has shown that the variant proteins have native folds, while data from other biophysical techniques have shown evidence for a reduction in the thermal stability of the protein. The

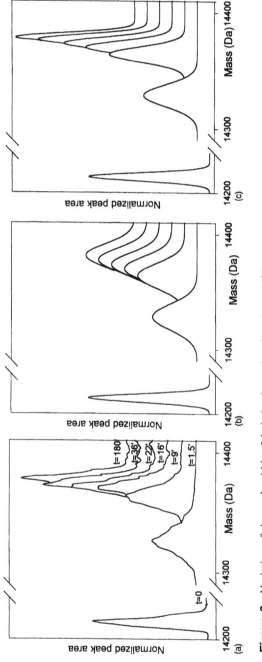

Figure 3 Variation of the peak width of *holo*-bovine α-lactalbumin at 15°C over the time course of the hydrogen exchange from solution at pH 3.8 containing a 1.7 molar excess of Ca^{2+}. The experimental peak widths (a) are compared with two simulations: (b) a random exchange model, in which exchange takes place via global unfolding and exchange from any of the 220 sites within the protein, and (c) a core protection model, where 158 sites are able to exchange. The similarity between the core protection model and the experimental data suggests that exchange takes place via local fluctuations rather than global unfolding. (Reprinted with permission from ref. 12. Copyright 1997 Cambridge University Press.)

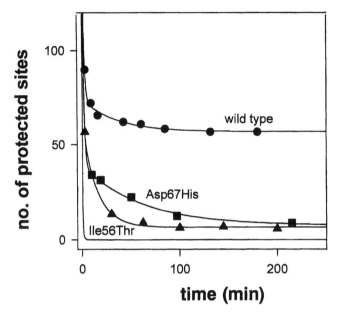

Figure 4   Kinetic profiles of hydrogen exchange at pH 5.0, 37°C, for wild type (●), Asp 67His (■), and Ile56Thr (▲) monitored by ESI MS. The plain black line is the simulated curve for exchange from completely unstructured human lysozyme at pH 5 and 37°C calculated from near neighbor inductive effects [27]. (Reprinted with permission from ref. 13. Copyright 1997 Macmillan Magazines Limited.)

properties of the wild-type and variant protein were investigated by HX-MS at 37°C and pH 5. In comparison with wild-type human lysozyme (Fig. 4), the resilience of the variant proteins to hydrogen exchange is considerably reduced with remarkably little protection observed. This lack of protection in the variants strongly suggests that unfolding events, exposing the hydrophobic core of the protein, occur more readily in the variant proteins [13]. The conformational dynamics of the variant proteins suggest a possible route to amyloid fibril formation via an intermediate that is substantially unfolded and hence not only exchanges more readily but is presumably able to associate with further protein molecules to assemble the amyloid fibril [13].

IV.   PROTEIN–LIGAND INTERACTIONS

Investigations of noncovalently bound protein–ligand complexes by ESI MS are of great interest because of their relevance to molecular recognition and to combinatorial library searching. The relatively small sample requirements and the rapidity of the method make MS particularly attractive to screening libraries of peptides and other ligands to examine binding interactions with proteins. In addition to measuring protein–ligand interactions by direct mass measurement, the application of HX-MS has the potential to provide additional detailed information regarding the structure and dynamics of the protein–ligand interaction. HX-MS has been used to monitor the conformation of myoglobin in the presence of its noncovalent heme [15] and to study the binding of peptides to an SH2 domain [16].

In our laboratory we have studied an acyl coenzyme A binding protein (ACBP) in the presence of a number of its coenzyme A (CoA) ligands with differing binding affinities [17]. ACBP is an 86-residue protein present in a wide range of eukaryotic organisms [18]. The protein binds long-chain acyl CoAs with high affinity and is thought to have a role in the transport and storage of acyl CoA molecules in the cell [19]. The protein molecule adopts a four-helix bundle structure, which forms a bowl shape with a polar rim and a predominantly apolar core. Protein and acyl CoA molecules associate by a combination of electrostatic, hydrophobic, and stacking interactions and have been the subject of detailed investigations by NMR [20] and MS [17]. Measurement of dissociation constants for the complexes has shown a distinct preference for long-chain acyl CoA esters (C_{14} to C_{22}). The dissociation constants of the complexes are $0.5 \pm 0.2 \times 10^{-10}$ M for the C_{16} CoA complex and $0.7 \pm 0.2 \times 10^{-8}$ M for C_{12} CoA complex [21]. The binding of free CoA to ACBP is much weaker, in the region of 1×10^{-3} M. Despite the high binding affinity of the C_{12}CoA ligand, it was not possible to obtain a spectrum in which total protein–ligand complex was observed from either conventional electrospray (ES) MS or from nanoflow ES MS [22]. Solution studies, however, have shown that at the concentration used for obtaining these spectra the protein exists entirely in its bound form. Some gas-phase dissociation of the complex must be occurring (Fig. 5).

Point mutations of the protein together with CoA ligands with varying acyl chain length were used to show that the nature of the interaction affects the survival of the complex in the gas phase [17]. The results of this study show that ionic and van der Waals interactions are well represented in the gas phase, but hydrophobic interactions, which play an important role in solution binding, are not well preserved in the gas phase.

HX-MS was used to examine the relative stability of the protein complexes formed with different CoA ligands (Fig. 6). For these measurements labile protons in the protein were exchanged for deuterium and the protein diluted into H_2O to initiate exchange. The results show that HX of the free protein is more rapid than that of the protein–CoA complexes. This finding is interpreted in terms of the increased stability of the ligand bound protein over

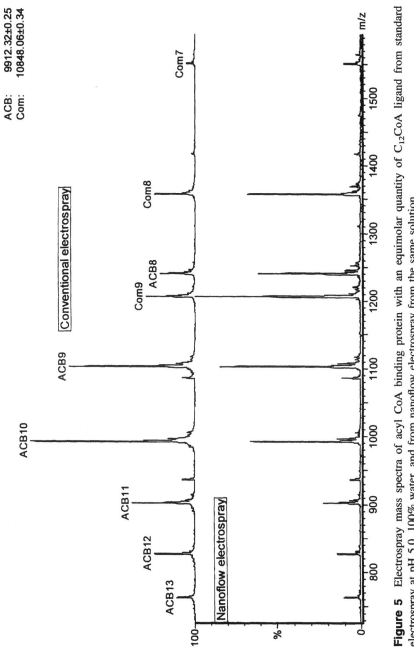

Figure 5 Electrospray mass spectra of acyl CoA binding protein with an equimolar quantity of $C_{12}CoA$ ligand from standard electrospray at pH 5.0, 100% water, and from nanoflow electrospray from the same solution.

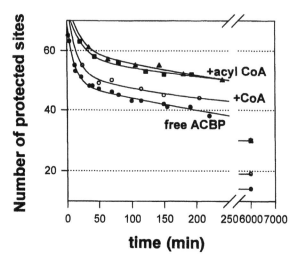

Figure 6 Comparison of hydrogen–deuterium exchange kinetics for unliganded ACBP (●) and ACBP ligated with C_{16} acyl CoA (▲), C_8 acyl CoA (■), and free CoA (○). (Reprinted with permission from ref. 17. Copyright 1996 American Chemical Society.)

that of the free protein and suggests that the free protein is able to unfold, either globally or locally, and exchange with solvent protons more readily than complexed protein. Interestingly, the weaker binding free CoA ligand is less protected against exchange than the more tightly bound protein–ligand complexes formed with the octanoyl and dodecanoyl CoA ligands. Thus, these hydrogen exchange measurements are sensitive to changes in the stability of the protein–ligand complex and provide a means of probing this stability in the presence of different ligands.

This hydrogen exchange study also highlighted another important feature of the HX-MS technique. Although the complex dissociates in the gas phase of the mass spectrometer, the dissociated protein and the complexed species carry the same number of deuterons (Fig. 7). Thus the solution history of the complex is carried by the free protein in the gas phase. This result is critical to the success of ligand binding investigations since it allows the study of even weakly bound protein–ligand interactions and does not rely on their preservation in the gas phase. Furthermore, it suggests a means for probing much larger protein–protein interactions by dissociation in the gas phase and measurement of hydrogen exchange properties.

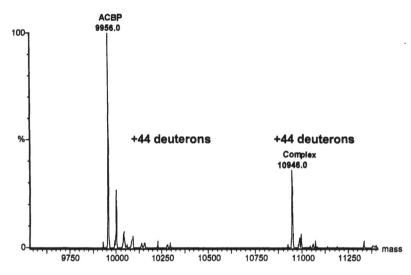

Figure 7 Electrospray mass spectrum of ACBP:C_{16}CoA complex transformed onto a mass scale. The mass spectrum of the complex shows the same number of deuterons retained in the complexed and dissociated protein, demonstrating that gas-phase dissociation protects hydrogen exchange.

V. PROTEIN–PROTEIN INTERACTIONS

Having established a method for studying large protein complexes by HX-MS and dissociation in the gas phase, this method was applied to study changes in protein conformation of proteins in multiprotein complexes. Using the techniques described earlier, we have monitored the conformation of a protein ligand bound in the central cavity of the chaperone GroEL [23]. These complexes, with molecular weight close to 1 MD, present a considerable challenge to many areas of structural biology because of their sheer size and complexity. A schematic of the experimental design is presented in Fig. 8. In these experiments protein ligand is denatured in ^{2}H-chemical denaturant and all labile sites in the protein ligand are exchanged for ^{2}H. The denaturant is diluted into D_2O containing the chaperone GroEL and the resulting GroEL–protein complex is washed in D_2O several times to remove all trace of salts and buffers. Hydrogen exchange is measured by dilution into H_2O at pH 5 and ESI mass spectra obtained from solution conditions in which the protein remains in its native state until solvent evaporation and dissociation of the protein complex in the gas phase. The resulting ES mass spectrum (Fig. 8) clearly shows the presence of both the chaperone, labeled GroEL, and the human dihydrofolate reductase (DHFR) protein ligand. The charge states arising from DHFR are readily identified in this spectrum since 50% of the protein carries an additional N-terminal methionine residue, and thus the protein ligand appears as doublet (Fig. 8).

In our first study using the approach presented here, the hydrogen exchange kinetics of a three-disulfide derivative of BLA bound to GroEL were investigated and compared with the Ca^{2+}-bound native state and a partially folded state of the protein in free solution. From hydrogen exchange measurements, deuterons in the GroEL-bound protein were found to be significantly protected from isotope exchange as compared with the simulated random coil exchange rate. These experiments demonstrated that the protein conformation bound to the molecular chaperone was closely similar to the partially folded state seen in free solution, suggesting that at least some regions of the protein ligand are structured [23].

In an extension to this work, the binding and release of protein ligands to the chaperone GroEL was investigated using the model substrate dihydrofolate

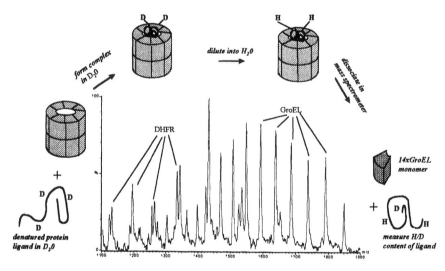

Figure 8 Schematic representation of the experiment designed to monitor hydrogen exchange of the protein ligand bound in the molecular chaperone GroEL. The protein ligand is unfolded in denaturant in which all labile hydrogen atoms are preexchanged for deuterium and added to a D_2O solution containing GroEL (method described in detail in ref. 28). The complex is then diluted into H_2O to initiate exchange and mass spectra are recorded after appropriate time intervals. The doublets arising from the protein ligand are labeled DHFR and represent protein with and without N-terminal methionine. Charge states labeled GroEL arise from GroEL monomers.

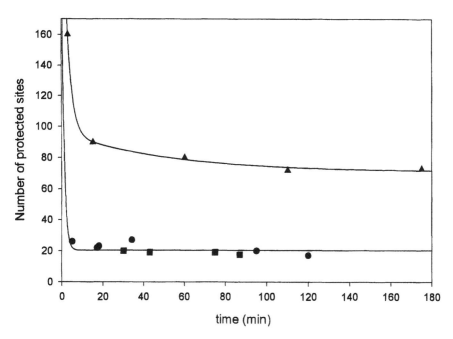

Figure 9   Kinetic profiles of hydrogen exchange at pH 5 and 20°C for fully folded DHFR (▲), the initial bound complex of DHFR with GroEL (●), and the bound state after multiple rounds of Mg^{2+}/ATP-induced binding and release to GroEL (■). (Reprinted with permission from ref. 24. Copyright 1996 Cambridge University Press.)

reductase [24]. In this experiment the protein ligand was allowed to undergo multiple rounds of binding and release to GroEL before analysis and compared with the initial bound form. From these experiments we conclude that the initial bound state and the state that has undergone multiple rounds of binding and release are similar (Fig. 9). This has important implications for the mechanism of GroEL-assisted protein folding, because rather than multiple rounds of substrate binding and release leading to a more folded state of the protein, the protein substrate bound to GroEL has the same hydrogen exchange protection irrespective of the number of rounds of binding and release.

The analyses of peak widths for protein ligands having undergone hydrogen exchange while bound to GroEL were not obtained under optimal resolution conditions. Thus the major contribution to the peak width is the resolving power of the mass spectrometer. However, these peaks can be analyzed in terms of contribution of different states. For example, two peaks recorded in the ES mass spectra of GroEL complexes at $t = 11$ min for BLA and much later for DHFR ($t = 92$ min) are shown in Fig. 10. These measurements discount the

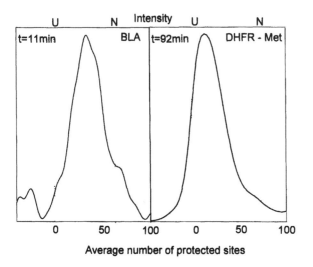

Figure 10 Comparison of the peak widths of chaperone bound protein ligands BLA and DHFR after 11 and 92 min of hydrogen exchange, respectively. The positions of the unfolded and native states under the same conditions and after the same HX times are shown for comparison. (Reprinted with permission from ref. 28. Copyright 1996 Cambridge University Press.)

possibility that a number of different conformations of the protein ligand are bound in the central cavity of GroEL.

VI. FUTURE PROSPECTS

One limitation of our current methodology is that although HX-MS provides valuable insight into populations of protein molecules in hydrogen exchange environments, the information is not specific for hydrogen sites within a given macromolecule. However, there are recent technological developments that bear promise of site-specific information for the near future. The fundamental principle is that the protein is fragmented prior to mass analysis such that changes in isotope content can be attributed to regions of the sequence. There are two methods for fragmentation and analysis: the well-established technique of proteolysis followed by coupled HPLC/MS [25] in which protein is digested at low pH and low temperature to conserve hydrogen exchange labeling information, or Fourier transform ion cyclotron resonance (FTICR) MS, and collision-induced dissociation in the gas phase followed by separation of the molecular fragments and analysis by FTICR MS [26]. The FTICR method holds great promise because enzymatic digestion and HPLC will destroy the chaperone complex and weakly protected sites may be lost. However, the possibility of hydrogen deuterium rearrangements occurring under the high-energy conditions required to fragment the chaperone complex in the gas phase cannot be ruled out at this stage. Other limitations arise from the nature of the proteins investigated. While the molecular weight range accessible to current quadrupole mass spectrometers allows analysis of charge states up to 4000 Da, many protein assemblies do not have charge states that fall within this mass range. The introduction of ESI sources coupled to time-of-flight analyzers may alleviate this current shortfall.

The recent introduction of nanoflow sample introduction offers increased sensitivity, a considerable reduction in the amount of sample required, and an increased tolerance to salts and buffers. Furthermore, the possibility exists of following a complete time course for exchange from 2 μl of a 5 μM protein solution contained in the nanoflow needle. This method opens the door to protein–protein interactions limited by sample availability, typically receptor–ligand complexes.

The rapid development of MS instrumentation capable of studying high-molecular-weight complexes under physiological conditions will enable these studies to be applied to increasingly challenging areas of structural biology. Moreover, the ability to dissociate large protein complexes and retain their hydrogen exchange history holds great promise for the study of protein–protein and protein–ligand interactions that are important in molecular recognition processes. These features, coupled with the inherent attributes of MS, namely, the generous molecular weight limits, tolerance of protein mixtures, and minimal sample requirements, coupled with the sensitivity of HX to changes in protein conformation, make HX-MS a powerful tool for modern structural biologists.

VII. EXPERIMENTAL

A. Mass Spectrometry

All mass spectra in this chapter were recorded on either a Platform I or II mass spectrometer (Micromass). Throughout the course of this work a number of different mass spectrometers have been evaluated. Instruments that employ heated capillary inlets for desolvation often lead to loss of hydrogen exchange information, through thermal unfolding and exchange within the electrospray droplet. The most consistent results were obtained from mass spectrometers capable of operating from 100% water in the absence of any heating. This mode of operation requires a high flow rate of both nebulizer and drying gases, typically 40 and 280 L/hr, respectively.

B. Preparation of Protein Samples

HX-MS may be carried out in two different ways: either using protein in which all labile hydrogens have been exchanged for deuterium (^{2}H-protein) or by dissolving protein containing natural-abundance isotopes (^{1}H-protein) in D_2O at the appropriate pH. ^{2}H-protein was prepared by heating protein solution above the T_m in D_2O, followed by lyophilization. The process was then repeated at least three times. It is possible to initiate exchange of proteins from lyophilized samples, either containing ^{2}H-protein or ^{1}H-protein; however, more consistent results were obtained when exchange was initiated by dilution from protein in solution. This method of sample dilution to initiate exchange gives rise to low levels of the original solvent in the exchanging solution. For example, a protein solution (concentration 100 pmol/μl) is diluted 10-fold into D_2O (final concentration 10 pmol/μl) and the isotopic content of the solvent is 90% D_2O/10% H_2O. Furthermore, the hygroscopic nature of D_2O reduces the percentage of deuterium in the exchange solvent. It is necessary, therefore, to adjust the calculation of protected sites to account for the isotopic composition of the solvent. The precise ratio of D_2O to H_2O is measured by heating the sample to 50–90°C, depending on the T_m of the protein, to accelerate the exchange and remove any residual protection from the 3D structure of the protein. This measurement effectively provides an endpoint for the exchange reaction as well as an accurate measure of the isotopic content of the solvent and is used to calculate the number of protected sites.

C. Sample Introduction

Samples were introduced via a Rheodyne injector and solvent delivery was by means of a fluid delivery module (Michrom Bioresources). The background solvent was matched isotopically and at the same pH as the protein sample solution. Nanoflow electrospray was carried out as described by Wilm and Mann [22] using pulled borosilicate needles coated in gold, in the absence of nebulizer gas and with a low flow rate of drying gas (20 L/hr). For HX-MS of partially folded labile states, samples were precooled on ice before introduction and the solvent delivery system was cooled using an ice salt bath. The temperature of the nebulizer gas was also reduced by passing through a copper coil immersed in an ice salt bath. These measures effectively quench hydrogen exchange during sample introduction and thus the time recorded for exchange was taken as the time at which exchange was quenched.

D. Analysis of Mass Spectra

All mass spectra represent the raw data with minimal smoothing. The average masses of protein samples were calculated from at least three charge states, and masses were calibrated using hen lysozyme. The estimated error in these HX-MS measurements, calculated from the standard deviation of the charge states, is ±2 Da. The number of protected sites was calculated by subtraction of the mass of the protein containing natural-abundance isotopes from that of the measured mass at each time point (^{2}H-protein into H_2O). Alternatively, the measured mass is subtracted from the mass of the protein plus the number of labile sites calculated from the amino acid sequence (^{1}H-protein into D_2O). The masses measured are the centroid values, which take into account the asymmetry of the peak arising from the natural isotopic distribution. The HX-MS kinetic data and normalization of peak areas were carried out using Sigma Plot (Jandell Scientific).

ACKNOWLEDGMENTS

I am particularly grateful to my students Evonne Chung, Ewan Nettleton, Dimitri Masselos, and Adam Rostom for their hard work and enthusiasm. I also acknowledge with thanks Dennis Benjamin, Andrew Miranker, Robin Aplin, Michael Groß, Sheena Radford, and Ulrich Hartl. Finally I thank Prof. C. M. Dobson for his continued help and encouragement throughout all this work. This program of research is supported by the Oxford Center for Molecular Sciences (funded by BBSRC, EPSRC, and MRC) and the Royal Society.

REFERENCES

1. UA Mirza, BT Chait. Effects of anions on the positive ion electrospray ionization mass spectra of peptides and proteins. Anal Chem 66:2898–2904, 1994.
2. YW Bai, JJ Englander, L Mayne, SW Englander. Thermodynamic parameters from hydrogen-exchange measurements. Methods Enzymol 259: 334–356, 1995.
3. A Hvidt. Hydrogen exchange in proteins. Adv Protein Chem 21:287–385, 1996.
4. CK Woodward. Hydrogen exchange rates and protein folding. Curr Opin Struct Biol 4:112–116, 1994.
5. Y Bai, JS Milne, L Mayne, SW Englander. Protein stability parameters measured by hydrogen exchange. Proteins Struct Funct Genet 20:4–14, 1994.
6. V Katta, BT Chait. Conformational changes in proteins probed by hydrogen exchange and electrospray mass spectrometry. Rapid Commun Mass Spectrom 5:214–217, 1993.
7. V Katta, BT Chait. Hydrogen/deuterium exchange electrospray ionization mass spectrometry: a method for probing protein conformational changes in solution. J Am Chem Soc 115: 6317–6321, 1993.
8. A Miranker, CV Robinson, SE Radford, RT Aplin, CM Dobson. Detection of transient protein folding populations by mass spectrometry. Science 262:896–899, 1993.

9. M Jaquinod, F Halgand, M Caffrey, C Saint-Pierre, J Gagnon, J Fitch, M Cusanovich, E Forest. Conformation properties of *Rhodobacter capsulatus* cytochrome c_2 wild-type and site-directed mutants using hydrogen-deuterium exchange monitored by electrospray ionization mass spectrometry. Rapid Commun Mass Spectrom 9:1135–1140, 1995.

10. RS Johnson, K Walsh. Mass spectrometric measurement of protein amide hydrogen exchange rates of *apo-* and *holo*-myoglobin. Protein Sci 3:2411–2418, 1994.

11. DS Wagner, LG Melton, Y Yan, BW Erickson, RJ Anderegg. Deuterium exchange of α-helices and β-sheets as monitored by electrospray ionization mass spectrometry. Protein Sci 3:1305–1314, 1994.

12. EW Chung, EJ Nettleton, CJ Morgan, M Groß, A Miranker, SE Radford, CM Dobson, CV Robinson. Hydrogen exchange properties of proteins in native and denatured states monitored by mass spectrometry and NMR. Protein Sci 6:1316–1324, 1997.

13. D Booth, M Sunde, V Bellotti, CV Robinson, WL Hutchnison, PE Fraser, PN Hawkins, CM Dobson, SE Radford, CCF Blake, MB Pepys. Instability, unfolding and aggregation of human lysozyme variants underlying amyloid fibrillogenesis. Nature 385:787–793, 1997.

14. MB Pepys, PN Hawkins, DR Booth, DM Vigushin, GA Tennent, AK Soutar, N Totty, O Nguyen, CCF Blake, CJ Terry, TG Feest, AM Zalin, JJ Hsuan. Human lysozyme gene mutations cause hereditary systemic amyloidosis. Nature 362:553–557, 1993.

15. DS Wagner, RJ Anderegg. Conformation of cytochrome *c* studied by deuterium exchange-electrospray ionization mass spectrometry. Anal Chem 66:706–711, 1994.

16. RJ Anderegg, DS Wagner. Mass spectrometric characterization of a protein-ligand interaction. J Am Chem Soc 117:1374–1377, 1995.

17. CV Robinson, EW Chung, BB Kragelund, J Knudsen, RT Aplin, FM Poulsen, CM Dobson. Probing the nature of non-covalent interactions by mass spectrometry. A study of protein-CoA ligand binding and assembly. J Am Chem Soc 118:8646–8653, 1996.

18. J Knudsen, S Mandrup, JT Rasmussen, PH Andreasen, F Poulsen, K Kristiansen. The function of acyl-CoA-binding protein (ACBP)/Diazepam binding inhibitor (DBI). Mol Cell Biochem 123:129–138, 1993.

19. J Knudsen, NJ Færgeman, H Skott, R Hummel, C Borsting, TM Rose, JS Andersen, P Roepstorff, K Kristiansen. Yeast acyl-CoA-binding protein: acyl-CoA-binding affinity and effect on intracellular acyl-CoA pool size. Biochem J 302:479–485, 1994.

20. BB Kragelund, KV Andersen, JC Madsen, J Knudsen, FM Poulsen. Three-dimensional structure of the complex between acyl-coenzyme A binding protein and palmitoyl-coenzyme A. J Mol Biol 230:1260–1270, 1993.

21. S Mandrup, P Horjup, K Kristiansen, J Knudsen. Gene synthesis, expression in *E. coli*, purification and characterization of the recombinant bovine acyl-CoA-binding protein. Biochem J 276:871–823, 1991.

22. M Wilm, M Mann. Electrospray and Taylor-cone theory, Dole's beam of macromolecules at last? Int J Mass Spectrom Ion Proc 136:167–180, 1994.

23. CV Robinson, M Groß, SJ Eyles, JJ Ewbank, M Mayhew, FU Hartl, CM Dobson, SE Radford. Conformation of GroEl-bound α-lactalbumin probed by mass spectrometry. Nature 372:646–651, 1994.

24. M Groß, CV Robinson, M Mayhew, FU Hartl, SE Radford. Significant hydrogen exchange protection in GroEL-bound DHFR is maintained during iterative rounds of substrate cycling. Protein Sci 5:2506–2513, 1996.

25. Z Zhang, DL Smith. Determination of amide hydrogen exchange by mass spectrometry: A new tool for protein structure elucidation. Protein Sci 2:522–531, 1993.

26. A Miranker, GH Kruppa, CV Robinson, RT Aplin, CM Dobon. Isotopic-labeling strategy for the assignment of protein fragments generated for mass spectrometry. J Am Chem Soc 118:7402–7403, 1996.

27. YW Bai, SJ Milne, L Mayne, SW Englander. Primary structure effects on peptide group hydrogen exchange. Proteins Struct Funct Genet 17:75–86, 1993.

28. CV Robinson, M Groß, SE Radford. Probing the conformation of GroEL-bound substrate proteins by mass spectrometry. Methods Enzymol, in press.

15

Characterization of Oligonucleotides by Electrospray Mass Spectrometry

Pamela F. Crain
University of Utah, Salt Lake City, Utah

I. INTRODUCTION

The development of electrospray ionization (ESI) and matrix-assisted laser desorption ionization (MALDI) has led to a rapid increase in their utilization for mass spectrometry for the study of large biopolymers. Analysis of nucleic acids is a major beneficiary of these new ionization methods [1,2], for which prior mass spectrometric methods were not especially useful for structural studies above about the tetramer level, substantially lagging in applications as compared to proteins. As applications of ESI mass spectrometry to nucleic acid analysis have been extensively reviewed recently [1,2], this account emphasizes practical aspects of ESI-based methodologies, especially sample handling, and provides a brief overview of how they are applied to problems of nucleic acid structure. References are not intended to be comprehensive, but rather are chosen to illustrate the points being made. Further information about nucleic acid structure can be obtained from Saenger's classic account [3]. A recent book covering principles and applications of ESI mass spectrometry [4] is recommended for an overview of the methodology.

II. SAMPLE PREPARATION

Nucleic acids are most commonly analyzed by detection of negatively charged ions. The presence of sodium (or other nonvolatile) counterions will therefore have two deleterious effects. First, the mass of the principal ion in the multiply charged cluster will include a contribution from the masses of an unknown number of cations, leading to inaccuracy in the determination of the true mass of the nucleic acid. Second, sensitivity will be reduced as the number of molecular species within a charge state is correspondingly increased, thereby attenuating the signal and leading to reduced spectral intensity. The key to successful oligonucleotide analyses is therefore careful sample preparation, such that the amount of salt (expecially nonvolatile cations) is reduced to a minimum. Although small amounts of cations can be compensated for by solution additives and sample inlet probe modifications, unnecessarily salty solutions will more quickly contaminate transfer lines and instrument surfaces. There are a number of methods that can be used to effect sample desalting, and the choice is dictated by the length of the nucleic acid, the amount of material present, and the method of introduction of the analyte into the mass spectrometer.

All salts and buffers used for sample preparation should be of the highest purity available. Sterile, nuclease-free water should always be used to make required solutions. Nucleases are present on the skin; they can be transferred to pipet tips and sample containers and retain activity. Disposable gloves should therefore be used when handling solutions of nucleic acids and any materials they contact. We use RNase-free pipet tips and 1.5-ml sample tubes (Ambion; Austin, TX) for sample manipulations. A decontaminating solution (RNase-ZAP; Ambion) can be used to rinse or wipe down other items.

A. Desalting

For short oligonucleotides, reversed-phase (RP) high-performance liquid chromatography (HPLC) using volatile eluents may be used, typically on C_{18} columns. The choice of solvent system is not critical; we routinely use a triethylammonium bicarbonate + acetonitrile binary solvent system [5], from which the oligonucleotide can be recovered by removal of the elution solvent in vacuo.

   For larger oligonucleotides, (e.g., approximately 15-mer and longer), precipitation from 2.5 M ammonium acetate with alcohol can be used to generate the ammonium salt [6,7]. One-third volume of 10 M ammonium acetate is added to the nucleic acid solution and thoroughly mixed to ensure homogeneity. The 10 M ammonium acetate stock solution should be filter sterilized. After several minutes, the alcohol can be added; ethanol is generally used (2.5 volumes). If initial recovery of the nucleic acid was from a phosphate-based buffer, however, isopropanol (1 volume) must be used to avoid coprecipitation of the phosphate salt. Phosphorothioate oligonucleotides bind cations strongly, and precipitation should be initiated from 10 M ammonium acetate [8]. The nucleic acid is recovered by centrifugation; if sufficient material is available, the process can be repeated for further cation removal. The pellet will contain excess ammonium acetate, so it should be washed by mixing with cold 80% (aq.) ethanol and respun. This precipitation method is simple to execute and sample recovery is essential quantitative. The extent of counterion conversion is generally greater than that which can be achieved using HPLC for large oligonucleotides (e.g., 50-mers or so).

   Ultrafiltration and size-exclusion microdevices are widely available from suppliers of molecular biological products and may also be used to effect desalting. We have used NENSORB 20 cartridges (NEN Life Sciences; Boston, MA) for preparation of salt-free nucleic acids [9]. These cartridges are intended to allow recovery of nucleic acids from solutions containing proteins and small molecules (e.g., unincorporated nucleotides from polymerase chain reactions). The stated capacity is 20 μg total, but when used for simple cleanup of pure nucleic acids, the load can be increased to 50 μg. The sample is eluted with an aqueous alcohol solution, with obvious advantage for subsequent mass spectral analysis.

B. Sample Storage

A nucleic acid pellet that has been obtained from alcohol precipitation is best stored as the wet pellet following decantation of supernatant. Just prior to use it can be dried and dissolved in water or buffer as desired. If the sample is to be used for multiple analyses it should be aliquoted into suitably small amounts and dried for storage in at $-20°C$. Solutions should be made up in and stored in plasticware, not glass. Chain cleavage and loss of terminal phosphate groups can occur if care is not taken with containers and storage solvents.

III. SAMPLE INTRODUCTION

Pure samples or simple mixtures can be dissolved in an appropriate solvent and introduced into the mass spectrometer using a syringe pump or a simple sample injector in-line with a flowing carrier solvent. More complex mixtures, as well as analytes present in limited amount, will benefit from chromatographic or electrophoretic sample introduction. In the latter instance, sensitivity is increased because the sample is delivered to the ion source as a bolus. (There are few examples of oligonucleotide analyses using electrophoresis [e.g., 10,11] and they are not discussed here, although they may prove to be very useful.) Very low flow rate inlet ("nanospray") probes are available from instrument manufacturers or may be constructed; they are likewise useful for introducing small sample amounts.

A. Infusion

Desalting by any of the methods discussed earlier will yield a nucleic acid sample that still contains sodium salts. Bottles in which organic solvents are supplied are a significant source of leached cations. In addition, certain liquid chromatographs utilize glass solvent delivery bottles that contribute substantial amounts of Na^+. We have substituted plastic bottles with 29/32 necks (number E-06039-80; Cole Parmer; Vernon Hills, IL) for the 29/35 bottles provided with our Hewlett Packard HP 1090 liquid chromatograph. Small amounts of plasticizers are leached, but the sodium content of the elution solvents is noticeably lower. It is undoubtedly impossible to generate a totally salt-free sample; nonetheless, choice of solvent used for sample infusion can compensate for this problem to a some extent. A modified inlet probe that incorporates an in-line microdialysis fiber has recently been described [12]; the extent of sodium removal from highly contaminated samples is impressive, but the flow rate and internal volume of the device may limit its application for trace analysis.

Various solvents have been described for analysis of oligonucleotides by direct infusion. There are no truly systematic studies reported in which a wide variety of buffer/organic modifier combinations have been studied. This may not prove to be especially useful, because the electrospray interfaces used by different mass spectrometer manufacturers differ in design and properties, and what may be the "optimal" solvent derived using one instrument may not be so for another. Nonetheless, certain basic conclusions have emerged. Higher amounts of organic modifier give generally more stable sprays [13]. Isopropanol has very favorable properties as an organic modifier for negative-ion work [14]. Generation of ions from a slightly basic solution is advantageous, and triethylamine-containing solvents have been used with success to suppress sodium cationization [7,15]. Solutions containing imidazole and either piperidine or triethylamine proved especially useful for analysis of phosphorothioate oligonucleotides [8]. Note that imidazole should not be used in infusion buffers for RNA, which it cleaves. It should be noted, however, that infusion of samples in 80% (aq.) methanol without organic modifier is successfully used in some laboratories [5,16]. The only safe advice to offer is to test the particular "cocktail" on the user's own instrument, with the warning that tuning parameters, especially interface temperature, may have to be adjusted. Nucleic acid concentrations of 5–10 pmol/μl or lower can be accommodated in a clean, well-tuned mass spectrometer.

Despite careful desalting and the use of adjuvants in the infusion solvent, there still remains the possibility for adduction of divalent cations too strongly bound to be removed by the treatments described above. During studies of natural transfer RNAs using ESI/MS we observed persistent cation adduction following standard cleanup protocols. Knowing that tRNA contains modified nucleosides that strongly bind divalent cations, especially Mg^{2+}, we devised a solvent for sample introduction that contained the chelating agent *trans*-1,2-diaminocyclohexane-N,N,N',N'-tetraacetic acid (CDTA), along with triethylamine and isopropanol as organic modifier [7]. Shown in Fig. 1 are spectra illustrating the effects of these adjuvants, singly and in combination, on the mass spectrum of a mixture of tRNAs (76-mers). Reduction of peak width with use of CDTA is evident, and masses were measured within an accuracy of 0.005%. CDTA should prove to be a generally useful additive for analyzing natural or enzymatically generated nucleic acids, where the possibility of contamination with divalent cations exists.

B. High-Performance Liquid Chromatography

Relatively few applications of LC/ESI-MS for oligonucleotide analysis have been reported [17–20], despite the requirement of a flowing liquid solution

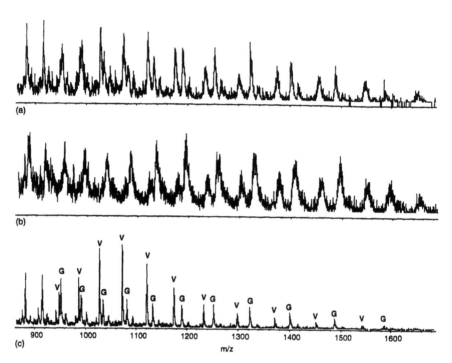

Figure 1 Electrospray mass spectra of an equimolar mixture of valyl (V) and glycyl (G) transfer RNAs from *E. coli.* Samples were precipitated once from 2.5 *M* ammonium acetate; see ref. 7 for details. Sample introduction by infusion from (a) 0.03% triethylamine; (b) 1 nmol CDTA; (c) 0.03% triethylamine + 1 nmol CDTA; all solutions + isopropanol, 1:2, v/v. (Reprinted by permission of Elsevier Science, Inc. from ref. 7. Copyright 1995 by American Society for Mass Spectrometry.)

to effect electrospray ionization, which should favor this method of sample introduction. A simple ammonium acetate + acetonitrile gradient has been described [17] and is currently being used in our laboratory. Low ionic strength volatile buffers (e.g., 5–10 m*M* ammonium acetate or triethylammonium acetate) that are required for compatibility with electrospray ionization do not generally yield good oligonucleotide separations, although interpretable mass spectra can generally be obtained from incompletely resolved peaks. A gradient system that utilizes 1,1,1,3,3,3-hexafluoro-2-propanol (HFIP) based solvents has recently been described for the liquid chromatography/mass spectrometry (LC/MS) analysis of a variety of nucleic acids, including polythymidylic acid mixtures, and normal backbone and phosphorothioate oligonucleotide mixtures [19]. Note: Goggles and gloves must be worn when handling this reagent. Separations were shown to be superior to those obtained with the commonly used triethylammonium acetate-based solvents for liquid chromatography of oligonucleotides.

A gradient system consisting of aqueous 0.4 M HFIP (adjusted to pH 7.0 with triethylamine) and 0.4 M HFIP (pH 7.0; in methanol, 1:1 v/v) was used for the LC/ESI-MS analysis of the porcine kidney metabolites of a phosphorothioate antisense oligonucleotide 20-mer [20]. Shown in Fig. 2 (top) are the total ion current chromatogram and the reconstructed ion chromatograms of the base peaks for five oligonucleotide metabolites. The electrospray mass spectrum acquired for the 12-mer metabolite (Fig. 2, bottom) shows excellent signal-to-noise properties, demonstrating that the very good chromatographic separations obtained for this closely related family of oligonucleotides were not achieved at the expense of mass spectral intensity. The administered 20-mer was found to have been degraded by removal of nucleotides from one or the other end, and also by simultaneous degradation from both termini. Interestingly, one metabolite was generated by addition of a dG residue to the 3′ terminus of the administered drug.

IV. INSTRUMENTATION ISSUES

In addition to issues of sample preparation and the mechanics of sample introduction, attention must be paid to instrument variables. Oligonucleotides have an unfortunate reputation for being difficult to analyze by ES/MS, which is not always deserved. Invariably, following a day's worth of protein analyses by infusion, if analysis of an oligonucleotide sample is then attempted, a poor (or no) signal is obtained. Most likely the sample has been adsorbed onto the protein-coated transfer line. We use capillary PEEK tubing instead of fused silica for the sample transfer line, as the latter tubing can become deactivated by cations carried along with previous samples, thus necessitating acid treatment and lengthy subsequent water flushing. Finally, we routinely clean the ion source surfaces before oligonucleotides are analyzed on an instrument that must be used for a variety of samples.

V. MOLECULAR WEIGHT DETERMINATION

Many problems in nucleic acid chemistry and molecular biology can be solved by a straightforward determination of molecular weight (reviewed in ref. 2). Quality control of synthetic oligonucleotides, mapping of modified residues within large natural RNAs [5,21], and mass analysis of polymerase chain reaction (PCR) products [22,23] constitute major areas of application to actual

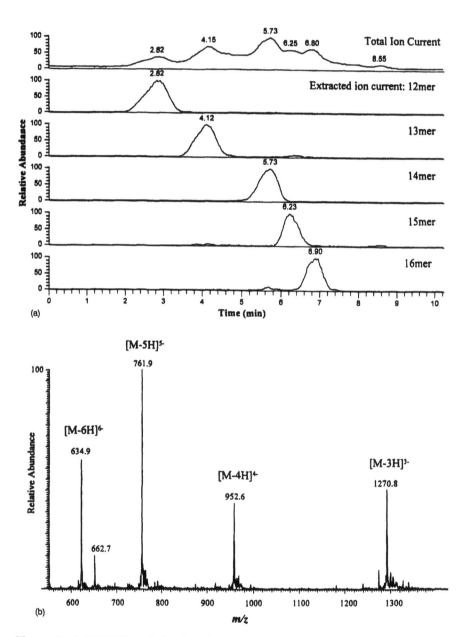

Figure 2 LC-MS/MS analysis of a mixture of in vivo metabolites of the phosphoro-thioate oligonucleotide GCCCAAGCTGGCATCCGTCA. (a) Total ion current and re-constructed ion chromatograms. (b) Electrospray mass spectrum obtained from the 12-mer oligonucleotide. (Reproduced with permission from ref. 20. Copyright 1997 John Wiley & Sons Limited.)

problems. Given proper attention to instrument calibration, and depending on the mass analyzer with which the electrospray ionization is coupled, molecular weights can be measured to about 0.005–0.01% or better with quadrupole instruments, somewhat less accurately using quadrupole ion traps, and more accurately using magnetic-sector or ion cyclotron resonance mass analyzers.

An extensive set of calculations were undertaken to determine the extent to which composition can be derived from mass measurement [24]. Where the sequence or composition are not known, compositions can be derived from measured mass only for oligonucleotide tetramers and smaller, in the absence of any molecular constraints. If the number of occurrences of any one residue can be determined, such as through selective chemical modification of one of the base types, or cleavage at specific residues (e.g., for RNase T1, after G, in which G is fixed at one), the number of allowable compositions can be significantly reduced. For example, mass measurement to the nearest integer mass permits determination of chain length to the 7-mer (DNA) or 8-mer (RNA) level. If mass can be measured to within 0.01%, then composition can uniquely be specified up to the 14-mer level (or up to the 25-mer level if chain length is known) for RNA. The concept of increasing the size of nucleic acid for which mass defines composition by constraining the allowable compositions was further extended by very accurate mass measurement (within 0.5 Da using an ICR mass analyzer) of double-stranded DNA [25]. In the latter situation, an important restriction arises from mass measurement of both of the DNA strands; since they are complementary, the number of A's in one strand is equal to the number of T's in the other, and likewise for the C and G content.

VI. OLIGONUCLEOTIDE SEQUENCE DETERMINATION

The sequence of a nucleic acid is one of its fundamental properties. Classical methods for sequence determination utilized nucleases that sequentially remove nucleotides from the termini. Typically, chromatographic methods were used to determine the identity of the residue removed at each step, to reconstruct the sequence. Greater certainty of identification of modified nucleotides within the sequence can be effected by determination of mass, a fundamental molecular property. Two general mass spectrometric approaches are available: mass analysis of exonuclease digestion products, and analysis of oligonucleotide fragment ions produced from collision-induced dissociation (CID) or "nozzle-skimmer" (NS) dissociation in the mass spectrometer (see Chapter 1). Direct sequencing by CID has the advantage of simplicity if sufficient material and an appropriate instrument are available. Fragmentation may be incomplete, however, and the exonuclease-based method can be used in a complementary fashion, particularly for determining the nucleotides at the termini.

A. Determination of Sequence by Collision-Induced Dissociation

The pioneering studies of McLuckey and colleagues, conducted using a quadru-pole ion trap mass analyzer, provided the initial fundamental framework for understanding the collision-induced dissociation (CID) of multiply charged oli-gonucleotide negative ions [16,26,27]. Further elucidation of fragmentation pathways and resolution of ambiguous ion identities was accomplished using Fourier transform MS (FTMS) for more accurate mass determination [28]. Fragmentation behavior of negatively charged oligonucleotides in the triple-quadrupole instrument has been detailed [29–31]; one of these reports [30] compares and contrasts the CID pathways of the DNA oligonucleotide CGAGCTCG in the triple-quadrupole mass analyzer with those reported using FTMS [28]. Comparative aspects of oligonucleotide CID in these various mass analyzers have been reviewed [2].

A typical product ion spectrum (quadrupole ion trap mass analyzer), from the LC/MS analysis shown in Fig. 2, is shown in Fig. 3. The nomenclature for

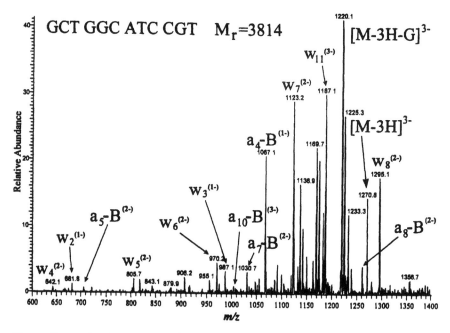

Figure 3 Product ion spectrum from collision-induced dissociation of m/z 1270.8, the $(M - 3H)^{3-}$ ion from the phosphorothioate oligonucleotide GCTGGCATCCGT, acquired during the LC/MS analysis shown in Fig. 2. (Reproduced with permission from ref. 20. Copyright 1997 John Wiley & Sons Limited.)

designation of the ion types and the location of the bond cleaved is that of McLuckey [16]. For example, w_7^{3-} designates a triply charged ion containing nucleotides 7–12; this ion type indicates the sequence in the $3' \rightarrow 5'$ direction. The $a_{10} - B_{10}$ ion indicates cleavage between nucleotides 10 and 11, with loss of the base from residue 10; this ion type indicates sequence in the $5' \rightarrow 3'$ direction. Product ion spectra of multiply charged oligonucleotides are inherently complex, but can generally be interpreted to yield at least partial, and often complete, sequence information, depending on the mass analyzer used. In this example [20], the oligonucleotide is one of several metabolites from in vivo nuclease degradation of a phosphorothioate oligonucleotide therapeutic agent. Although complete sequence information could not be derived via CID using the ion trap, the sequence of the starting material was known, and measured molecular weights and partial sequence information were sufficient to permit structures of the metabolites to be derived.

The CID spectra of multiply charged oligonucleotides are sufficiently complex that even the interpretation of the spectrum of a known compound is very tedious. If CID is to be used for verifying a known sequence, it is a trivial task to computer-generate a list of calculated fragment ions based on the known oligonucleotide cleavage reactions. Allowance must be made, however, for the existence of the same cleavage ion at different charge states. The mass spectral data can then be compared with the calculated mass values, taking into account simple abundance rules [31] used to avoid misassignment of minor ions resulting from backbone double-cuts.

Most fundamental studies have consisted of interpretation of the CID spectra of reference compounds whose sequence is known. An approach to de novo oligonucleotide sequence determination has been described that relies on computer-assisted sequence derivation [31]. An algorithm was described that was developed through iterative testing with model oligonucleotides designed to explore ambiguities in fragment ion assignment. In addition to mass spectral data, other input parameters are M_r, precursor ion mass and charge, error tolerance for calculated mass values, nature of the polynucleotide (RNA or DNA), structures of the termini (OH or PO_4H; identities will be known from the analyte origin, e.g., synthetic or enzymatically derived, etc.), and mass increments for any modifications known or suspected. There is no a priori sequence information required. For derivation of sequence in the $3' \rightarrow 5'$ direction, the most abundant ion is assigned as the w_1 ion, and candidate values are calculated for the next ion in the series, w_2. If there is more than one match in the mass list, the most abundant one is flagged as the most probable candidate. Likewise, the sequence in the $5' \rightarrow 3'$ direction is derived using the $a_n - b_n$ ion series. "Trees" are iteratively generated for all combinations for which a calculated ion is present in the spectrum, and the calculation is continued until a sequence is derived that is consistent with the measured molecular weight of the oligonucleotide.

In an early report of ESI of oligonucleotides, it was observed that the molecules fragmented in a complex way at elevated declustering voltages [6]. In fact, these NS fragment ions are now known to be many of the same ion types resulting from CID. NS dissociation has, in fact, been used to generate precursor ions for subsequent CID [32]. If the sample is introduced by direct infusion, high purity is required because the precursor ion is not isolated. Although not reported to date, NS dissociation in conjunction with LC/MS might prove to be useful to generate partial sequence information from a single quadrupole instrument, and to simulate an MS^3 measurement in a tandem mass spectrometer by subsequent CID of an NS-generated ion.

The fragmentation behavior of positively charged oligonucleotides has been reported [33]. Backbone cleavage pathways that indicate the sequence are generally analogous to those described for negatively charged oligonucleotide ions. The generally lower level of charging, typical of positively charged polynucleotide ions [34], may limit the number of residues that can be sequenced, as more complete fragmentation is effected from high- versus low-charge-state ions [29]. It was noted that owing to the low proton affinity of U and T derivatives, gaps in the mass sequence ladder are observed at their locations [33].

B. Mass Analysis of Exonuclease Digestion Products

Detection of the masses of the residual oligonucleotides following exonucleolytic hydrolysis has been reported using MALDI [35] and ESI [36]. Several advantages accrue ' he ESI-based approach. Sample manipulation is minimal because hydrolysis proceeds directly in the syringe used for sample infusion into the mass spectrometer (a volatile buffer with acetonitrile as the organic component is used). Progress of the hydrolysis can be monitored in real time, and, if needed, the reaction rate can be directly controlled by manipulating the temperature of the solution during the course of data acquisition. Finally, the masses of both the mononucleotide removed and the residual oligonucleotide can be determined. Shown in Fig. 4 is the sequence analysis of an oligonucleotide derived from valyl transfer RNA from *Escherichia coli* [36]. The tRNA was digested with RNase T_1, which yields a mixture of oligonucleotides terminating with Gp at the 3' terminus. After ion exchange chromatography, an oligonucleotide was recovered whose measured mass [24] was consistent with the composition $CU_2Gp + CH_3$. During in situ digestion with phosphodiesterase I, which sequentially removes nucleotides from the 3' terminus, spectra were summed over the time ranges shown, allowing the methyl group to be assigned to the 5'-terminal U residue, and the complete sequence to be assigned as m^5UUCGp [36].

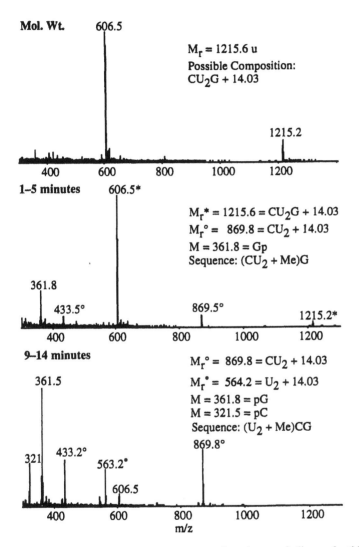

Figure 4 Electrospray mass spectrum of a mixture of oligonculeotides produced from phosphodiesterase I digestion of an oligonucleotide derived from valyl transfer RNA from *E. coli*. (From ref. 36 by permission of Oxford University Press.)

VII. CONCLUSIONS

The value of electrospray ionization mass spectrometry for analysis of the (formerly) intractable oligonucleotides and small nucleic acids cannot be overestimated; they are the principal beneficiaries of advances in mass spectrometry instrumentation. A wide variety of useful applications is beginning to emerge, driven by the increased availability of instruments and advances in methods development. Electrospray has been used to ionize megadalton size DNA molecules [37], and the size limitation for nucleic acid ionization is not yet apparent.

ACKNOWLEDGMENT

Support from National Institutes of Health grant GM 29812 is gratefully acknowledged.

REFERENCES

1. E Nordhoff, F Kirpekar, P Roepstorff. Mass spectrometry of nucleic acids. Mass Spectrom Rev 15:67–138, 1996.
2. PF Crain. Electrospray ionization mass spectrometry of nucleic acids and their constituents. In: RB Cole, ed. Electrospray Ionization Mass Spectrometry. New York: John Wiley & Sons, 1997, pp 421–457.
3. W Saenger. Principles of Nucleic Acid Structure. Springer Advanced Texts in Chemistry, ed. CR Cantor. New York: Springer-Verlag, 1984.
4. RB Cole. Electrospray Ionization Mass Spectrometry. New York: John Wiley & Sons, 1997, p 577.
5. JA Kowalak, SC Pomerantz, PF Crain, JA McCloskey. A novel method for the determination of post-transcriptional modification in RNA by mass spectrometry. Nucleic Acids Res 21:4577–4585, 1993.
6. JT Stults, JC Marsters. Improved electrospray ionization of synthetic oligodeoxynucleotides. Rapid Commun Mass Spectrom 5:359–363, 1991.
7. PA Limbach, PF Crain, JA McCloskey. Molecular mass measurement of intact ribonucleic acids via electrospray ionization quadrupole mass spectrometry. J Am Soc Mass Spectrom 6:27–39, 1995.
8. M Greig, RH Griffey. Utility of organic bases for improved electrospray mass spectrometry of oligonucleotides. Rapid Commun Mass Spectrom 9:97–102, 1995.
9. PF Crain. Analysis of 5-methylcytosine in DNA by isotope dilution gas chromatography-mass spectrometry. Methods Enzymol 193:857–865, 1990.
10. W Schrader, M Linscheid. Determination of styrene oxide adducts in DNA and DNA components. J Chromatogr A 717:117–125, 1995.
11. JP Barry, J Muth, S-J Law, BL Karger, P Vouros. Analysis of modified oligonucleo-

tides by capillary electrophoresis in a polyvinylpyrrolidone matrix coupled with electrospray mass spectrometry. J Chromatogr A 732:159–166, 1996.

12. C Liu, Q Wu, AC Harms, RD Smith. On-line microdialysis sample cleanup for electrospray ionization mass spectrometry of nucleic acid samples. Anal Chem 68:3295–3299, 1996.

13. K Bleicher, E Bayer. Various factors influencing the signal intensity of oligonucleotides in electrospray mass spectrometry. Biol Mass Spectrom 23:320–322, 1994.

14. RF Straub, RD Voyksner. Negative ion formation in electrospray mass spectrometry. J Am Soc Mass Spectrom 4:578–587, 1993.

15. N Potier, A Van Dorsselaer, Y Cordier, O Roch, R Bischoff. Negative electrospray ionization mass spectrometry of synthetic and chemically modified oligonucleotides. Nucleic Acids Res 22:3895–3903, 1994.

16. SA McLuckey, GJ Van Berkel, GL Glish. Tandem mass spectrometry of small, multiply charged oligonucleotides. J Am Soc Mass Spectrom 3:60–70, 1992.

17. K Bleicher, E Bayer. Analysis of oligonucleotides using coupled high performance liquid chromatography-electrospray mass spectrometry. Chromatographia 39:405–408, 1994.

18. B Bothner, K Chatman, M Sarkisian, G Siuzdak. Liquid chromatography mass spectrometry of antisense oligonucleotides. Bioorg Med Chem Lett 5:2863–2868, 1995.

19. A Apfel, JA Chakel, S Fischer, K Lichtenwalter, WS Hancock. Analysis of oligonucleotides by HPLC-electrospray ionization mass spectrometry. Anal Chem 69:1320–1325, 1997.

20. RH Griffey, MJ Greig, HJ Gaus, K Liu, D Monteith, M Winniman, LL Cummins. Characterization of oligonucleotide metabolism in vivo via liquid chromatography/electrospray tandem mass spectrometry with a quadrupole ion trap mass spectrometer. J Mass Spectrom 32:305–313, 1997.

21. E Bruenger, JA Kowalak, Y Kuchino, JA McCloskey, H Mizushima, KO Stetter, PF Crain. 5S rRNA modification in the hyperthermophilic archaea *Sulfolobus solfataricus* and *Pyrodictium occultum*. FASEB J 7:196–200, 1993.

22. Y Naito, K Ishikawa, Y Koga, T Tsuneyoshi, H Terunuma, R Arakawa. Molecular mass measurement of polymerase chain reaction products amplified from human blood DNA by electrospray ionization mass spectrometry. Rapid Commun Mass Spectrom 9:1484–1486, 1995.

23. DC Muddiman, GA Anderson, SA Hofstadler, RD Smith. Length and base composition of PCR-amplified nucleic acids using mass measurements from electrospray ionization mass spectrometry. Anal Chem 69:1543–1549, 1997.

24. SC Pomerantz, JA Kowalak, JA McCloskey. Determination of oligonucleotide composition from mass spectrometrically measured molecular weight. J Am Soc Mass Spectrom 4:204–209, 1993.

25. DJ Aaserud, NL Kelleher, DP Little, FW McLafferty. Accurate base composition of double-strand DNA by mass spectrometry. J Am Soc Mass Spectrom 7:1266–1269, 1996.

26. SA McLuckey, S Habibi-Goudarzi. Decompositions of multiply charged oligonucleotide anions. J Am Chem Soc 115:12085–12095, 1993.

27. SA McLuckey, S Habibi-Goudarzi. Ion trap tandem mass spectrometry applied to

small multiply charged oligonucleotides with a modified base. J Am Soc Mass Spectrom 5:740–747, 1994.

28. DP Little, RA Chorush, JP Speir, MW Senko, NL Kelleher, FW McLafferty. Rapid sequencing of oligonucleotides by high-resolution mass spectrometry. J Am Chem Soc 116:4893–4897, 1994.

29. JP Barry, P Vouros, A Van Schepdael, S-J Law. Mass and sequence verification of modified oligonucleotides using electrospray tandem mass spectrometry. J Mass Spectrom 30:993–1006, 1995.

30. PF Crain, JM Gregson, JA McCloskey, CC Nelson, JM Peltier, DR Phillips, SC Pomerantz, DM Reddy. Characterization of posttranscriptional modification in nucleic acids by tandem mass spectrometry. In: AL Burlingame, SA Carr, eds. Mass Spectrometry in the Biological Sciences. Totawa, NJ: Humana Press, 1996, pp 497–517.

31. J Ni, SC Pomerantz, J Rozenski, Y Zhang, JA McCloskey. Interpretation of oligonucleotide mass spectra for determination of sequence using electrospray ionization and tandem mass spectrometry. Anal Chem 68:1989–1999, 1996.

32. DP Little, FW McLafferty. Sequencing 50-mer DNAs using electrospray tandem mass spectrometry and complementary fragmentation methods. J Am Chem Soc 117:6783–6784, 1995.

33. J Ni, MAA Mathews, JA McCloskey. Collision-induced dissociation of polyprotonated oligonucleotides produced by electrospray ionization. Rapid Commun Mass Spectrom 11:535–540, 1997.

34. KA Sannes-Lowery, DP Mack, P Hu, H-Y Mei, JA Loo. Positive ion electrospray ionization mass spectrometry of oligonucleotides. J Am Soc Mass Spectrom 8:90–95, 1997.

35. U Pieles, W Zürcher, M Schär, HE Moser. Matrix assisted laser desorption ionization time-of-flight mass spectrometry: a powerful tool for the mass and sequence analysis of natural and modified oligonucleotides. Nucleic Acids Res 21:3193–3196, 1993.

36. PA Limbach, JA McCloskey, PF Crain. Enzymatic sequencing of oligonucleotides with electrospray mass spectrometry. Nucleic Acids Res Symp Ser 31:127–128, 1994.

37. R Chen, X Cheng, DW Mitchell, SA Hofstadler, Q Wu, AL Rockwood, MG Sherman, RD Smith. Trapping, detection, and mass determination of coliphage T4 DNA ions of 10^8 Da by electrospray ionization Fourier transform ion cyclotron resonance mass spectrometry. Anal Chem 67:1159–1163, 1995.

16

Methods and Applications of Sizing DNA by Matrix-Assisted Laser Desorption/Ionization Mass Spectrometry

David M. Lubman, Jian Bai*, Yan-Hui Liu[†], Jannavi R. Srinivasan, Yongdong Zhu, and David Siemieniak
University of Michigan, Ann Arbor, Michigan

Patrick J. Venta
Michigan State University, East Lansing, Michigan

Current affiliations:
*Hewlett-Packard Company, Palo Alto, California.
†Schering-Plough Research Institute, Kenilworth, New Jersey.

I. INTRODUCTION

The advent of the Human Genome Project, which plans to sequence a large portion of the human genome over the next 10 years, will require ever more rapid and cost-effective methods for sequencing DNA. Further, advances in molecular biology that allow the manipulation and processing of DNA have resulted in the identification of genes and specific mutations linked with a variety of human diseases [1–13]. The use of large-scale DNA sequencing and various methods for rapid genetic screening will become increasingly important as tools for genetic diagnosis, which is expected to play an important role in the fields of molecular pathology/genetics. The speed of analysis for such large-scale sequencing/screening projects is limited by the analytical tools available for sizing DNA fragments. The size of DNA fragments is generally determined by gel electrophoresis, which presently forms the basis for DNA sequencing, mapping, and screening. However, electrophoresis is time-consuming, not easily automated, and occasionally susceptible to errors due to possible irregularities of both the gel and the migration behavior of certain fragments. In addition, DNA-mers <50 base pairs (b.p.) long are often difficult to detect and size using current gel methods. The development of new rapid, accurate, and cost-effective methods for large-scale characterization of both synthetic and natural genetic materials has become an important area of research for the human genome initiative in terms of mapping genes, detection of genetic defects and mutations, and drug development.

A potential alternative method for enhanced speed for DNA sequencing [13,14] and screening is mass spectrometry [15–22]. The advent of matrix-assisted laser desorption/ionization (MALDI) mass spectrometry (MS) has extended the capabilities of mass spectrometry for detection of high mass and for analyzing mixtures of biolgoical molecules with minimal sample preparation. MALDI/MS thus has the potential to replace gel electrophoresis for direct analysis of DNA sequencing reaction mixtures, enzymatic digests of DNA, and polymerase chain reaction (PCR) products for sequencing, mapping, and screening.

The problems associated with applying MALDI to sizing DNA include extending the range of detection to sufficiently large mass, achieving high sensitivity, and reaching a resolution capable of detecting a single base-pair difference. In order to be viable for DNA sequencing or screening, a detection range of at least 300 b.p. and preferably 500 b.p. is required. The latter corresponds to a molecular weight of approximately 175,000 Da for single-stranded DNA. In early studies in MALDI/MS of DNA only small oligonucleotides could be detected intact with commonly used matrices, where fragmentation of DNA was the major problem. However, the introduction of 3-hydroxypicolinic acid (3-HPA) as a matrix for MALDI/MS of DNA by Becker and co-workers [23,24] and later mixtures of 3-HPA and picolinic acid (PA) by Chen et al. [25] provided a major breakthrough, where large mixed-base DNA-mers could be ionized and detected. Indeed, using mixtures of 3-HPA and PA as matrix, Chen et al. [26] have been able to detect DNA-mers as large as 500 b.p. by MALDI/MS. The problems of achieving sufficient resolution at high mass and also sensitivity still remain as important issues. In addition, the reproducibility of the MALDI process for large DNA-mers and the ability to analyze complicated mixtures of DNA-mers also remain to be solved.

A major issue limiting the mass range, resolution, and sensitivity of MALDI/MS analysis of DNA is the presence of impurities [16,27–33]. Indeed the presence of certain proteins or other additives in the PCR procedure or in the Sanger method can completely suppress the MALDI process, and some prepurification method must be used. Even in the case of synthesized single-stranded DNA, impurities can cause reduced sensitivity and poor resolution. In particular, cations such as Na^+ and K^+ in buffering salts can lead to cation attachment product formation, which results in peak broadening and poor resolution. Other impurities may affect the matrix crystallization process of the analyte/matrix mixture, which is a critical aspect of the MALDI process. In the case of DNA this may be particularly problematic since DNA molecules contain strong acidic phosphate groups, and their high polarity and strong interactions among the DNA molecules may explain the general difficulty in desorbing these species intact. It appears for the MALDI results that the negative charges of DNA molecules are counterbalanced by positively charged counterions often present in biological samples. Generally Na^+ and K^+ form the counterions, but Fe^{2+} has also been observed [30,32]. The formation of these adduct ions not only degrades the sensitivity by distributing the signal over several ions but also causes the resolution to deteriorate if those multiple peaks cannot be resolved. The counterions and their interactions with DNA may affect the desorption/ionization of DNA in the MALDI process and may account for the fact that the sensitivity for DNA drops as the size of the DNA increases. Thus, some method of purification is needed for removal of buffers, reagents, and counterions for enhancement of mass range, resolution, and sensitivity.

There are a number of purification schemes that have been used to purify the DNA, especially for the Na^+ and K^+ ions. These methods include the use of ion exchange beads, addition of ammonium base such as diammonium citrate or tartrate, and complexation reagents [27,28,31,33]. Indeed, it appears that DNA molecules present in the $M[H]_n$ or $M[NH_4]_n$ forms often result in enhanced sensitivity in mass spectrometric analysis. In MALDI/MS, improved signal has been obtained by converting DNA-mers into their ammonium salts using ion-exchange beads. Other work by Chen and co-workers [27] has shown that diammonium citrate can improve the MALDI/MS of DNA by ammoniating the DNA and also provides enhanced ionization efficiency even with some of the common MALDI matrices other than 3-HPA or PA. Other purification procedures have also been used that are quite effective but more complicated, such as the use of biotin tagged onto streptavidin-coated magnetic beads [33]. These procedures have been successfully used to purify small DNA-mers for analysis in genetic screening procedures.

In this chapter, we discuss various detailed experimental aspects of sample preparation and purification for MALDI/MS of large DNA-mers. In particular, we discuss the use of modified films as active substrates for the detection of DNA by MALDI/MS. The detection of mixtures of DNA-mers that range to over 600 b.p. is demonstrated. The application of these methods for detection of DNA is demonstrated for a number of important problems in genetic screening. Various problems associated with the presence of metastable decay and matrix background that limit the resolution and sensitivity are also discussed.

II. DESCRIPTION OF EXPERIMENTAL APPARATUS

The system used for MALDI/MS of DNA-mers is shown in Fig. 1 and consists of a linear time-of-flight (TOF) mass spectrometer modified for high-voltage operation constructed of parts purchased from R. M. Jordan Co. (Grass Valley, CA). The system consists of four parts, which include ion generation, ion acceleration, ion separation, and ion detection and data collection.

The laser source used to produce MALDI was a DCR-11 Nd:YAG laser system (Spectraphysics, Mt. View, CA) which produces radiation at 355 nm or 266 nm. In this work for MALDI/MS of DNA, generally 355-nm radiation was used with 3-HPA or 3-HPA/PA matrices. The laser beam impinged on the probe tip at a 45° angle to the probe surface with a single 12.5-in focal length quartz lens to a spot size of ~0.2 mm×0.5 mm. The power density at this spot was generally between ~5×10^6 and 1×10^7 W/cm^2 at 355 nm in these experiments,

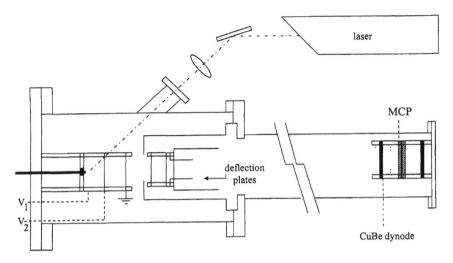

Figure 1   Schematic of MALDI/mass spectrometer.

although sometimes defocusing the beam was needed to obtain optimized signals. Desorption/ionization was performed from a stainless steel probe tip (12
mm^2) onto which the matrix/analyte was applied. The probe tip was inserted
into the ion source of the TOFMS through a vacuum lock and its surface was
positioned flush with the backing plate.

The ion source of the TOF mass spectrometer consists of a modified two-
stage acceleration source (see Fig. 1). There is an acceleration plate, which also
acts as the sample probe backing plate such that the front end of the sample
probe tip was flush with the first acceleration plate. The first acceleration plate
is followed by a second acceleration plate with a 90% fine mesh grid and then
by a third plate also with a grid that is held at ground potential. High voltages
are used on the plates to accelerate the heavy MALDI-produced ions. Either
+20 kV or −17 kV is placed on the first plate and ±10 kV is placed on the
second plate depending on whether the positive or negative mode is used. Generally the negative mode is used in the DNA work. The dimensions of the two
acceleration regions are both 1.27 cm in length. This dimension is much larger
than that normally used in MALDI/MS TOF instruments where an ion source
spacing of ∼0.5 cm is commonly used. This spacing, though, appears to be a
key issue in detecting molecular ions in the MALDI/MS of large DNA-mers.
As the matrix plume expands in the MALDI process, the large ions collide with
the plume and may undergo extensive fragmentation at this point. Extensive
fragmentation of large DNA-mers is often observed with the 0.5-cm plate spac-

ings where much higher field strength is experienced by the ions. The larger plate spacings allow the plume to expand further, thus minimizing large-ion/matrix collisions in the expansion and thus fragmentation.

The ions are accelerated from the ion source into a field-free flight tube and are separated according to their m/z ratios. The flight tube in our instrument is 1 m. In addition, there is a split liner inside the drift tube with one half at ground and a high-voltage pulsed potential applied to the other half. This pulsed voltage can be used to eliminate the ion matrix background before it strikes the detector and saturates the detector response.

The detector used in this work was a triple microchannel plate detector (MCP) with a Cu-Be conversion dynode with postacceleration capability up to ±15 kV. The triple MCP detector (R. M. Jordan Co., model C701/ZG) provides a gain of nearly 10^8 times. The top plate can be floated as high as -5 kV with a maximum of 1 kV bias voltage across each plate. However, it was observed that above 30 kD the efficiency for detection declined rapidly for these large ions. The problem becomes even more severe for large negative ions where there is a negative bias on the front plate that will repel the ions. A Cu-Be dynode postacceleration stage with a suppression grid was placed in front of the MCP as shown in Fig. 2. There is a grounded grid in front of the conversion dynode and a second grounded grid between the dynode and the detector. The first grid is grounded to the liner. The spacing between the first

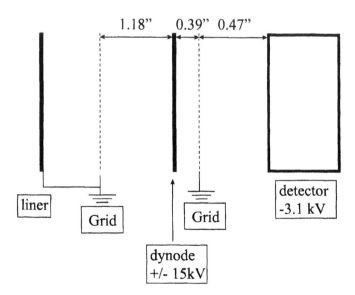

Figure 2 Postacceleration detection scheme.

grid and the dynode is 3 cm, while the spacing between the dynode and second grid is 1 cm. In the case of positive ions, the dynode is floated at -15 kV so that there is a total of 32 kV across the instrument. The ions impinge upon the dynode, where it has been shown that Cu^+ ions and other small ions are sputtered from the surface and then accelerated into the MCP detector [35]. Alternatively, in the negative mode the large negative DNA-mers are accelerated into the dynode at $+15$ kV and produce sputtered species that are then detected by the MCP. The large DNA-mers could only be detected by the use of the postacceleration stage with the MCP detector. Although the use of the triple MCP detector using the postacceleration stage enhanced the efficiency for detection of heavy species significantly, there was some loss in resolution as a result. It should also be noted that although a total of >30 kV could be placed on the front end of the TOF, it was found that bias voltages above 20 kV were very much prone to arcing, especially when the ion source became contaminated. It was thus found to be much more convenient to split the bias voltage over the acceleration and detector regions.

Data were recorded using a LeCroy 9350 M digital oscilloscope and subsequently transferred to an IBM 586 compatible PC for processing. Each MALDI mass spectrum was generally averaged over 50–100 single-shot spectra. The averaged data from the oscilloscope were processed using Microsoft Origin 4.0.

III. USE AND PREPARATION OF POLYMER MEMBRANES FOR MALDI/MS OF LARGE DNA-MERS

In early work, it was observed that large DNA-mers could only be detected with difficulty and often nonreproducibly in MALDI/MS. The key to the detection of large DNA-mers in MALDI/MS appears to be the use of active polymer substrates. Several different polymer substrates have been explored in our work, where the optimal polymers appeared to be nitrocellulose (NC) and Nafion [16,17,36,38]. The use of these polymer films appears to be particularly effective in the MALDI/MS of impure samples and has been used to enhance the signal in the MALDI/MS analysis of proteins in bacterial cell analyses [37,38], in dairy milk [38], in chicken egg white samples [38], and in protein digest analysis [38], as well as in the DNA detection problem. Indeed, in many of the samples mentioned earlier, only marginal signals could be observed without the use of the polymer membrane assist. In this section, we describe the preparation of the polymer surfaces, the MALDI matrix, and the sample for analysis. We also describe some of the possible mechanisms responsible for the effect of these membranes.

A. Preparation of Nafion Surface and Matrix/Sample

The exact procedure for preparation of the Nafion film and the application of the matrix/sample is critical to the success of the MALDI/MS of large DNA-mers [31]. The Nafion was obtained as a Nafion solution (5% in a mixture of lower aliphatic alcohols and water) from Aldrich Chemical Co. The substrate was formed by air drying ~1.5 μl of the Nafion solution on the stainless steel probe tip to form a film ~12 mm^2 in size. Further modifications involved applying 3 μl of concentrated ammonium hydroxide solution over the dried Nafion film. Due to the hydrophobic nature of the prepared Nafion film, it was necessary to make sure that the NH$_4$OH solution applied to the surface covered a sufficient area of the Nafion film. This was done by spreading the NH$_4$OH solution using a polypropylene pipette tip. After allowing the NH$_4$OH solution to spread on the surface, the solution was allowed to stand briefly and the excesss liquid was removed with a paper tissue. The treated Nafion film with residual ammonia was allowed to dry.

The actual sample preparation must be performed such that the sample is deposited first as a film and then the matrix applied as a separate layer above it. In this procedure aliquots of 2.0–2.5 μl of the DNA samples were applied to the completely dried Nafion or modified Nafion film. This film should be hydrophobic and resistant to the type of organic solvents or compounds in-volved in the experiment. Again due to the hydrophobicity of the film prepared, a pipette tip was used to spread or disturb the small sample drop to a size of ~3–3.5 mm in diameter. The sample drop was then allowed to air dry; 3 μl of the matrix was then applied to the dried sample spot followed by agitation with a pipette tip to spread the sample. The matrix was allowed to dry before MALDI-MS was performed. The agitation of the matrix solution in the above procedure was intended to allow sufficient contact of the impurities with the substrate surface and the matrix with the analytes, since consistent and repro-ducible results were obtained from such sample preparation.

B. Preparation of Nitrocellulose Substrate

Two procedures were used for the preparation of the NC substrate in these experiments [16]. One involves the application of NC solution in appropriate solvents onto the stainless steel probe tip followed by air-drying. The nitrocellu-lose solution was obtained by dissolving a piece of Immobilon-NC pure mem-brane in organic solvents, such as acetone, methanol, or acetonitrile, at concen-trations of 10–32 mg/ml. Usually a concentration of ~16 mg/ml was used for these experiments. Approximately 5 μl of this solution was applied to a stain-less steel probe tip using a glass pipette. After the initial application, the excess liquid (~4 μl of the solution) was taken away by the same pipette and the remaining part was allowed to air-dry. A thin and homogeneous layer of NC film covering an area of ~12 mm^2 was obtained. Then 2 μl of the DNA sample was applied to the hydrophobic surface followed by air-drying. A 4-μl aliquot of the matrix was then applied to the sample spot and air-dried.

A second method tested for preparation of the NC substrate involved the direct use of an untreated NC blotting membrane. In the case of the untreated NC membrane substrate, 3 μl of DNA sample was applied to the center of the membrane (~2×2 mm^2). After drying, 6 μl of the matrix was deposited onto the same side of the membrane as where the sample was applied. The dried membrane was then attached to the stainless steel probe tip using double-sided tape. In either case the sample is deposited on the substrate as described previously and the matrix is deposited above the sample.

It should be noted that, as in the case of the Nafion film, the success of using NC as a substrate for MALDI/MS analysis of double-stranded DNA is highly dependent on the preparation of the NC film. Among the important factors are the concentration of the original NC solution and the solvent used. A wide range of concentrations of the NC solution was tested in this work. NC solutions of ~10–32 mg/ml generate a colorless, thin, and homogeneous film. This range of concentration provides the most consistent results for molecular ion yields of DNA using the substrate preparation method described herein. A concentration of ~16 mg/ml was generally used for the solution in this work, which is similar to that in PDMS where usually a 25–50 μl NC solution with a concentration of 2 mg/ml is electrosprayed onto a target [39]. NC solutions with a concentration that is too high or too low usually provide less satisfactory results. When the concentration of the solution was too high, the application of NC solution by the glass pipette was difficult and usually resulted in a wrinkled, thick film from which quality spectra could not be obtained. If the NC concentration was too low, the capacity of the NC film substrate appeared to be insufficient and no effect was observed. Acetone, methanol, and acetonitrile were good solvents for dissolving the NC membrane. However, acetone, which has the lowest boiling point of the solvents, was ultimately used as the solvent for the NC solution because it has the fastest evaporation rate and produces a homogeneous film.

The choice of the solvent system for the 3-HPA matrix is another factor that needs to be considered because of its potential to damage the nitrocellulose film substrate upon matrix application. The conventional solvent for the matrix is a mixture of acetonitrile/water (v/v, 1:1). However, this solvent composition can partially degrade the NC film. By examining the dissolvation of the NC blotting membrane by matrix solutions in different solvent systems ranging from 10 to 50% of acetonitrile in water and comparing the quality of the spectra obtained, ~36% acetonitrile in water was selected as the solvent for the matrix in this work. A further decrease in the proportion of acetonitrile will eventually limit the solubility of the matrix materials.

IV. EFFECTS OF NITROCELLULOSE AND NAFION SUBSTRATES ON MALDI/MS OF DNA

An example of the effect of the nitrocellulose substrate on Na^+ and K^+ adducts in the MALDI/MS of DNA is shown in Fig. 3, which is a comparison of the negative-ion mass spectra of a 5-mer single-stranded oligonucleotide obtained on both an NC film and a stainless steel probe tip using 3-HPA as the matrix. It can be seen that adduct ions due to sodium and potassium are greatly reduced with the use of the NC film substrate. It appears that NC effectively binds the inorganic cation contaminants to its "negative" site and limits/eliminates their interference during the MALDI process for oligonucleotides even though the deposited sample was not washed. The procedure used in this work for applying the DNA sample prior to the matrix may further enhance this effect, since the analyte sample is in direct contact with the active NC surface and the cation impurities can access the "negative" sites. DNA molecules are highly hydrophilic and negatively charged, so that their interactions with the NC surface are minimal. It should be noted that in more recent work [26,27,30] it has been shown that the addition of diammonium citrate and other such ammonium salts can further reduce the cation impurity distribution and enhance the resolution of small DNA-mers.

V. DNA SAMPLES FROM POLYMERASE CHAIN REACTION

The samples for analysis by MALDI MS were prepared by amplifying a selected region of the genome. This procedure, called polymerase chain reaction (PCR), is a technique for the detection of polymorphisms in DNA. It allows the amplification of selected regions of DNA extracted from a variety of sources to a detectable level [40]. Several PCR products derived from the human genome ranging from 100 to 450 b.p. were analyzed by MALDI/MS using an NC film substrate, a Nafion film substrate, and a stainless steel probe tip. It should be noted that no signal could be obtained even under the best conditions directly from the PCR reaction without some purification before MALDI/MS. Apparently the polymerases and other reagents in the PCR process suppress the MALDI/MS signal. We have found that for larger PCR products, between 100 and 500 b.p., the amplified products were quickly and adequately purified (following the manufacturer's protocol) using a commercial PCR purification kit,

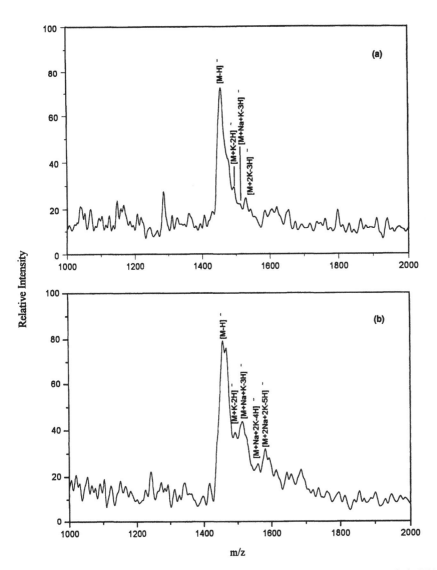

Figure 3 Negative-ion MALDI mass spectra of a 5-mer single-stranded DNA ($5' - TTTTT - 3'$) with 3-HPA as the matrix on (a) NC film substrate and (b) stainless steel probe. (Reprinted with permission from ref. 16. Copyright 1995 American Chemical Society.)

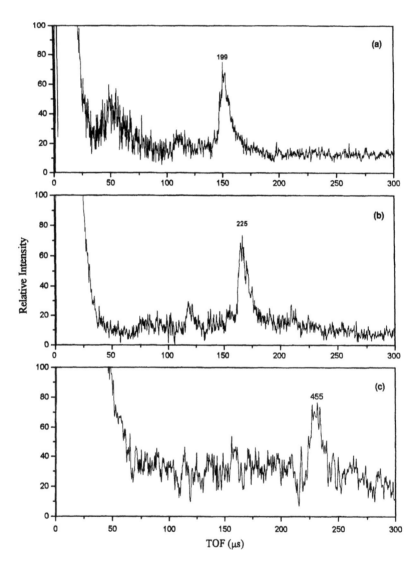

Figure 4 Negative-ion MALDI mass spectra of double-stranded DNA of human genome on NC film substrate: (a) 199 b.p., (b) 225 b.p., and (c) 455 b.p. Matrix, 3-HPA/PA with molar ratio of 4:1. (Reprinted with permission from ref. 16. Copyright 1995 American Chemical Society.)

the Qiagen spin column (Qiagen, Chatsworth, CA). For PCR products <100 b.p. a different Qiagen column was required in order not to lose the DNA sample. For significantly smaller PCR products, purification is difficult and other methods may be required.

Figure 4 shows the MALDI/MS spectra of a 199-, 225-, and 455-b.p. double-stranded DNA from the human genome obtained by using a NC film substrate. Only signals corresponding to single-stranded DNA were detected for all of the PCR products. Indeed, we have never observed signals from double-stranded DNA in any of our experiments. Very similar results were obtained with the use of a Nafion film. Generally, no discernible signal could be observed by MALDI from a stainless steel surface. The use of the polymer film substrates greatly enhance the ion yields, the reproducibility of the signal, and the signal persistence. In these MALDI/MS spectra the matrix was either a 60–70 mg/ml 3-HPA solution or a mixture of 3-HPA and PA with a molar ratio of 4:1 (60–70 mg/ml 3-HPA) dissolved in 10–50% acetonitrile and water mixture solvent.

VI. ANALYSIS OF MIXTURES OF DOUBLE-STRANDED DNA BY MALDI/MS

Using MALDI/MS for DNA sequencing and mapping requires the ability to analyze mixtures containing DNA fragments from several base pairs to several hundred base pairs [16,18,31]. It should be noted that the digest mixture did require some purification before MALDI/MS analysis and no signal could be observed without purification. The digest sample used in these figures was subjected to chloroform/isoamyl alcohol and phenol extraction. After extraction, the DNA was precipitated with the addition of 0.1 volume of 3 M sodium acetate and 2 volumes of ice-cold ethanol by incubating in dry ice overnight. The long incubation time was used to avoid the loss of smaller fragments. The DNA was run in a microcentrifuge at 17,000 rpm for 15 min, washed with 70% ethanol, and hydrophilized. The dried DNA was resuspended in 150 μl of deionized water for MALDI/MS analysis.

The MALDI/MS methodology for detecting mixtures of large bases has been demonstrated using our procedures as shown in Figs. 5 and 6. In these experiments MALDI/MS on a NC film has been used to analyze the double-stranded fragments from the DNA pBR322 after MspI digestion. The MspI digest of pBR 322 is a commercially available molecular marker (New England Biolabs, Beverly, MA). This digest sample is used in routine gel electrophoresis for analysis of amplified DNA (600 b.p. $\leq$ DNA $\leq$ 30 b.p.). The MspI digest of pBR 322 DNA yields 26 fragments [41,42]. Using MALDI/MS fragments ranging from 9 to 622 b.p. are detected. Due to the relatively low resolution, mix-

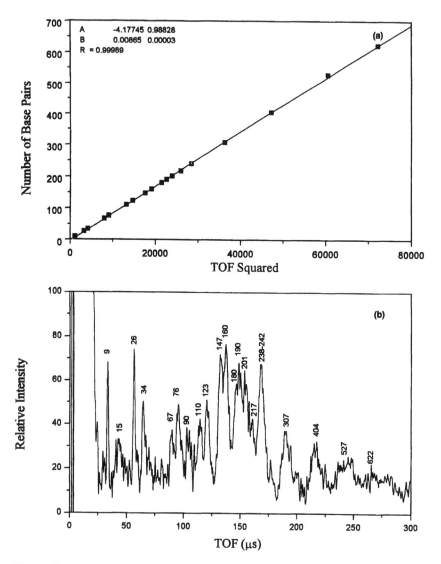

Figure 5 (a) Calibration curve of number of bases versus the TOF squared and (b) negative-ion MALDI mass spectrum of Msp I digest of DNA pBR322 on NC film substrate. Matrix, 3-HPA/PA with molar ratio of 4:1. (Reprinted with permission from ref. 16. Copyright 1995 American Chemical Society.)

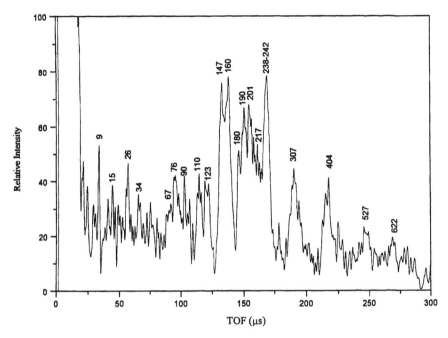

Figure 6 Negative-ion MALDI mass spectrum of Msp I digest of DNA pBR322 on NC film substrate with laser power increased by 10%. Matrix, 3-HPA/PA with molar ratio of 4:1. (Reprinted with permission from ref. 16. Copyright 1995 American Chemical Society.)

tures of different fragments with the same number of base pairs, such as 9, 26, 34, 147, and 160 b.p., cannot be resolved. However the intensities of these peaks are considerably higher than those of the other peaks with only one single fragment as expected. In Fig. 5, the DNA fragments of up to 404 b.p. are clearly detectable, whereas larger fragments may be lost due to the S/N. However, an enhancement of the signal intensity of higher mass fragments could be achieved by increasing the laser power by ~10% as shown in Fig. 6, where fragments at 527 and 622 b.p. are observed. The accompanying calibration of the spectrum as a function of the number of base pairs is shown in Fig. 5a. MALDI/MS of DNA mixtures with the NC film substrate was found to be highly reproducible and uniform in desorption of all of the fragments in the mixture. Similar results were obtained using the Nafion substrate, although improved results were obtained using the NC. In comparison, the use of a stainless steel substrate resulted in a much degraded spectrum, where DNA fragments of >100 b.p. could not be detected. This may be due to interference by salts introduced during the precipitation, such as sodium acetate and residue buffer ingredients in the sample, which can disturb the crystallization process.

VII. SCREENING OF DNA POLYMORPHISMS BY MALDI/MS USING BUCCAL CELL COLLECTION

A. Sample Preparation

A key element in using rapid MALDI/MS methods for screening of genetic diseases will be the ability to conveniently acquire small samples of materials from individuals for analysis. The collection of DNA for diagnostic analysis based on DNA obtained from blood samples, hair roots [43,44], urine, saline rinses, and so forth may be rather inconvenient. Most of these methods involve extensive sample handling, which is time-consuming and expensive for use in high-volume testing. We have chosen a method of PCR amplification using DNA prepared from buccal cells collected on cytology brushes as has been reported by Richards et al. [45]. This method is a fairly simple method for DNA sample preparation. The DNA obtained using this method has also proved to be relatively stable and can be stored for 2–3 weeks at 4°C.

In this study, DNA extracted from buccal cells collected on a cytology brush has been employed as the primary source for PCR amplification. Buccal epithelial cells were collected by twirling a sterile cytology brush (Scientific Products, Romulus, MI) on the inner cheek for less than 30 sec. DNA was extracted immediately by immersing the cytology brush in 400 μl of 50 mM NaOH in a microfuge tube. After vortexing, the microfuge tube still containing the brush was heated at 95°C for 5 min and the brush was removed. The liquid solution containing DNA was then neutralized with the addition of 40 μl of 1 M Tris and vortexed again before being stored as a stock solution at ~4°C. Five microliters of the stock solution was used in a 100-μl PCR reaction.

For PCR amplification of some of the variants described herein, we used primers as described next. Table 1 lists the primers used in this work for PCR amplification on CA II and CA III genes, and exon 6a and exon 12 genes of cystic fibrosis transmembrane conductance regulator (CFTR). For normal PCR amplification, a 100-μl reaction consists of 5 μl of the buccal cell DNA stock solution, 0.1 μl of each of the primers, 10 μl of 2 mM dNTP (Sigma Chemical Co., St. Louis, MO), 6μl of 25 mM of MgCl$_2$, 2.5 units of Taq DNA polymerase, and 10 μl of 10× PCR reaction buffer (Promega, Madison, WI). All of the reactions were performed for 40 cycles. No optimization of the cycling conditions was attempted throughout the experiments. Direct restriction enzyme digestion of PCR reaction products was used to simplify the digestion process. Restriction enzyme of 10–20 units [Acc I and Rsa 1 (Boehringer Mannheim GmbH, Germany), BstN I and Hinc II (New England Biolabs, Beverly, MA)]

Table 1 Primers for PCR Amplification

Genes	Primers (5' → 3')	PCR product (b.p.)
CA II	TGT GTC TGC TGC TCT CCT ACC (1)	212
	GGG TTC CTT GAG CAC AAT (2)	
CA III	GGT CCT GAC CAC TGG CAT GA (1)	
	TGC CTG ATG TCT TTA GTA TGC AGG[a] TCA A (2)	86
	CGG CAG GTC TTC CCA TTA TT[b] (3)	167
CFTR exon 6a	TTA GTG TGC TCA GAA CCA CG (1)	385
	CTA TGC ATA GAG CAG TCC TG (2)	
CFTR exon 12	GTG AAT CGA TGT GGT GAC CA (1)	426
	CTG GTT TAG CAT GAG GCG GT (2)	

[a] The base G in the primer 2 of CA III is the mismatched one.
[b] The primer is used for the heminesting process, and is also used to perform multiplex amplification in this work.

was added directly to the cycled PCR reaction and incubated at an appropriate temperature for 1–4 h. The detailed digestion conditions and enzymes used for each gene are demonstrated in Table 2.

The PCR products and their digests were purified using a commercially available purification kit (Qiagen, Chatsworth, CA, USA) following the manufacturer's procedures, except that the final elution of the DNA fragments was

Table 2 Restriction Endonuclease Digestion of PCR Products

PCR product	Enzyme	Fragments produced (b.p.)	Recognition sequence	Incubation conditions
CA II	*Bst*N I	86 normal	5'-CC/(A,T)GG-3'	1 hr/60°C
		60 mut		
CA III	*Hinc* II	152 normal	5'-GTPy/PuAC-3'	4 hr/37°C
		133 mut		
CFTR exon 6a	*Rsa* I	94, 291 normal	5'-GT/AC-3'	4 hr/37°C
		385 mut		
CFTR exon 12	*Acc* I	125, 301 normal	5'-GT/(A,C)(T,G)AC-3'	4 hr/37°C
		426 mut		

Note: 10–20 units of the enzyme was added directly to each 100 μl PCR reaction product.

performed using a 1 mM Tris buffer. The final eluent was dried in a SpeedVac, and the DNA contents were resuspended in 8 μl of deionized water; 2.0–2.5 μl of this solution obtained was used for MALDI/MS analysis.

B. Screening of DNA Polymorphisms by MALDI/MS

A biological method employing restriction fragment length polymorphisms (RFLP) for the detection of DNA polymorphism involves the use of a restriction enzyme to cut a specific sequence bearing the polymorphisms in DNA to yield fragments of defined lengths. The resulting restriction fragments are then separated by gel electrophoresis according to their molecular sizes. In this study, MALDI/MS was used in place of gel electrophoresis as a rapid means for separating and measuring the restriction fragment length. The reliability of MALDI/MS as a method to analyze DNA mixtures has been greatly improved by incorporating active substrates into the sample preparation step. In addition, the samples shown here were collected from individuals using the buccal cell swipe method.

Tables 3–5 summarize the polymorphic fragments generated from PCR products of CA II, CA III, and CFTR on exon 6a and exon 12. The related endonuclease sequences responsible for the digestion and the corresponding restriction enzyme are also shown. The polymorphic fragments after digestion with corresponding enzymes are fragments of 133 (allele A) and 152 (allele B) b.p. for the CA II gene; 60 (allele A) and 86 (allele B) b.p. for the CA III gene; 94, 291 (normal) and 385 b.p. (mutated) for the CFTR exon 6a gene; and 125, 301, (normal) and 426 b.p. (mutated) for the CFTR exon 12 gene.

Figures 7 and 8 show the MALDI/MS spectra of the restriction enzyme digests of the PCR products obtained by buccal-cell collection and processing for CA genes. Results from three representative individuals are shown here among a total of six individuals randomly sampled, two of which are homozygous and one heterozygous. Homozygosity is detected by the presence of a single polymorphic fragment after digestion, while heterozygosity is indicated by two fragments. The single fragment of 133 b.p. (individual A) or 152 b.p. (individual B) is produced for homozygous individuals in which both alleles are either of one type or the other for the CA II gene. For the heterozygotes, both fragments of 133 and 152 b.p. for the CA II gene (individual C) are generated, representing both alleles (Fig. 7). Similar explanations can be applied to the CA III gene, where individuals A and B are homozygous, and C is heterozygous (Fig. 8).

Figure 9 shows the MALDI/MS spectra of restriction enzyme digests of the PCR products from CFTR genes at exon 6a and 12. In this case, PCR products of mutated alleles are not cleaved by the respective restriction enzymes. Only samples from individuals A and B were analyzed, and both of

Table 3 Sequence of Part of the CA II Gene and the PCR Amplified Region for the Typing of the *Bst*N I Polymorphism

ATTTCAAGTG	AAGCTCGTAA	TTCTTTTATT	TGTTGCCAGT	GATATAGAAC	50
CCCGTTTTTT	AAAAAGTGTT	TTTGACCATC	AGAGGGGAGT	ATACCTATT<u>T</u>	100
<u>GTGTCTGCTG</u> <u>CTCTCCTACC</u> primer 1		TTCCTCCTAC	TCTGTCAATG	TGATTGTTTG	150
AAGCTGCGTA	TTTGCCTTGT	TCTAGGGCAA	GAGTGCTGAC	TTCACTAACT	200
TCGATCCTCG	TGGCCTCTTT	CCTGAATC░░ ░░░░ATTACTG	.GACCTAC░░░		250
░░CTCACTGA	CCACCCCTCC	TCTTCTGGAA	TGTGTGA░░░ ░░░ATTGTGCT primer 2		300
<u>CAAGGAACCC</u>	ATCAGCGTCA	GCAGCGAGCA	GGTTTGTTTT	GTAATGACAG	350
GTCTGTTTAC	GGGTGGAGCA	TTTAGTCAAG	GCAGAAGACC	TTGGCCTCCA	400
GAGTGAGAGA	GACCTGAGAT	TTAATCCTTC	TCC		433

Note: The sequences underlined are primer regions for PCR. Asterisk at position 230 indicates polymorphism site, where T is replaced by C in another allele, in which case the resulting sequence CCTGG (shaded box) can be cleaved by *Bst*N I. Sequence in shaded box can be cleaved by *Bst*N I.

them are homozygous, representing a population with both alleles normal (unmutated). This is indicated by the presence of fragments of 94 and 291 b.p. and the absence of the 385-b.p. fragments for the exon 6a gene, and the presence of fragments of 125 and 301 b.p. and the absence of the 426-b.p. fragment for the exon 12 gene. This observation was also verified by gel analysis. Although the CFTR disorder is found with a relatively high frequency, the gene mutations studied here are relatively rare. PCR amplification of these particular genes were chosen as compared to other mutated forms of the CFTR gene since, in this case, larger PCR fragments, which are suitable for the evaluation of the effectiveness of the MALDI/MS detection, were generated.

Another type of polymorphism is due to the variation of the copy numbers of a segment of DNA in the gene. In this work, the copy number of a hexamer repeat sequence (AGGATT) in intron 40 of the canine von Willebrand's factor gene was detected by screening the PCR products directly [16].

Table 4 Sequence of Part of the CA III Gene and the PCR Amplified Regions for the Typing of the *Hinc* II Polymorphism

GGGAAGAGAA	AGCAGGAGCC	GTCCAGCACG	GAGGAAGGCA	GACCATGGCC	50

AAGGAGTGGG	GCTACGCCAG	TCACAACGGT	CCTGACCACT	GGCATGAACT	100

primer 1

TTTCCCAAAT	GCCAAGGGGG	AAAACCAGTC	GCCC▓▓▓▓▓	CTGCATACTA	150

primer 2

AAGACATCAG	GCATGACCCT	TCTCTGCAGC	CATGGTCTGT	GTCTTATGAT	200

GGTGGCTCTG	CCAAGACCAT	CCTGAATAAT	GGGAAGACCT	GCCGAGTTGT	250

primer 3

ATTTGATGAT	ACTTATGATA	GGTCAATGCT	GAGAGGGGGT	CCTCTCCCTG	300

GACCCTACCG	ACTTCGCCAG	TTTCATCTTC	ACTGGGGCTC	TTCGGATGAT	350

CATGGCTCTG	360

Note: The sequences underlined are primer regions for PCR. Asterisk at position 135 indicate polymorphic site, where G can be replaced by A in another allele. Asterisk at position 140 indicate the mismatched base for primer 2, where it is replaced by C, so the resulting sequence GTTGAC (shaded box) can be cleaved by *Hinc* II.

Figure 10 shows the spectra of the PCR products obtained from the genes of a Siberian huskey and a mixed-breed dog. Figure 10a was obtained from the Siberian huskey, which is homozygous containing nine repeats of AGGAAT, and the PCR fragment is 147 b.p. in length. Figure 10b is from the mixed-breed dog, which is heterozygous containing both 9 and 14 repeats of AGGAAT, and the PCR fragments obtained are 147 and 177 b.p. in length.

The data presented herein from several model genes, along with the data in the previous work [16,17,38], demonstrate that the MALDI/MS methodology with the application of a nitrocellulose film substrate can be successfully used for the detection of DNA polymorphisms or mutations involving DNA fragments up to 400 bp or more. The MALDI/MS technique, while still limited in resolution, is much more rapid in screening PCR products and restriction enzyme digests than gel electrophoresis. Even though the detection of only RFLP is demonstrated in this work, the MALDI MS method presented herein can also be used to detect other types of polymorphisms based on the detection of DNA fragment length, such as variable number of tandem repeats (VNTR) polymor-

Table 5 Sequence of Exon 6a and Exon 12 Genes of CFTR and PCR Amplification Region for the Typing of *Rsa* I and *Acc* I Polymorphism

Exon 6a

TTAGTGTGCT <u>CAGAACCACG</u> AAGTGTTTGA TCATATAAGC TCCTTTTACT
<u>Primer 1</u>
TGCTTTCTTT CATATATGAT TGTTAGTTTC TAGGGGTGGA AGATACAATG

|→ exon 6a

ACACCTGTTT TTGCTGTGCT TTTATTTTCC AG GGA CTT GCA TTG GCA CAT
TTC GTG TGG ATC GCT CCT TTG CAA GTG GCA CTC CTC ATG GGG CTA
ATC TGG GAG TTG TTA CAG GCG TCT GCC TTC TGT GGA CTT GGT TTC
CTG ATA GTC CTT GCC CTT TTT CAG GCT GGG CTA

                     *

GGG AGA ATG ATG ATG **AAG TAC** AG GTAGCAACCT ATTTTCATAA
                    exon 6a ←|

CTTGAAAGTT TTAAAAATTA TGTTTTCAAA AAGCCCACTT TAGTAAAA<u>CC</u>
<u>AGGACTGCTC TATGCATAG</u>
      <u>Primer 2</u>

Exon 12

<u>GTGAATCGAT GTGGTGACCA</u> TATTGTAATG CATGTAGGAA CTGTTTAAGG
  <u>Primer 1</u>
CAAATCATCT ACACTAGATG ACCAGGAAAT AGAGAGGAAA TGTAATTTAA

|→ exon 12
                *

TTTCCATTTT CTTTTTAG A GCA **GTA TAC** AAA GAT GCT GAT TTG TAT
TTA TTA GAC TCT CCT TTT GGA TAC CTA GAT GTT TTA ACA GAA
AAA GAA ATA TTT GAA AG GTATGTTCTT TGAATACCTT ACTTATAATG
              exon 12 ←|

CTCATGCTAA AATAAAAGAA AGACAGACTG TCCCATCATA GATTGCATTT
TACCTCTTGA GAAATATGTT CACCATTGTT GGTATGGCAG AATGTAGCAT
GGTATTAACT CAAATCTGAT CTGCCCTACT GGGCCAGGAT TCAAGATTAC
TTCCATTAAA ACCTTTCTC <u>ACCGCCTCAT GCTAAACCAG</u>
                 <u>Primer 2</u>

Note: The sequences underlined are primer regions for PCR. Asterisk indicates polymorphic site; for exon 6a, it is where C can be replaced by T in the mutated population, and in exon 12, it is where G can be replaced by A in the mutated population. In exon 6a, the sequence <u>GTAC</u> in the bracket can be cleaved by *Rsa* I. In exon 12, the sequence <u>GTATAC</u> can be cleaved by *Acc* I.

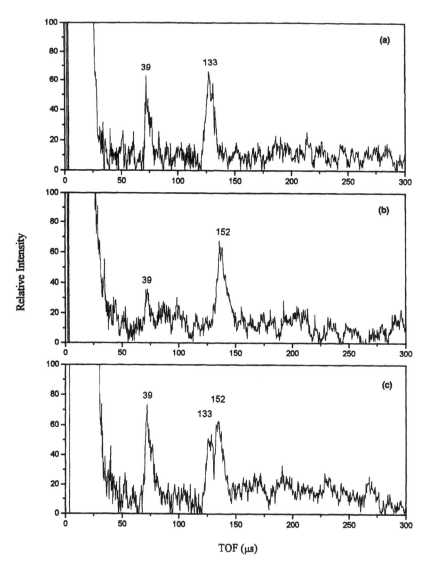

Figure 7 Negative-ion MALDI mass spectra of *Bst* NI digestion of CA II genes of (a) individual A (133 b.p.), (b) individual B (152 b.p.), and (c) individual C (133 b.p. and 152 b.p.). Nitrocellulose film substrate was used. (From ref. 17. Copyright 1995 John Wiley & Sons, Ltd.)

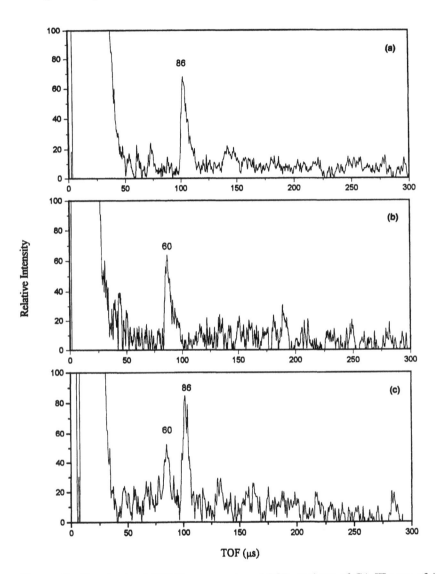

Figure 8 Negative-ion MALDI mass spectra of PCR products of CA III gene of (a) individual A (86 b.p.), (b) individual B (60 b.p.), (c) individual C (60 b.p. and 86 b.p.). Nitrocellulose film substrate was used. (From ref. 17. Copyright 1995 John Wiley & Sons, Ltd.)

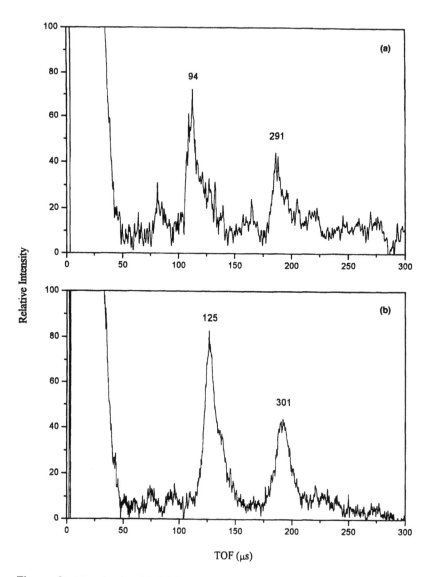

Figure 9 Negative-ion MALDI mass spectra of restriction enzyme digests of CFTR genes on exon 6a and exon 12: (a) Rsa I digest of CFTR gene on exon 6a (94 b.p. and 291 b.p.) and (b) Acc I digest of CFTR gene on exon 12 (125 b.p. and 301 b.p.). Nitrocellulose film substrate was used. (From ref. 17. Copyright 1995 John Wiley & Sons, Ltd.).

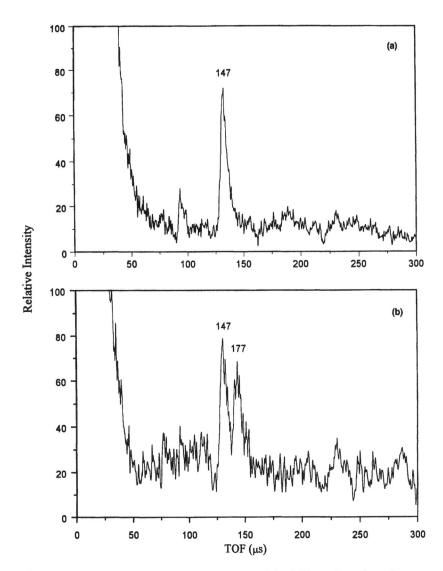

Figure 10 Negative-ion MALDI mass spectra of the PCR products from the genes of (a) Siberian huskey (homozygous) and (b) mixed-breed dog (heterozygous) with NC film substrate. Matrix, 3-HPA/PA with molar ratio of 4:1. (Reprinted with permission from ref. 16. Copyright 1995 American Chemical Society.)

phism, and the allele discrimination by primer length [41] (ADPL) method proposed recently.

VIII. FUTURE DIRECTIONS

The success of MALDI/MS as a method for screening of genetic polymorphisms and ultimately for sequencing will depend on further advances in several critical areas. An important issue continues to be the limited resolution and sensitivity of the method, especially for increasingly larger DNA-mers. A number of methods have been used to improve the resolution in MALDI/MS, most recently pulsed-delayed extraction (PDE) [26,46–48]. Several groups have used PDE to achieve a resolution of >1000 for a mixed base single-stranded 35-mer at the time of this writing [26,45,46]. In more recent work, our own group has demonstrated a resolution of >500 for a single-stranded 60-mer using extensive purification prior to PDE MALDI/MS [49]. The use of PDE also appears to enhance sensitivity by eliminating much of the low-mass matrix that saturates the detector and by narrowing the peak profile of the ions detected. Efforts are now underway to improve the mass range of this experiment and to simplify the high-voltage circuitry required of this work.

A second promising technology is the use of ion trapping methods for enhancement of resolution and sensitivity. It has recently been shown that small DNA-mers (<15-mers) can be vaporized by MALDI/MS and externally injected into an ion trap [50]. In this configuration, the DNA can be introduced into the trap without fragmentation and stored for an extended period of time (>50 msec) intact. The injection process is such that the fields can be adjusted to eliminate ionic matrix background and small neutral products. These species are not trapped and never reach the detector. Also, the MALDI process is performed at relatively low voltage and the ions are floated into the trap using an Einzel lens, which reduces the metastable decay process. In addition, using a high-repetition-rate laser, ions produced by many laser pulses can be stored in the trap, integrated, and the sensitivity enhanced by this method. Also, the use of a buffer gas provides cooling of the MALDI-produced ions in the trap so that resolution can be improved using a time-of-flight device as the detector [50]. Initial results appear promising and are presently being extended to higher mass. Such improvements in resolution will be especially critical in the case of screening DNA polymorphisms, where for a particular genetic disease (such as cystic fibrosis, CF [51]) several different polymorphisms exist. In order to accomplish rapid screening in such cases, multiplexing of the PCR procedure is required and simultaneous detection of several different DNA-mers corresponding to these polymorphisms is necessary. If 10–12 such polymorphisms must be screened together as in the case of CF, then improved mass resolution will be required.

The other critical area where advances will be necessary for transforming MALDI/MS of DNA into a practical method is in sample preparation. The development of several matrix materials such as 3-HPA, PA, and others have allowed the detection of intact DNA in MALDI/MS. However, new matrices and combinations of matrix materials will be required to further enhance the detection limit in terms of size and sensitivity for DNA. Improvements in sample purification will be particularly important in the MALDI/MS process. The presence of impurities in DNA samples may totally suppress the MALDI/MS signal. As discussed earlier in this chapter, various methods such as Qiagen spin columns, polymer membrane substrates, and affinity methods have been used for purification. However, even after initial purification, adducts may still be observed in the MALDI/MS spectra of DNA, which may limit the resolution and sensitivity, especially for increasingly larger DNA-mers. The future viability of MALDI/MS as a high-resolution and highly sensitive method for detection of DNA will depend upon the development of new rapid and relatively simple methods for purification of DNA prior to MS analysis.

ACKNOWLEDGMENTS

We gratefully acknowledge support of this work by the National Institutes of Health under the National Center for Human Genome Research under grants 1R21 HG0068501A2 and 1R01 HG0068503. We also thank Dr. A. Killeen of the University of Michigan, Medical Center, Clinical Diagnostic Laboratory, for helpful suggestions during this work.

REFERENCES

1. U Landegren, R Kaiser, CT Caskey, L Hood. DNA diagnostics— Molecular techniques and automation. Science 242:229–242, 1988.
2. JS Bell. Identifying defective and controlling genes. Gazette July:82–88, 1993.
3. JL Weber, PE May. Abundant class of human DNA polymorphisms which can be typed using the polymerase chain reaction. Am J Hum Genet 44:388–396, 1989.
4. K Kravitz, M Skolnick, C Cannings, D Carmelli, B Baty, B Amos, A Johnson, N Mendell, C Edwards, G Cartwright. Genetic linkage between hereditary hemochromatosis and HLA. Am J Hum Genet 31:601–619, 1979.
5. D Botstein, RL White, M Skolnick, RW Davis. Construction of a genetic linkage map in man using restriction fragment length polymorphisms. Am J Hum Genet 32:314–331, 1980.
6. JL Davies, Y Kawaguchi, ST Bennett, JB Coperman, HJ Cordell, LE Pritchard, PW Reed, SCL Gough, SC Jenkins, SM Palmer, KM Balfour, BR Rowe, M Farrall, AH Barnett, SC Bain, JA Todd. A genome-wide search for human type 1 diabetics susceptibility genetics. Nature 371:130–135, 1994.

7. Y Nakamura, M Leppert, P O'Connell, R Wolfe, T Holm, M Culver, C Martin, E Fujimoto, M Hoff, E Kumlin, R White. Variable number of tandem repeat (VNTR) markers for human gene mapping. Science 235:1616–1622, 1987.

8. SE Antonarakis. Diagnosis of genetic disorders at the DNA level. N Engl J Med 320:153–161, 1989.

9. JM Rommens, MC Iannuzzi, BS Kerem, ML Drumm, G Melmer, M Dean, R Rozmahel, JL Cole, D Kennedy, N Hidaka, M Zsiga, M Buchwald, JR Riordan, LC Tsui, FS Collins. Identification of the cystic fibrosis gene: chromosome walking and jumping. Science 245:1059–1065, 1989.

10. FS Collins, P O'Connell, BAT Ponder, BR Seizinger. Progress towards identifying the neurofibromatosis (NF1) gene. Trends Genet 5:217–221, 1989.

11. L Chadderdon. The canine genome project: making a mark in the fight against disease. Report of the Michigan State University College of Veterinary Medicine, Fall:17–22, 1991.

12. G Siest, T Pillot, A Regis-Bailly, B Leninger-Muller, J Steinmetz, M-M Galteau, S Visvikis. Apolipoprotein E: an important gene and protein to follow in laboratory medicine. Clin Chem 4(1/8):1068–1086, 1995.

13. TA Shaler, Y Tan, JN Wickham, KJ Wu, CH Becker. Analysis of enzymatic DNA sequencing reactions by matrix-assisted laser desorption/ionization time-of-flight mass spectrometry. Rapid Commun Mass Spectrom 9:942–947, 1995.

14. S Mouradian, DR Rank, LM Smith. Analyzing sequencing reactions from bacteriophage M13 by matrix-assisted laser desorption/ionization mass spectrometry. Rapid Commun Mass Spectrom 10:1475–1478, 1996.

15. LY Chang, K Tang, M Schell, C Ringelberg, KJ Matteson, SL Allman, CH Chen. Detection of ΔF580 mutation of the cystic fibrosis gene by matrix-assisted laser desorption/ionization mass spectrometry. Rapid Commun Mass Spectrom 9:772–774, 1995.

16. YH Liu, J Bai, X Liang, DM Lubman, PJ Venta. Use of a nitrocellulose film substrate in matrix-assisted laser desorption/ionization mass spectrometry for DNA mapping and screening. Anal Chem 67:3482–3490, 1995.

17. YH Liu, J Bai, Y Zhu, X Liang, D Siemieniak, PJ Venta, DM Lubman. Rapid screening of genetic polymorphisms using buccal cell DNA with detection by matrix-assisted laser desorption/ionization mass spectrometry. Rapid Commun Mass Spectrom 9:735–743, 1995.

18. GB Hurst, MJ Doktycz, AA Vass, MV Buchanan. Detection of bacterial DNA polymerase chain reaction products by matrix-assisted laser desorption/ionization mass spectrometry. Rapid Commun Mass Spectrom 10:377–382, 1996.

19. C Jurinke, D vande Boom, A Jacob, K Tang, R Worl, H Koster. Analysis of ligase chain reaction products via matrix-assisted laser desorption/ionization time-of-flight mass spectrometry. Anal Biochem 237:174–181, 1996.

20. DC Muddiman, DS Wunschel, C Liu, L Pasa-Tolic, KF Fox, A Fox, GA Anderson, RD Smith. Characterization of PCR products from bacilli using electrospray ionization FTICR mass spectrometry. Anal Chem 68:3705–3712, 1996.

21. NI Taranenko, CM Chung, YF Zhu, SL Allman, VV Golovler, MR Isola, SA Martin, LA Haff, CH Chen. Matrix-assisted laser desorption/ionization for sequencing single-stranded and double-stranded DNA. Rapid Commun Mass Spectrom 11:386–392, 1997.

22. S Mouradian, DR Rank, LM Smith. Analyzing sequencing reactions from bacterio-phage M13 by matrix-assisted laser desorption/ionization mass spectrometry. Rapid Commun Mass Spectrom 10:1475–1478, 1996.

23. KJ Wu, A Steding, CH Becker. Matrix-assisted laser desorption time-of-flight mass spectrometry of oligonucleotides using 3-hydroxypicolinic acid as an ultraviolet-sensitive matrix. Rapid Commun Mass Spectrom 7:142, 1993.

24. KJ Wu, TA Shaler, CH Becker. Time-of-flight mass spectrometry of underivatized single-stranded DNA oligomers by matrix-assisted laser desorption. Anal Chem 66:1637, 1994.

25. K Tang, NI Taranenko, SL Allman, CH Chen, LY Chang and KB Jacobson. Picoli-nic acid as a matrix for laser mass spectrometry of nucleic acids and proteins. Rapid Commun Mass Spectrom 8:673–677, 1994.

26. K Tang, CH Chen. Detection of 500 base single stranded DNA by MALDI mass spectrometry. Rapid Commun Mass Spectrom 8:727–730, 1994.

27. YF Zhu, NI Taranenko, SL Allman, SA Martin, L Huff, CH Chen. The effect of ammonium salt and matrix in the detection of DNA by matrix-assisted laser de-sorption/ionization time-of-flight mass spectrometry. Rapid Commun Mass Spec-trom 10:1591–1596, 1996.

28. NP Christian, SM Colby, L Giver, CT Houston, RJ Arnold, AD Ellington, JP Reilly. High resolution matrix-assisted laser desorption/ionization time-of-flight analysis of single-stranded DNA of 27 to 68 nucleotides in length. Rapid Commun Mass Spectrom 9:1061–1066, 1995.

29. E Nordhoff, R Cramer, M Karas, F Hillenkamp, F Kirpekar, K Kristiansen, P Roepstorff. Ion stability of nucleic acids in infrared matrix-assisted laser desorp-tion/ionization mass spectrometry. Nucleic Acids Res 21:3347–3357, 1993.

30. TA Shaler, JN Wickham, KA Sonnes, KJ Wu, CH Becker. Effect of impurities on the matrix-assisted laser desorption mass spectra of single-stranded oligodeoxy-nucleotides. Anal Chem 68:576, 1996.

31. J Bai, YH Liu, X Liang, Y Zhu, DM Lubman. Procedures for detection of DNA by matrix-assisted laser desorption/ionization mass spectrometry using a modified Nafion film substrate. Rapid Commun Mass Spectrom 9:1172–1176, 1995.

32. NP Christian, L Giver, AD Ellington, JP Reilly. Effects of matrix variations and the presence of iron on matrix-assisted laser desorption/ionization mass spectra of DNA. Rapid Commun Mass Spectrom 10:1980–1986, 1996.

33. K Tang, D Fu, S Kotter, RJ Cotter, CR Cantor, H Koster. Matrix-assisted laser desorption/ionization mass spectrometry of immobilized duplex DNA probes. Nu-cleic Acids Res 23:3126–3131, 1995.

34. C-W Chou, SE Bingham, P Williams. Affinity methods for purification of DNA sequencing reaction products for mass spectrometric analysis. Rapid Commun Mass Spectrom 10:1410–1414, 1996.

35. SR Weinberger, E Donlon, Y Kaphun, R Kornfeld, L Li, R Whittal, L Russon. Further resolution improvements in linear matrix-assisted laser desorption/ioniza-tion TOF-MS. Proceedings of the 44th ASMS Conference on Mass Spectrometry and Allied Topics, May 12–16, 1996, p 269.

36. J Bai, YH Liu, DM Lubman, D Siemieniak. Matrix-assisted laser desorption/ion-ization mass spectrometry of restriction enzyme-digested plasmid DNA using an active Nafion substrate. Rapid Commun Mass Spectrom 8:687–691, 1994.

37. TC Cain, W Weber, DM Lubman. Differentiation of bacteria using protein profiles from matrix-assisted laser desorption/ionization time-of-flight mass spectrometry. Rapid Commun Mass Spectrom 8:1026–1030, 1994.

38. J Bai, YH Liu, TC Cain, DM Lubman. Matrix-assisted laser desorption/ionization using an active perfluorosulfonated ionomer film substrate. Anal Chem 66:3423–3430, 1994.

39. GP Jonsson, AB Hedin, PL Hakansson, BUR Sundquist, BGS Sane, PF Nielson, P Roepstroff, K-E Johansson, I Ramensky, MSL Lindberg. Plasma desorption mass spectrometry of peptides and proteins adsorbed on nitrocellulose. Anal Chem 58:1084–1087, 1986.

40. KB Mullis, FA Faloona. Specific synthesis of DNA in vitro via a polymerase-catalyzed chain reaction. Methods Enzymol 155:335, 1987.

41. JG Sutcliffe. Complete nucleotide sequence of the *Escherichia coli* plasmid pBR 322. Cold Spring Harbor Symp Quant Biol 43:77–90, 1978.

42. KWC Peden. Revised sequence of the tetracycline-resistance gene of pBR 322. Gene 22:277–280, 1983.

43. R Higuchi, CH von Beroldinger, GF Sensabaugh, HA Erlich. DNA typing from single hairs. Nature 2:543–546, 1988.

44. PCR analysis of hair root specimens to detect Tay-Sachs' disease carriers in Ashkenazi Jews. (Tech Brief). Clin Chem 41:321–322, 1995.

45. B Richards, J Skoletsky, AP Shuber, R Balfour, RC Stern, HL Doskin, RB Parad, D Witt, KW Klinger. Multiplex PCR amplification from the CFTR gene using DNA prepared from buccal brushes/swabs. Hum Mol Genet 2:159, 1993.

46. Y Dai, RM Whittal, L Li, SR Weinberger. Accurate mass measurement of oligonucleotides using a time-lag focusing matrix-assisted laser desorption/ionization time-of-flight mass spectrometer. Rapid Commun Mass Spectrom 10:1792–1796, 1996.

47. P Juhasz, MT Roskey, IP Smirnov, LA Huff, ML Vestal, SA Martin. Applications of delayed extraction matrix-assisted laser desorption ionization time-of-flight mass spectrometry to oligonucleotide analysis. Anal Chem 68:941–946, 1996.

48. RS Brown, JJ Lennon. Mass resolution improvement by incorporation of pulsed ion extraction in a matrix-assisted laser desorption/ionization linear time-of-flight mass spectrometer. Anal Chem 67:1998–2003, 1995.

49. Y Zhu, L He, J Srinivasan, DM Lubman. Improved resolution in the detection of oligonucleotides up to 50 b.p. in matrix-assisted laser desorption/ionization time-of-flight mass spectrometry using pulsed-delay extraction with a simple high voltage transistor switch. Rapid Commun Mass Spectrom, 11:987–992, 1997.

50. L He, Y Zhu, Y Liu, DM Lubman. Detection of oligonucleotides by external injection into an ion trap storage/reflection time-of-flight device. Rapid Commun Mass Spectrom 11:1440–1448, 1997.

51. NL Taranenko, KJ Matteson, LN Chung, Y Zhu, LY Chano, SL Allman, L Haff, SA Martin, CH Chen. Laser desorption mass spectrometry for point mutation detection. Genet Anal Biomol Eng 1–10, 1996.

17

Network-Based Bioinformatics in Protein Mass Spectrometry

David Fenyö
Rockefeller University, New York, New York

Ronald C. Beavis*
The Skirball Institute of Biomedical Research, New York University Medical School, New York, New York

---

Current affiliation: Eli Lilly & Company, Indianapolis, Indiana.

I. INTRODUCTION

Proteins are directly responsible for almost all of the metabolic processes that occur within a cell. Specialized proteins are used for the generation of structural elements in a cell or an entire organism. The elaborate structure of DNA in a cell, now referred to as its genome, is meant to record the structure of an organism's proteins, both to serve as a template for the construction of copies of those proteins and to pass that crucial information on to the next generation. The proteins encoded by a genome can be organized into larger structures that are used as cellular machinery for performing specific tasks in a cell, such as transcribing DNA into more protein (which requires several discrete protein-based machines), transporting other molecules through membranes, or automatically repairing damage to cellular subsystems. When genetic material is passed from one generation to the next, it is also necessary to pass a complete working set of protein-based machinery for the reading of the DNA strands and performing all other necessary steps in cell metabolism.

Individual proteins are composed of discrete polypeptide chains, called subunits, that may be linked together either covalently or by the local structure of the surrounding solvent molecules. The linear sequence of individual amino acid residues in a particular polypeptide chain is referred to as the "primary structure" of the polypeptide. This primary structure is what is directly encoded by an organism's genome. Also encoded in the genome are the instructions to make protein-based machines for the express purpose of modifying particular residues in other proteins. These modifications occur after a polypeptide has been and are therefore called "posttranslational" modifications.

It should be noted that the word "protein" is normally used very loosely to describe a number of different physical or conceptual objects. This confusing state of affairs exists because the word protein was used originally to describe materials present in living organisms that performed particular tasks or catalyzed specific reactions, without reference to what these materials were chemically. Now that the fundamental structure of the molecules that carry out these tasks is known, the functionally derived names for proteins may be either inappropriate or slightly misleading, but they carry with them the weight of history. The realization that a particular protein may be a small part of a much larger machine makes the functional definition of a protein even more difficult. Regardless of shifts in philosophy about the true name and function of a protein, the component subunit polypeptide chains are real physical objects that are amenable to study and analysis.

This chapter describes the application of mass spectrometry to the analysis of the subunit polypeptide chains that are the building blocks of a protein (or protein-based machine). These polypeptide chains are the direct products of genome transcription and therefore their original structure can be predicted with a knowledge of an organism's genome. Until recently, mass spectrometry was seen as a possible method for the determination of the primary structure of a polypeptide without reference to genomic information, in a manner analogous to Edman sequence analysis. The task has proved to be difficult, because of the unpredictable gas-phase cleavage behavior of peptides. Fortunately, genomic sequencing methods have made this task unnecessary in a growing list of species. The complete nucleic acid sequences that in turn correspond to all of the polypeptide sequences produced by those organisms are now known with a fair degree of reliability.

The existence of a truly huge amount of nucleic acid sequence information has opened up the field of protein analysis by mass spectrometry in unforeseen directions. It has, however, introduced a new problem: How can mass spectroscopists gain access to this information in a format that is appropriate for their work? Sequence databases were originally distributed in book form, but the paper format soon became too difficult to use because there is no possibility of properly indexing and searching the sequences for particular patterns. The compact disk read-only memory (CD-ROM) format became the standard method in the early 1990s, but this computerized format has become cumbersome because of the rapid rate of sequence generation. To stay current, these CD-ROMs needed to be replaced on a monthly basis, which is too costly. In the latter half of the 1990s, the distribution of databases has been almost completely replaced by electron-based formats (communications network transport). Using the currently popular Internet protocol (discussed later), it is possible to easily send information from one computer network to another, allowing the exchange of gigabytes of information with the same ease as making a telephone call.

Philosophically, all of the new applications of mass spectrometry can be viewed in terms of a simple hypothesis–confirmation model of experimental design [1]. If a hypothesis can be made about the structure of a polypeptide, an experiment can be designed in such a way that a mass spectrometric measurement will either confirm or disprove that hypothesis. A straightforward case (see Fig. 1) would be a polypeptide sample in which the amino acid sequence is thought to be known, such as a peptide produced by recombinant DNA methods. The known sequence can then be treated as a hypothesis: If this hypothetical structure is true, then cleaving that polypeptide with a sequence-specific protease should result in a collection of peptides with a certain set of allowed molecular masses. Performing the experimental cleavage and measuring the masses of the resulting peptides either will confirm the hypothetical structure, or will suggest that some parts of that hypothetical structure are incorrect. Hypotheses about those incorrect portions of the structure can then be formulated and tested by further experimentation. This type of experimental design leads to either the determination of the posttranslational modifications that have been made to the polypeptide or to the discovery of errors in the nucleic acid sequence used as the basis of the original hypothesis.

The use of the combination of computers, analytical chemistry, and databases has been christened "bioinformatics." The following sections discuss the fundamental aspects of network-based resources that can be used to help analyze protein mass spectra, using bioinformatics methods, rather than manual calculations and analysis. The structure and use of the protein sequence databases currently available are reviewed, along with the current methods of accessing these data. These databases are the core of current protein chemistry network-based applications, so understanding the differences between databases is very important. These databases are currently available in the form of network-connected resources. Some discussion of the state-of-the-art of network information retrieval technology will also be presented. Details of hardware and software construction are fleeting; however, the general framework of server–client relationships and the principals of network-based computing are sufficiently important that they are discussed in some detail.

II. PROTEIN SEQUENCE DATABASES

Databases have become the preferred method for storing both polypeptide amino acid sequences and the nucleic acid sequences that code for these polypeptides. The databases come in a variety of different types that have advantages and disadvantages when viewed as the hypothesis for a polypeptide identification experiment. The properties of the most common databases currently in use are listed in Table 1.

While the "database entry" for an amino acid sequence may appear to be a simple text file to a user browsing for a particular polypeptide, many databases are being organized into very flexible, complicated structures [2]. The detailed implementation of the database on a particular system may be based on a collection of simple text files (a "flat-file" database), a collection of tables (a "relational" database), or it may be organized around concepts that stem from the idea of a protein, gene, or organism (an "object-oriented" database). The organization of the database is of more concern to programmers than it is to the user, and it does not directly affect the usefulness of the database for protein analysis and identification.

Any protein sequence database contains a collection of amino acid sequences represented by a string of single-letter codes for the residues in a polypeptide, starting at the N-terminus of the sequence. The letter codes may be either upper case or lower case, depending on the database interface. These codes may contain nonstandard characters to indicate ambiguity at a particular site (such as "B" indicating that the residue may be "D" or "N"). All of the sequences have a unique number–letter combination associated with them that

Table 1　Sequence Databases Currently Used for Protein Bioinformatics

Database	Peptide sequences	Nucleic acid sequences	Peptide annotation	Redundancy	Sequence reliability	Entries (/1000)
SWISSPROT [3]	Yes	No	Complete	Low	Excellent	59
PIR [4]	Yes	No	Complete	Moderate	Excellent	95
EMBL [5]	No	Yes	Some	High	Good	1438
TREMBL [5]	Yes	No	Some	High	Good	137
GENBANK [6]	No	Yes	Some	High	Good	1766
GENPEPT [6]	Yes	No	Some	High	Good	262
OWL [7]	Yes	No	Some	Low	Good	210
dbEST [8]	No	Yes	None	Very high	Low	1317

Note: The sequence reliability for GENBANK and EMBL refers to the normal nucleic acid sequences stored in those databases. These databases also include expressed sequence tag data, which are not as reliable.

is used internally by the database to identify the sequence, usually referred to as the accession number for the sequence (more technically referred to as a "key attribute" for the database entry).

A.　Redundancy

A sequence database may contain multiple copies of a particular sequence (redundant entries) or it may have been constructed so that there are no multiple copies—that is, it is nonredundant. The presence of redundant sequences in a database may have several origins. The cause may be a historical accident, such as the entry of the same sequence under more than one "protein" name because of confusion about the function of the protein, or it may be deliberate—for example, multiple entries for a polypeptide sequences that has been spliced together after translation. The presence of many copies of the same sequence in a database has the effect of making the database larger, resulting in slower searches. It also may result in multiple hits in a protein identification experiment, while all of the hits are actually the same molecule. Database managers are currently engaged in the process of removing as many redundant entries as they can find from their databases. Some databases exist only to be nonredundant collections of sequence entries from other redundant databases.

B.　Annotation

Databases may contain a combination of amino acid sequences, comments, literature references, and notes on known posttranslational modifications to the sequence. A database that contains all of these elements is referred to as "annotated." Other databases only contain the sequence, an accession number, and a descriptive title. Annotation of each entry is obviously very time-consuming and difficult to maintain without errors; therefore annotated databases usually have many fewer sequence entries than nonannotated ones. Annotation also implies that some functional or structural information is known about the mature protein, as opposed to a sequence that is known only from the translation of a stretch of nucleic acid sequence. Even the best annotated databases now include large numbers of entries that have very little real information about the mature protein other than some reference to who sequenced and translated the nucleic acid sequence.

Annotated databases are technically superior for performing protein identification searches, because they contain information about the true form of the mature protein, making search misses caused by posttranslational modifications less likely. They also can screen out spurious hits that could occur if parts of the sequence that are not present in the mature protein are included in the search. The nucleic acid sequences of many polypeptides contain stretches of sequence that are removed either immediately following translation of the pre-protein (signal peptides) or when an inactive proprotein is activated by the removal of the corresponding propeptide. Including these pre- and propeptides in the search may lead to errors.

Nonannotated databases have the tremendous advantage of being simpler to maintain. This simplicity means that new sequences are incorporated more quickly and the effort necessary to verify all new entries is not required. Several very large amino acid sequence databases (GENPEPT and TREMBL) are simply translations of corresponding large nucleic acid databases (GENBANK and EMBL). The number of entries in these translated databases makes them very attractive for protein identification searches because the chance of finding positive hits is much greater than in the much smaller annotated databases. It is necessary to be much more careful about interpreting the results from searching this type of database because no information about the mature polypeptide is available. Even with this caveat, the large number of polypeptide sequences available in nonannotated databases and the current ascendancy of molecular biology have resulted in their use as the hypothetical data set for comparison with experiment in protein identification searches.

C. Sequence Reliability

It is usually assumed that if something is published, it has been proofread and it is letter perfect. Protein sequence databases are generally not proofread and they are far from perfect. The problems stem from several sources. A major problem is that the data is difficult to proof, once it is manually entered into the database. The sequences are a string of random-looking capital letters and it is difficult to pick out typing errors in this type of text. Database creators have different schemes in place to screen out typing errors, but they remain a consistent problem in all databases and they tend to migrate from database to database. If the detailed sequence of a protein is necessary for comparison with the primary structure of a sample, it is always necessary to check the original source of the data (frequently a journal article) to verify the sequence. Relying too heavily on the accuracy of protein sequence database has led to misunderstandings and confusion between groups examining the same protein.

The automated calling and transmission of sequences into a database directly from nucleic acid sequencing machines has practically eliminated human error in the data entry in some types of sequences, such as expressed sequence tags. This type of entry introduces a new class of error into a database, however. In order to achieve high-throughput sequencing, it is not possible to investigate portions of a particular sequence that cannot be called reliably. The current generation of sequence calling software will place an N at any position where the sequence cannot be called with confidence. Nucleic acid databases are accumulating sequence containing N's faster than they are accumulating sequences that are completely characterized. Automated sequencing software is also written with the idea that it will make errors in calling the sequence at some rate. This error rate will be unevenly distributed in particular sequences. Sequences called from good, strong signals will be very nearly perfect, while sequences called from poor quality data will have a very high error rate. Both sequences will appear the same in the database; no measure of signal strength or residue assignment confidence exists in current databases. It is therefore wise to treat all sequences that are not derived from journal publications as raw data, that is, with a certain amount of skepticism.

Another source of "reliability" problems arises even in cases where peer-reviewed journal articles describing a sequence exist. Most complete sequences of mature proteins are the results of several publications that may disagree in detail. Many "conflicts" exist between different references on the same protein. Annotated databases usually include some indication of these conflicts between sequences, but the sequence actually entered in the database represents one or the other of the sequences published. Database searching and sequence matching software does not take these conflicts into account: It considers the sequence as entered in the database as true. This situation is frequently the case for proteins where both protein sequencing and nucleic acid sequencing has been performed. Generally, the nucleic acid sequence seems to be more reliable, but nucleic acid sequencing gives no indication of posttranslational modifications or sequence conflicts that arise because of real variation in protein sequence within a population. The proportion of database entries with sequence conflicts is decreasing, because most sequencing is now only done once by nucleic acid sequencing from homogeneous cell populations. This style of sequence determination eliminates the possibility of the discovery and reporting of conflicts.

D. Protein Identification

Frequently, the sequence of a particular polypeptide is not known, even though the complete genomic sequence of the organism is available. This problem exists because there is no simple, direct method of connecting an experimentally observed polypeptide with the genomic sequence; that is, a band on a sodium dodecyl sulfate polyacrylamide gel electrophoresis (SDS-PAGE) gel cannot be connected to a gene without considerable effort. It is this particular problem that has attracted the attention of protein chemists and their mass spectrometers. A number of mass spectrometer-based experiments have been formulated to try to connect a stained band on a gel with genomic information. All of these experiments are the result of the realization that the sum total of the genomic information from an organism can be treated as one big hypothesis. If one is confronted with an unknown polypeptide, it is reasonable to make the hypothesis that this polypeptide sequence is encoded by the known nucleic acid sequence, along with thousands of other polypeptides. If it is possible to generate a set of mass spectrometric data points that can be tested against this hypothetical nucleic acid sequence by some algorithm, then there is the possibility of linking the real polypeptide with the portion of genome that produced it.

The process of linking molecular mass data with a translated genomic sequence has come to be called mass spectrometric "protein identification"—a rather inaccurate but evocative name [9–13]. Various clever strategies have been formulated for producing mass spectrometric information that is sufficiently unique to identify a particular stretch of genomic sequence. While there is no a priori reason to believe that such strategies can in fact be devised, the sequences of real polypeptide subunits greatly aid the process. The sequences of real subunits are extremely varied, even in molecules with similar three-dimensional shapes. The fact that amino acid residues with very similar physical properties have different molecular masses means that sequences that perform similar but discrete tasks in a cell will produce mass spectrometric data that can distinguish between the two molecular species.

The general pattern of the experiments used to identify proteins is similar to that illustrated in Fig. 1, but with the addition of bioinformatic database searching. Some combination of specific and/or nonspecific peptide bond cleavage experiments is performed and the experiments are mass analyzed. The determined masses are then compared against the collection of hypothetical polypeptide sequences in a genome. A match between the experimental results and the hypothetical sequences is then made on the basis of some reasonable algorithm that scores the probability that a particular sequence could give rise to

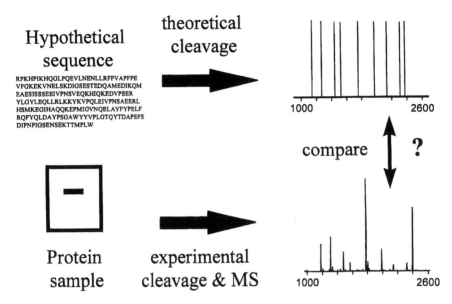

Figure 1   Typical hypothesis-driven, protein mass spectrometry experiment. Hypothetical sequences are screened against an observed protein cleavage pattern. A good match between the two patterns constitutes "identification"—that is, the hypothetical sequence is assumed to be the measured protein's primary structure.

the experimental results. The original polypeptide is then "identified" with the hypothetical sequence that produces the best score.

Any information about a polypeptide's partial sequence can be used in combination with the measured masses of peptides generated by proteolysis to constrain the search for a polypeptide. Several strategies using multidimensional mass spectrometry (MS/MS) have been demonstrated. Figure 2 illustrates this type of bioinformatics experiment. A single peptide from a protein digest is subjected to MS/MS measurement and the observed pattern of fragment ions is compared to the patterns of fragment ions predicted from database sequences. This type of comparison can be very constraining, resulting in high-confidence identification of the peptide with a known sequence. The approach is very attractive, requiring only one or two peptides to identify a protein sequence [14–20].

Figure 3a shows a MALDI ion trap mass spectrometry (ITMS) spectrum of a tryptic digest of an unknown protein from *Saccharomyces cerevisiae* that was observed as a spot on a gel. The spectrum has more than a dozen major peaks. If the corresponding masses are used to search all *S. cerevisiae* sequences in OWL with ProFound, a list of the proteins that are most likely to

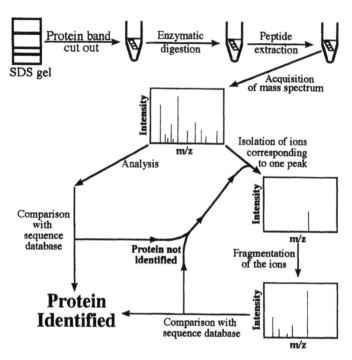

Figure 2 Flow chart showing the steps in a complete protein identification MS/MS experiment.

give the observed tryptic map is obtained (Table 2). In this example, subunit P130 of eukaryotic initiation factor 4F (IF42 __YEAST) is the most probable protein. To further increase the confidence in the identification, the ions with $m/z = 2596$ were isolated and fragmented (Fig. 3b). The spectrum contains three major fragment ions. The peak at $m/z = 2468$ is loss of the C-terminal lysine and contains little information. The two other fragment peaks, on the other hand, correspond to fragmentation at the C-terminal side of acidic amino acids. If a database is searched for proteins that have tryptic peptides with mass 2595 Da that fragment at the C-terminal side of acidic amino acids to give rise to b or y ions with mass 1984 and 2337 Da there is only one yeast protein (IF42 __YEAST) in the public databases that agree with this information. The tryptic peptide is AQPISDIYEFAYPENVERPDIK and the two fragment ions correspond to fragmentation on the C-terminal side of the aspartic acids at residues 6 and 20, respectively. If a theoretical trypsin digest of IF42 __YEAST is compared with the peptide map, all major peaks can be assigned to tryptic peptides from IF42 __YEAST or to peptides from autolysis of trypsin.

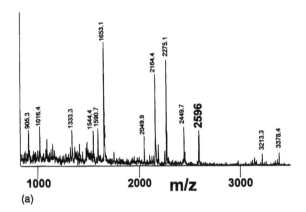

(a)

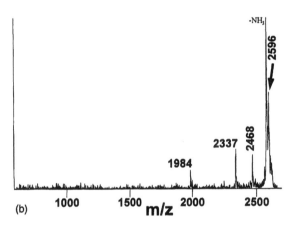

(b)

Figure 3 (a) MALDI-ITMS spectrum of a tryptic digest of an unknown protein from *S. cerevisiae*. (b) MALDI-ITMS/MS spectrum if peptides at $m/z = 2596$ (unpublished results, Jun Qin, NIH).

The observed molecular mass or isoelectric point of a polypeptide can also be used to constrain a mass spectrum pattern-matching search, although care must be taken not to apply this type of constraint too firmly. When nonannotated nucleotide sequence databases are used (such as TREMBL or GEN-PEPT), subsequent processing can greatly alter the pI or molecular mass of a protein, so much so that no identification can be made. For example, the small, highly conserved protein ubiquitin (SWISSPROT accession number P02248) has a molecular mass of 8.6 kD, which is the mass that would be measured by

Table 2   Result from a ProFound Search of All *S. Cerevisiae* Proteins in OWL with Masses from Fig. 3a

1. IF42_YEAST (probability = 7.9e-01) EUKARYOTIC INITIATION FACTOR 4F SUBUNIT P130
2. NPR1_YEAST (probability = 2.8e-02) NITROGEN PERMEASE REACTIVATOR PROTEIN
3. S63138 (probability = 1.9e-02) PROBABLE PROTEIN KINASE NPR1
4. RL3E_YEAST (probability = 1.8e-02) 60S RIBOSOMAL PROTEIN L30E
5. SCYOR206W (probability = 1.5e-02) HYPOTHETICAL PROTEIN SCYOR206W
6. CBF5_YEAST (probability = 1.4e-02) CENTROMERE/MICROTUBULE BINDING PROTEIN CBF5
7. S52893 (probability = 9.5e-03) HYPOTHETICAL PROTEIN YMR044W

Note: m/z = 2164 and 2275 were not used in the search, because they are from autolysis of trypsin.

a mass spectrometer or a gel. A simple keyword search of the translated-nucleotide database GENPEPT results in several sequences for the same protein [accession numbers M26880 (77 kD), U49869 (25.8 kD) and X63237 (17.9 kD)]. None of these nucleotide-translated sequences give the correct molecular mass or pI, so using those parameters to limit a search would result in missing the database sequence altogether. Only annotated databases that fully outline known modifications can be used when the properties of the mature protein are being used to constrain a search.

All of the protein identification strategies just outlined are currently available as CGI programs (discussed later) that can be accessed using a browser. Table 3 is a list of the programs available, as well as some of the relevant characteristics of the programs. The list of available programs will change with time, as will their features, but the table should remain a reasonable starting place for several years.

E.   Software Strategies for Protein Identification

Because there is a range of CGI software available for protein identification, this application can serve as an example of the types of choices that must be made to create a bioinformatics application for mass spectrometry. Some protein identification software takes the original protein sequence database as input and calculates all the proteolytic masses every time an identification is performed. This has the advantage that the software can be easily modified to take into account different kinds of additional information (see previous section). Performing the mass calculations "on the fly" works well when the search is performed using mass spectrometric fragmentation information. However, for

Table 3 CGI Programs Currently Used for Protein Identification

CGI program	Data type	Databases	Cleavage chemistry	Additional search parameters	Help	AI	Report format
PepFrag [21]	MS/MS	10 (aa+na)	7 Enzymes + CNBr	Protein mass, peptide masses, taxonomy, missed cleavage sites, daughter ion types, mass accuracy	Pages + examples	No	Active + tools
ProFound [22]	Masses	1 (aa)	5 Enzymes + CNBr	Protein mass, taxonomy, missed cleavage sites, mass accuracy	Pages + examples	Yes	Active + tools
MS-Fit [23]	Masses	5 (aa+na)	9 Enzymes + CNBr	Protein mass, taxonomy, missed cleavage sites, mass accuracy	1 Example	No	Active
MS-Tag [24]	MS/MS	5 (aa+na)	9 Enzymes + CNBr	Protein mass, taxonomy, missed cleavage sites, daughter ion types, mass accuracy	1 Page	No	Active
PeptideSearch [25]	Masses	1 (na)	7 Enzymes + CNBr	Protein mass, mass accuracy	None	No	Active
PeptideSearch [26]	MS/MS	1 (na)	7 Enzymes + CNBr	Protein mass, daughter ion types, mass accuracy	None	No	Active
Mass Search [27]	Masses	2 (aa+na)	11 Enzymes + 7 chemical	Protein mass	1 example	No	Dead
Mowse [28]	Masses	1 (aa)	6 Enzymes + CNBr	Protein mass, amino acid composition, mass accuracy	1 Page	Yes	Active

Note: Abbreviations: aa, amino acid sequence database; na, nucleic acid sequence database; CNBr. cyanogen bromide; and AI, any type of artificial intelligence/expert system used to rate results. An "active" report format has links that you can follow to get the sequence of a matched protein, and "tools" refer to additional programs to help make use of that sequence information. A "dead" report just lists the matching proteins' names and the user is expected to find the sequence information and pass judgment on the results without any assistance.

peptide mapping speed becomes an issue. It is important that the search is fast (<1 min) so that after obtaining some experimental data a search can immediately be performed and evaluated, so that a decision can be made whether the information is enough or if more experiments are necessary. One way of increasing the speed for peptide mapping is to first make a secondary database that contains the masses of all possible proteolytic peptides that can arise from enzymatic digestion of all the proteins in the original database. This secondary database is then used for the search. Unfortunately, the secondary proteolytic peptide mass databases usually take much more space than the original protein sequence database.

The simplest way to present the results of a database search with mass spectrometric information is to list all proteins that match the constraints given. This type of search is possible when the experimental data does not contain much noise (e.g., when searching with mass spectrometric fragmentation information). Since it is not always possible to separate proteins well, peptide maps often contain more than one protein; therefore, peptide mapping algorithms usually rank the proteins in the sequence database according the how well they fit the list of masses provided in the search. This ranking can in the simplest case be according to the number of masses that match proteolyic peptides in the protein. More sophisticated algorithms take into account the size of the protein, the spread of the proteolytic peptide mass errors, and the relative position of the proteolytic peptides in the protein.

For mass spectrometric protein identification the ideal situation occurs when the DNA sequence of the complete genome of the organism of interest is available and all protein coding regions have been predicted correctly. In this case, the search can be performed on a database of the translated protein sequences. However, even with organisms having their genomes completed it can happen that no known proteins match the experimentally obtained mass spectrometric constraints. In this instance, all the six reading frames of the complete DNA sequence must be translated without regard to whether a region of DNA is considered to code for proteins or not. If a nucleic acid sequence database is searched in this way, it is possible to find proteins that have not been predicted from the DNA sequence or when a frame shift has occurred in a predicted protein. To be able to find a protein in this way more information is necessary than when the protein database is searched because a lot of noise will be introduced by the six-reading-frame translation of both coding and noncoding regions.

If the protein of interest is posttranslationally modified, the identification problem becomes more complicated. Most protein identification algorithms can handle modifications that are present at all occurrences of an amino acids (e.g., alkylation of cysteines where the chemistry can be tuned so every cysteine residue is modified). Other modifications, like phosphorylation, that give a known mass shift can be handled but increase the noise level, because many more peptide masses have to be considered. Modifications like glycosylation, on the other hand, cannot be handled by the search algorithm because there are many possibilities for the mass shifts. Usually posttranslational modifications do not interfere with the identification because in most cases only a few of the proteolytic peptides bear modifications; however, they do become a significant issue when the detailed structure of a molecule is to be determined.

III. NETWORK-BASED ACCESS TO INFORMATION

A. Computer Networks

Computers have been connected together with communication lines for many years. Originally, these lines allowed the transfer of data from one computer to another, or from a terminal to a particular computer. The network consisted of a continuous wire connection between the computers and terminals, taking the name "network" from the electrical engineering word that refers of a set of interconnected wires.

Modern usage of the term "computer network" refers to a much more complicated arrangement of computers, wires, and/or fiber-optic cables and software [29]. The wires themselves are no longer "the network": They are now the method for "connecting to the network." In a typical, modern analytical laboratory, the same set of wire cabling will carry AppleTalk, Microsoft, and Internet protocol network messages. Telephone lines can be used for the same purpose. The word "network" no longer has a precise meaning out of context; it is used in conjunction with other words to fully describe what type of technology is being used to communicate between computers. For the purposes of this discussion, "network" designates the combination of cables, software, and computers necessary to send a message from one place to another.

B. The Internet Protocol

At the same time that protein sequence information proliferated, a revolutionary development in computerized network information access made it possible for everyone to have nearly instant access to that information. This development was the creation of a simple, standardized protocol for individual computer networks to communicate with each other. With the widespread use of this protocol, any computer that can be attached to a telephone line can connect to a local network that communicates with any other publicly accessible network. The concurrent development of powerful, inexpensive computers and mass data storage devices has lead to a rapid rate of change in information storage and access technologies. To someone working with these rapidly changing technologies, William Gibson's description of using a future global information access system as the projection of disembodied consciousness into a consensual hallucination seems to have come true.

The protocol for internetwork communication that has given rise to this consensual hallucination is called the Internet Protocol (IP) [30]. While the popular media frequently discusses the "Internet" as though it is a physical object that can be examined and judged, it does not exist at any level other than the imagination. The Internet Protocol was designed to allow any computer network to exchange information with any other computer network, regardless of the architecture of the networks communicating with each other. Therefore, using IP, network software companies are free to maintain the proprietary nature of their local-area network (LAN) or wide-area network (WAN) systems without affecting IP communication. IP deals strictly with networks; the characteristics of individual computer operating systems are not included in the protocol specification. Each computer must have a layer of translation software that mediates its communication with IP-compliant message. This layer is called the transport control protocol (TCP): it intercepts all IP messages and reformats them in such a way that a computer's operating system can interpret the message. The combination of TCP and IP software on a computer is called a "TCP/IP stack." The standard nature of the inputs and outputs from this stack makes it much simpler for programmers to write software that deals with network communication, rather than having to write interfaces for a large number of proprietary network specifications.

The Internet Protocol's success is based on three things. First, it is free. It was developed for the U.S. government and the use of the protocol does not require the payment of royalties. Second, it is straightforward to implement on any computer or network. As any user of UNIX will attest, the combination of free availability and straightforward implementation leads to software popularity, regardless of the obvious shortcomings of the system [31]. Finally, IP addresses the problem of transporting information between two networks over unreliable physical connections (such as telephone lines) that may be slower or faster than the networks that are trying to communicate. IP does this by breaking up any data into a set of small, formatted blocks that are referred to as "packets." These packets contain the information about where the information is going and how they are to be reassembled into the whole message when they are received. These packets are sent out from the source computer, through a local-area network to a special computer that is called a "router." The router computer examines each packet that arrives and reads its destination address. The routing computer will be able to send the packet to the correct computer if it is within its own LAN; otherwise, it will pass the packet on to another router. The packet continues to pass from router to router until one is found that has the destination computer attached to its LAN. The destination computer's TCP/IP stack then reassembles the message from the packets it receives and passes that message up to the computer's operating system. The process of passing packets through routers that have lookup tables that allow them to guess where to send the message next means that individual routers do not need to keep an exhaustive list of all of the computers in the world on IP-compliant networks, which greatly simplifies the maintenance of the system. Packets are free to take any path available between two networks that is currently available; routers are free to make decisions about which path will be fastest or the most reliable on a packet by packet basis. Therefore, even if an intermediate pathway is interrupted or busy during the transmission of a message, subsequent packets are free to find other ways to their destination without loosing data. Because individual data packets can be switched to different routes based on load, it is possible to squeeze much more information down a physical component (such as a fiber optic) than can be done using a protocol that requires a continuous connection, such as is made with a conventional telephone call or FAX message.

C. Client–Server Interactions

Internet Protocol data packets can be used by a variety of software that communicates with the TCP/IP stack on a computer. By convention, part of the address for a computer specifies a "port" number, which determines what piece of software receives the information from the TCP/IP stack [10]. The file transfer protocol (FTP), hypertext transfer protocol (HTTP) and the simple mail transfer protocol (SMTP) refer to popular methods for sending information to and from the TCP/IP stack. When information is being dispensed using these protocols, even though many router computers are necessary for the message to be passed along, only the sending and the receiving computer are important. These computers are referred as either being a "client" or a "server." A client software on the "client" computer requests information or action from the server software on the "server" computer, which responds appropriately to the request. Figures 4 and 5 illustrate typical configurations for hypertext and e-mail transmission using the appropriate combination of client–server protocols and IP networking.

It must be stressed that the client and server designations refer to software running on a particular machine rather than to any special characteristics of a particular computer or operating system. A single computer may be acting as a client and a server simultaneously, depending on what software it is running. Because requests for service can occur at any time, server software usually runs continuously in the background on a computer. Operating systems that can run programs simultaneously (referred to as "multithreaded" systems) are appropriate for running server software because they can respond immediately to requests. Operating systems that allow multithreading, such as UNIX or Microsoft's Windows NT, are therefore the most common choice to run server software. Single-threaded operating systems, such as the Macintosh OS, can be used to run server software but the performance of the server can be seriously compromised by the operating system.

The most flexible of these client–server software combinations is the hypertext transfer protocol (HTTP). The protocol evolved around the idea of

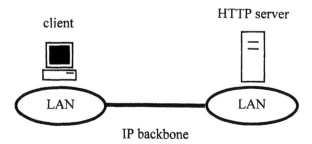

Figure 4 Typical network arrangement for the World Wide Web client–server relationship. The client computer is attached to a local-area network (LAN) of some configuration. The client's browser software makes a request using the HTTP format, which is routed out of the LAN, through the IP backbone (which can be local or international in scope). The HTTP request is finally received by the server, which is attached to its own LAN. The server then responds to the client, sending its message back to the client over the same IP backbone, even though the physical routing of the message may be very different.

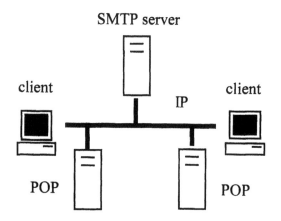

Figure 5 More complicated client–server configuration, commonly used to transmit e-mail. To send e-mail, a client machine will send a request to a simple mail transfer protocol (SMTP) server. The request contains the e-mail message and the address of the recipient's post office protocol (POP) server. The STMP server then sends the message to the POP server. When the recipient "checks" its e-mail, the client machine makes a request of the messages currently stored by the POP server, which responds to the request by sending the messages to the client.

being able to transfer elaborately formatted text pages to client software and to use highlighted portions of that text on the browser screen to request other pages of text from a specified server. A page with this property is referred to as "hypertext." The IP address of the new page and the location of that page on the server computer's mass storage device are easily specified using a simple text formatting language (the hypertext markup language, HTML), so anyone can create pages and place them as files on a computer running server software. HTTP has become ubiquitous because the protocol is royalty-free, it is straightforward to implement, and the client software (called a browser) is easy to use and either free or very inexpensive.

D. CGI, Databases, and Protein Analysis

In addition to being able to transfer pages of hypertext, HTTP has been extended to allow the client to request the server to run another piece of software that will accept some parameters from the client, perform some operation using those parameters, and respond to the client using HTML-formatted hypertext as output. The mechanism that the server uses for communicating with this additional software is called the common gateway interface (CGI), frequently referred to as "cgi-bin" because of a quirk in the conventional way for naming UNIX directories. The original HTTP specification has been extended to allow the inclusion of pictures along with text, and it also allows the server to respond to a request in such a way that the client will start a piece of software (called a "plug-in" or "helper" application) to receive and interpret the information that the server pushes back to the client. The clever combination of server-side CGI software and client-side browsers and plug-ins allows for very sophisticated interaction between the two computers in ways that were extremely difficult in the past. Rather than the clumsy process of manually "logging on" to a remote computer, running a remote program, downloading the results of that program, and reformatting it in such a way that it could be displayed on a local computer, HTTP hides the entire interaction behind a secure, easy-to-use interface that produces attractive displays.

The use of CGI programs to manipulate databases held on a remote server is potentially the most powerful application of network-based computing to protein analysis. Maintaining accurate protein sequence databases and developing sequence analysis tools is a time-consuming specialized task that is beyond the average computer user. Therefore, databases are held on servers that can be queried by users from anywhere that they can obtain a network connection and any computer capable of running a browser, including the newest generation of television receivers. Some programs exist that still attempt to perform database searches locally; however, this type of computing is a temporary transition step between isolated computers and network-aware software. Even the mention this type of database computing will hopefully seem quaint by the time this book is published.

E. Helpers and Plug-Ins

The CGI programs that have dominated the discussion so far revolve around programs that run on a remote server, but that receive input and "push" output back to a browser running on the user's local computer. Programs also exist that run on the local computer too, but are linked to the browser in such a way that when certain types of data is pushed at the browser, the browser responds by starting these other programs to deal with it. Applications of this type are called "helpers" if they run as separate, stand-alone programs, or they are called "plug-ins" if they require the browser to run. Many of these applications exist for multimedia data types, such as compressed audio and video formats. Many common programs can be used as helper applications, because the mechanism used by current browsers to start a helper is to download the data to be read into a file and then start up the helper with a command line parameter that points it to the appropriate downloaded temporary file. Therefore, any program that creates a specific type of file can be used as a helper application to display and manage downloaded files of compatible types. For example, a word processing application can be used as a helper to view documents created with any compatible word processor.

Unfortunately, mass spectrometry has no commercially available plug-ins or helper applications made specifically for that purpose. The necessary multipurpose Internet mail extension (MIME) codes necessary to start these programs in response to server data pushes have not been agreed upon. Each mass spectrometer manufacturer uses its own data format, and the manufacturers do not seem to have grasped the importance of people viewing data at remote sites. These data formats do not generally economize on data storage, resulting in very large files that are impractical to send over current networks without unacceptible delays. The programs that are supplied for the analysis of mass spectra in relation to known sequences are also based around the idea of an isolated computer running only the manufacturer's software.

One group (R. J. Lancashire, C. Muir, and H. Reichgelthas) has developed a helper application around the JCAMP-DX data format [32], which is available for Win16 and Win32 operating systems [33]. The program was designed for mass spectra obtained from small molecules, but it will certainly deal with protein mass spectra. Unfortunately, mass spectrometer manufacturers have specifically chosen not to support the JCAMP-DX format, so converting spectra into this format may not be easy. Instead of JCAMP-DX, manufacturers have opted for the ANSI standard OFS file format. This file format does not support any type of data compression, resulting in extremely large data files for many types of protein and peptide mass spectra. These files are not suitable for network exchange because of their size, but at least the data format is platform independent (i.e., the byte order of the data is the same regardless of the operating system that was used to create the file).

The authors of this chapter have also developed helper applications for dealing with protein mass spectra and sequences, the Protein Analysis Worksheet (PAWS) [34] and "m/z" [35], and they are available for Win16, Win32, and Macintosh operating systems. "m/z" is a mass spectrum viewing program that natively uses a highly compressed data format that was designed for data warehousing of large numbers of complex protein mass spectra [1]. It can accept files in the native formats used by most time-of-flight mass spectrometer manufacturers. PAWS is a protein sequence analysis tool that uses fuzzy-logic-based rules to interpret the results of protein mass spectroscopy experiments. PAWS can read data from any of the commonly used protein sequence database entry formats.

IV. RESOURCES OF PROTEIN MS INTERPRETATION

A. General Considerations

Software produced by scientific laboratories is almost always written by graduate students or postdoctoral fellows with little or no formal training in software engineering principles. The programs involved are written to solve particular problems associated with the research in a laboratory and are normally written to be used by the author. The code that results is usually very specific for a particular operating system (OS), because the authors use the details of the operating system to optimize the performance of the software. Distributing this type of software to other laboratories is very difficult: Even slight differences between the setup of superficially identical computers will cause unexpected behavior from the program. This problem is exaggerated because most student software is written for the UNIX OS because the operating system itself has the character of a student project. Different versions (or flavors) of UNIX may appear to be the same, but the small real difference may be so difficult to track down that it is a waste of time for the average users. Most users are also very intolerant of unexpected behavior (such as the program flashing unintelligibly worded error messages), and most student authors are not interested in supplying detailed support for their software (ascribing all errors to problems between the chair and the keyboard rather than their software). Distributing and maintaining a piece of software that is easily used and installed requires the dedication of a many people, ranging from small groups to huge corporate entities.

The CGI coupled with cheap, standardized World Wide Web browsers has alleviated almost all of the difficulties associated with amateur software.

Rather than sending code out to a user to install on the user's own computer (which must be of the same type as the program was developed upon), the program is written so that it can accept input using CGI and produce output that is compatible with HTML. The program can run on the computer it was developed and debugged on, even though a user's input and output was performed on a remote computer of any type or configuration. Therefore, in order for the authors to use the program themselves, they will eliminate all of the bugs that other users will see, rather than relying on the authors' altruism to remove bugs that they may not be able to properly reproduce on their own systems. HTML is a fairly simple page formatting language, making it easy for authors writing programs to produce elegantly formatted output pages quickly. The combination of the complicated nature of the tools for producing visually pleasing output in any OS and the almost magical ability to use the program from any place on earth (at least theoretically) makes CGI a very attractive interface.

The prevalence of "amateur" software is a great strength of network-based computing for scientists. The software tends to address problems associated with the cutting edge of scientific research and it need not address a problem that has any commercial motivation. If the piece of software is truly useful, then the student will obtain gratifying positive reactions from people all over the world and be encouraged to improve the program and to remove as many of the bugs that existed in the initial release as possible.

The prevalence of "amateur" software is also a tremendous weakness of network-based computing for scientists. The programmers rarely have any input from the nonprogrammers that actually use the software, and they have no particular reason to listen to any customer feedback that they receive. Consequently, the user interfaces that allow access to CGI programs tend to be an afterthought rather than the result of considered design. Consequently, the interface tends to be ideal for debugging a program (which is of great concern to the developer) rather than use of a program by an inexperienced user [11]. Documentation for the program is usually some combination of incomprehensible and incomplete, mainly because there is no conceivable justification for having a student or postdoctoral fellow spending time writing program documentation when the person could be working on the software.

The following passage is a good example of the curious type of documentation that currently exists for CGI programs:

EXPECT
    The statistical significance threshold for reporting matches against database sequences; the default value is 10, such that 10 matches are expected to be found merely by chance, according to the stochastic model of Karlin and Altschul (1990). If the statistical significance ascribed to a match is greater than the EXPECT threshold, the match will not be re-

ported. Lower EXPECT thresholds are more stringent, leading to fewer chance matches being reported. Fractional values are acceptable. (See parameter E in the BLAST Manual).

This clearly incomprehensible passage was taken from the online documentation for BLAST (Basic Local Alignment Search Tool) and it explains the most important numerical parameter for performing a search with the program. An inappropriate selection for the EXPECT parameter will result in a search that will not find sequence similarities that exist, depending on the value of other parameters and the length of the sequence fragment used to perform the search. It is not our intent to single out this piece of text out as an example of bad program documentation; it is a typical example. BLAST is probably the most used piece of biopolymer sequence searching software, and it is used every day by hundreds of molecular biologists who depend on the accuracy of the results obtained from its searches. With this said, almost no one who uses the program really knows what values to use for BLAST's parameters to optimize their particular search. Typically, they use the default parameters and hope for the best. Note: We have been unable to find the reference to "parameter E" in any of the documents associated with BLAST, either through HTTP- or FTP-accessible documents.

An underappreciated feature of professional software is "version-control"; that is, different versions of the software are released at long time intervals and the particular version of the software is easily determined by the user. When the output of a program is reported in a publication, it is possible to write down the version number of the software used, so that it can be examined in the future in light of any bugs that are subsequently discovered in that program. CGI programs almost never carry any version information with them, and they are changed often without warning by the developer. Therefore, there is no definitive way to determine whether a known bug has affected your calculation or not. Even when a CGI program does report a version number, it is difficult to determine with certainty whether that version number is correct or whether changing the version number was forgotten during the last build process.

B. Link Lists

Hypertext pages that are lists of other hypertext pages on a particular subject are the simplest type of network-accessible resources. These pages allow users to rapidly obtain information about a particular subject, relying on the judgment and diligence of a particular list's author. Many lists exist, although the most comprehensive one currently available is maintained by Kermit Murray at Emory University [36]. This list is so extensive that it has its own "search engine"—a CGI program that searches through all of the links available to find the ones that most closely match some search phrase.

V. THE FUTURE—DISTRIBUTED COMPUTING

The model for a desktop computer operating system that was current in 1995 was an easy-to-use, easy-to-configure graphical user interface that allowed users to run single copies of locally held applications. That is, the only data storage/data retrieval operations that an application would participate in would be using its own mass storage devices (magnetic disks, CD-ROM, etc.) or through a network that made remote disks behave as though they were local disks. All computer applications were written to take advantage of that model.

The 1997 model for a desktop computer operating system is completely different. It is now thought that the operating system on any desktop computer should expect to interact with an IP network in a nearly transparent manner. The physical location of disks (i.e., their location on a particular LAN configuration) will no longer be of much consequence. In fact, applications will interact with servers, rather than at the hardware level of dealing with actual devices. Using common, standard methods for addressing these servers, it will be possible to break monolithic computer applications up into much smaller parts, referred to as "objects," which can be assembled into a wide variety of software built from a common set of components. This approach is currently used in software development, but all the applications are meant to run under one operating system. In the new scheme, all programs built of these objects should be compatible with all computers, resulting in a great time saving for the development of applications. These software objects will be available to your local computer in a variety of ways: Some will be on a local hard disk, while others will be downloaded to your computer and used as needed.

Several different software and hardware development groups are vying for the lead in this new style of operating system. An operating system independent computer language called Java is the most favored candidate for producing these objects, while a scheme called JavaBeans will allow Java objects to communicate with each other in a straightforward manner. Java objects (also called "applets") are currently available and they run inside of the current generation of World Wide Web browser software. JavaBeans has just been proposed and it is not yet fully implemented in widely available platforms.

It is anticipated, however, that the combination of IP-network-aware operating systems, simple client–server interactions, and applications built of reusable and easily available objects will revolutionize data-intensive undertakings, such as bioinformatics [37]. Traditional bioinformatics, with centralized databases and large expensive computers, will probably be replaced with a large number of specialized servers that are much less expensive and that can be tuned for specific tasks. Applications that can take advantage of this new diversity of options and opportunities will hopefully make it possible to finally get an overall view of the vexing problems in protein chemistry, such as predicting folding, selective proteolysis, and understanding the role of sequence divergence in molecular evolution.

REFERENCES

Note: All of the universal resource locators (URLs) listed here were tested and operational in March 1997.

1. D Fenyö, W Zhang, BT Chait, RC Beavis. Internet-based analytical chemistry resources: a model project. Anal Chem 68:721A–726A, 1996.
2. R Elmasri, SB Navathe. Fundamentals of Database Systems, 2nd ed. Menlo Park, CA: Addison-Wesley, 1994, pp 611–700.
3. URL: http://expasy.hcuge.ch/sprot/sp-docu.html.
4. URL: http://www.gdb.org/Dan/proteins/pirusersdoc.html.
5. URL: http://www.ebi.ac.uk/ebi_docs/embl_db/embl_db.html.
6. URL: http://www.ncbi.nlm.nih.gov/Web/Genbank/index.html.
7. URL: http://www.biochem.ucl.ac.uk/bsm/dbbrowser/OWL/OWL.html.
8. URL: http://www.ncbi.nlm.nih.gov/dbEST/index.html.
9. WJ Henzel, TM Billeci, JT Stultz, SC Wong, C Grimley, C Watanbe. Identifying proteins from two-dimensional gels by molecular mass searching of peptide fragments in protein sequence databases. Proc Natl Acad Sci USA 90:5011–5015, 1993.
10. M Mann, P Højrup, P Roepstorff. Use of mass spectrometric molecular weight information to identify proteins in sequence databases. Biol Mass Spectrom 22:388–392, 1993.
11. DDJ Pappin, P Højrup, AJ Bleasby. Rapid identification of proteins by peptide-mass fingerprinting. Current Biol 3:327–332, 1993.
12. JR Yates III, S Speichner, PR Griffin, T Hunkapiller. Peptide mass maps: a highly informative approach to protein identification. Anal Biochem 214:397–408, 1993.
13. P James, M Quadroni, E Carafoli, G Gonnet, Protein identification by mass profile fingerprinting. Biochem Biophys Res Commun 195:58–64, 1993.
14. HH Rasmussen, E Mortz, M Mann, P Roepstorff, JE Celis. Identification of transformation sensitive proteins recorded in human two-dimensional gel protein databases by mass spectrometric peptide mapping alone and in combination with microsequencing, Electrophoresis 15:406–416, 1994.
15. M Mann, M Wilm. Error-tolerant identification of peptides in sequence databases by peptide sequence tag. Anal Chem 66:4390–4399, 1994.
16. JR Yates III, JK Eng, AL McCormack. Mining genomes: correlating tandem mass spectra of modified and unmodified peptides to sequences in nucleotide databases. Anal Chem 67:3202–3210, 1995.
17. SD Patterson. Matrix-assisted laser-desorption/ionization mass spectrometric approaches to the identification of gel-separated proteins in the 5–50 pmol range. Electrophoresis 16:1104–1114, 1995.
18. SJ Cordell, MR Wilkins, A Cerpa-Poljak, AA Gooley, M Duncan, KL Williams, I Humprey-Smith. Cross-species identification of proteins separated by two-dimen-

sional gel electromphoresis using matrix-assisted laser desorption/ionization time-of-flight mass spectrometry. Electrophoresis 16:438–443, 1995.

19. KR Clauser, SC Hall, DM Smith, JW Webb, LE Andrews, HM Tran, LB Epstein, AL Burlingame. Rapid mass spectrometric peptide sequencing and mass matching for characterization of human melanoma proteins isolated by two-dimensional PAGE. Proc Natl Acad Sci USA 92:5072–5076, 1995.

20. A Shevchenko, ON Jensen, AV Podtelejnikov, F Sagliocco, O Vorm, P Mortensen, H Boucherie, M Mann. Linking genome and proteome by mass spectrometry—Large-scale identification of yeast proteins from two dimensional gels. Proc Natl Acad Sci USA 93:14440–14445, 1996.

21. URL: http://prowl.rockefeller.edu/PROWL/pepfrag.html.

22. URL: http://prowl.rockefeller.edu/cgi-bin/ProFound.

23. URL: http://rafael.ucsf.edu/MS-Fit.html.

24. URL: http://rafael.ucsf.edu/mstag.html.

25. URL: http://www.mann.embl-heidelberg.de/Services/PeptideSearch/FR_Peptide-SearchForm.html.

26. URL: http://www.mann.embl-heidelberg.de/Services/PeptideSearch/FR_Peptide-PatternForm.html.

27. URL: http://cbrg.inf.ethz.ch/subsection3_1_3.html.

28. URL: http://gserv1.dl.ac.uk/SEQNET/mowse.html.

29. W Hioki. Telecommunications, 2nd ed. Englewood Cliffs, NJ: Prentice-Hall, 1995, pp 330–374.

30. C Hunt. TCP/IP Network Administration. Sebastopol, CA: O'Reilly and Associates, Inc., 1992, pp 6–49.

31. S Garfinkel, D Weise, S Strassmann. The UNIX-Haters Handbook. Indianapolis, IN: IDG Books Worldwide, 1994, pp 43–61.

32. URL: http://members.aol.com/rmcdjcamp/faq.htm.

33. URL: http://wwwchem.uwimona.edu.jm:1104/software/jcampdx.html.

34. URL: http://www.proteometrics.com/software/paws.htm.

35. URL: http://www.proteometrics.com/software/mz.htm.

36. URL: http://tswww.cc.emory.edu/~kmurray/mslist.html.

37. A Eccleston. Java needs to put bio into bioinformatics. Nature Biotech 15:315, 1997.

Index